The Retinal Müller Cell

PERSPECTIVES IN VISION RESEARCH

Series Editor: Colin Blakemore
University of Oxford
Oxford, England

Biochemistry of the Eye
Elaine R. Berman

Development of the Vertebrate Retina
Edited by Barbara L. Finlay and Dale R. Sengelaub

Parallel Processing in the Visual System
THE CLASSIFICATION OF RETINAL GANGLION CELLS AND ITS IMPACT ON THE NEUROBIOLOGY OF VISION
Jonathan Stone

Presbyopia Research
FROM MOLECULAR BIOLOGY TO VISUAL ADAPTATION
Edited by Gérard Obrecht and Lawrence W. Stark

The Retinal Müller Cell
STRUCTURE AND FUNCTION
Vijay Sarthy and Harris Ripps

Visual Development
Nigel W. Daw

A Continuation Order Plan is available for this series. A continuation order will bring delivery of each new volume immediately upon publication. Volumes are billed only upon actual shipment. For further information please contact the publisher.

The Retinal Müller Cell
Structure and Function

Vijay Sarthy
Northwestern University Medical School
Chicago, Illinois

and

Harris Ripps
University of Illinois College of Medicine
Chicago, Illinois

Kluwer Academic / Plenum Publishers
New York, Boston, Dordrecht, London, Moscow

Library of Congress Cataloging-in-Publication Data

Sarthy, Vijay.
The retinal Müller cell: structure and function/authored by Vijay Sarthy and Harris Ripps.
p. ; cm. — (Perspectives in vision research)
Includes bibliographical references and index.
ISBN 0-306-46470-5
1. Retina—Cytology. 2. Neuroglia. I. Ripps, Harris. II. Title. III. Series
[DNLM: 1. Retina-cytology. 2. Neuroglia—physiology. 3. Neurons—physiology. WW 270 S249r 2000]
QP479 .S275 2000
612.8′43—dc21

00-058749

Cover figure: Scanning electron micrograph of an isolated Müller cell from the turtle retina. Vijay Sarthy and Ann Bunt. The ultrastructure of isolated glial (Müller) cells from the turtle retina. *The Anatomical Record*. (Copyright 1982 Wiley-Liss, Inc., a subsidiary of John Wiley & Sons, Inc., reprinted with permission.)

ISBN 0-306-46470-5

233 Spring Street, New York, N.Y. 10013

http://www.wkap.nl/

10 9 8 7 6 5 4 3 2 1

A C.I.P. record for this book is available from the Library of Congress

Printed in the United States of America

To Mary and Jeanne

Preface

The human brain contains more than a billion neurons which interconnect to form networks that process, store, and recall sensory information. These neuronal activities are supported by a group of accessory brain cells collectively known as *neuroglia*. Surprisingly, glial cells are ten times more numerous than neurons, and occupy more than half the brain volume (Hydén, 1961). Although long considered a passive, albeit necessary, component of the nervous system, many interesting and unusual functional properties of glial cells are only now being brought to light. As a result, the status of these cellular elements is approaching parity with nerve cells as a subject for experimental study.

The term *glia* (or glue) seems today to be a misnomer in view of the diverse functions attributed to glial cells. Experimental studies in the last three decades have clearly established that the behavior of glial cells is far from passive, and that they are at least as complex as neurons with regard to their membrane properties. In addition, glial cells are of importance in signal processing, cellular metabolism, nervous system development, and the pathophysiology of neurological diseases. The Müller cell of the vertebrate retina provides a splendid example of an accessory cell that exhibits features illustrating every aspect of the complex behavior now associated with glial cells.

HISTORICAL PERSPECTIVE

It was more than a century ago that the generic term *neuroglia* was given to the non-neuronal elements of the nervous system. The notion that neural elements are embedded in an unusual connective tissue-like matrix (Bindegewebe) can be traced to the early writings of the eminent German pathologist Rudolph Virchow (1846), but the concept of neuroglia (Nervenkitt) and a description of its histological form did not appear until ten years later (Virchow, 1856; for a historical review, see Somjen, 1988). Without detracting from Virchow's enormous prescience, it is worth noting that five years earlier Heinrich Müller (1851), having studied a variety of species, provided a detailed description of the radial fibers which we now know to be the principal glial cell of the vertebrate retina. (The text of Müller's landmark paper, together with a translation, is provided in the Appendix.) Shortly thereafter, Kolliker (1854) observed similar structures in the human retina and ascribed to them eponymously the name by which they have come to be known: the Müller cells.

It is also interesting to note that whereas glial cells of the central nervous system (CNS) have been classified into a number of subtypes based on such distinguishing features as morphology, antigenicity, and functional properties (Raff, 1989), Müller cells are usually treated as belonging to a unique but functionally uniform class of glial cell. This is clearly not the case. Depending upon the species, there is a striking heterogeneity in Müller cell structure, antigenic properties, and responses to neurotransmitters. However, the molecular determinants of these differences are often unknown, and no rational basis for subclassification of Müller cells has emerged. The observed differences between species may reflect the different metabolic or functional demands on these cells, or the influence of different environmental factors; a comparative study of Müller cells from this perspective is sorely needed.

In focusing on the Müller cell, we cannot ignore the wealth of information available from the study of glial cells in other regions of the CNS. Indeed, many of the defining characteristics of Müller cells, such as their electrical properties, immunochemical features, metabolic activities, and cytoplasmic content, display similarities to those found in protoplasmic astrocytes (Kettenman and Ransom, 1995). Nevertheless, the Müller cell has become highly adapted in form and function to its retinal environment, and there is little to be gained by attempting to force it into one or another of the conventional categories of neuroglia.

SCOPE OF THE BOOK

Chapter 1 introduces the glial cells of the retina and describes their structural features and intercellular relationships. The morphology of Müller cells is considered in detail, with special attention paid to junctional contacts, membrane characteristics, and other cytological features. Chapter 2 deals with the role of Müller cells in retinal development and also with features of Müller cell development itself. The lineage, birth, and determination of Müller cells are reviewed. Potential roles of Müller cells in neuronal development are also discussed. Chapter 3 looks at the metabolic interactions between Müller cells and retinal neurons. The involvement of Müller cells in energy metabolism, transmitter inactivation, pH regulation, and retinoid metabolism are discussed. Chapter 4 focuses on the signaling pathways between retinal neurons and Müller cells. The properties of neurotransmitter transporters and receptors on Müller cells are described. Chapter 5 discusses ionic properties of the Müller cell membranes, the role of Müller cells in potassium homeostasis, and the involvement of Müller cells in generating electroretinographic (ERG) potentials. Last, Chapter 6 presents an overview of the involvement of Müller cells in retinal pathology. This chapter describes putative roles of Müller cells in excitotoxicity, reactive gliosis, and retinal diseases.

It will be clear from the information presented in these chapters that Müller cells perform a variety of tasks in the retina. Perhaps their best understood functions are in potassium homeostasis and in neurotransmitter uptake and metabolism. These activities support a number of vital processes in the normal retina. Müller cells play an active role in pathological conditions as well. For instance, in ischemic retina, Müller cells are likely to alleviate glutamate excitotoxicity by removing excess extracellular glutamate. The strong gliotic response of Müller cells also suggests a role in the phagocytosis of cell debris and scar formation. Moreover, it has recently been discovered that reactive Müller cells produce neuroprotective cytokines and neurotrophic factors that alleviate neuronal damage and degeneration. In contrast, our current knowledge of the mechanisms that determine Müller cell fate and the roles of Müller cells in retinal development is still in its infancy.

In writing this book, we have attempted to survey the current status of our knowledge of Müller cell structure and function with the hope that this effort will point out areas of Müller cell biology that need to be addressed in the future. Because of space constraints, we have been selective in our choice of topics and illustrations. Many similar or related examples have not been presented, and we apologize if we overlooked any important studies. Clearly, the content and emphasis of the book reflects our personal view of Müller cell biology. Although many details are still lacking, it is evident now that Müller cells and retinal neurons have evolved together to fashion an intricate cellular network, the retina, whose ultimate goal is to present a clear view of the surrounding world.

ACKNOWLEDGMENTS

We gratefully acknowledge the generosity of our colleagues who provided illustrations and data from their published work, and the publishers who very graciously permitted us to reproduce figures. We are especially thankful to Paul Witkovsky, Jack Saari, Tom Reh, Don Puro, Lee Jampol,

Steve Fisher, and Ruben Adler for helpful comments and suggestions on individual chapters. This book could not have been completed without the enthusiastic assistance of Jane Zakevicius, Mary Winneke of the UIC library, and the library staffs of the Marine Biological Laboratory, Woods Hole, Massachusetts, and Friday Harbor Laboratory, Friday Harbor, Washington. We are grateful to Lisa Birmingham who meticulously prepared all the illustrations, and we thank Tom Porter, Michael Schütte, Etha Schuette, and Rolf Behrmann for undertaking the challenging task of translating Müller's original report.

Contents

Structural Organization of Retinal Glia

1

The vertebrate retina contains four types of glia, which exhibit distinct morphological, developmental, and antigenic characteristics (Fig. 1.1). The Müller cell is the predominant glial element, comprising 90% of the retinal glia. Müller cell processes interdigitate with the perikarya, axons, and dendrites of neurons throughout the retina, a feature clearly consistent with a symbiotic relationship. Considering their strategic location, Müller cells are in a position to influence and be influenced by neuronal activity throughout the tissue. In addition to Müller cells, astrocytes and microglia are the two glial types seen most frequently in retina (cf. Vrabec, 1970; Boycott and Hopkins, 1981; Robinson and Dreher, 1989). These cells have different embryological origins and are found predominantly in species with vascularized retinas. The fourth type of glia, the oligodendrocyte, is seen occasionally in the retina, but only when myelinated ganglion cell axons are present in the nerve fiber layer, e.g., the medullary rays of the rabbit retina (cf. Schnitzer, 1987a; Ehinger et al., 1994). Although our focus is primarily on the Müller cell, the chapter also includes a brief description of other retinal glia, their distribution, and their interactions with blood vessels, Müller cells, and neurons.

1.1. MÜLLER CELLS

As already mentioned, Müller cells are radially oriented cells that traverse the retina from its inner (vitreal) border to the distal end of the outer nuclear layer (Fig. 1.2). Along their course, Müller cells extend branches that interdigitate with every class of retinal neuron, with other types of glia, and with the blood vessels of vascularized retinas.

This general description belies the remarkable morphological variability among species, which is prominently seen in the pattern of branching processes that project to the inner limiting membrane (Fig. 1.3A).

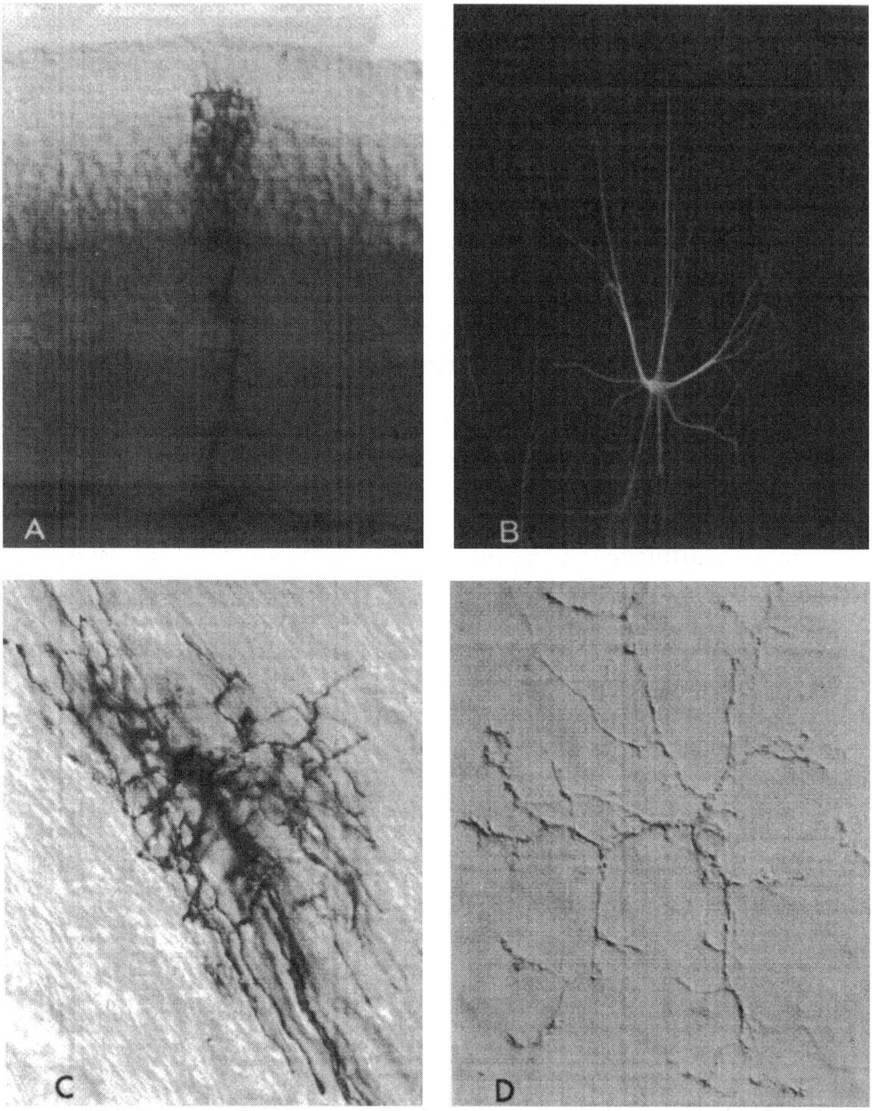

Figure 1.1. The four types of glial cells in the rabbit retina. A. Müller cell filled by intracellular injection with horseradish peroxidase. B. Astrocyte labeled with a monoclonal (MA1) antibody. C. An oligodendrocyte filled with biocytin extends processes that ensheathe the bundles of medullated nerve fibers that enter the optic nerve. D. A lectin-labeled microglial cell shows the multiple small branches that project from the cell body (Robinson, 1992). (Copyright 1992 *Today's Life Science*, reprinted with permission.) For a color representation of this figure, see color insert facing this page.

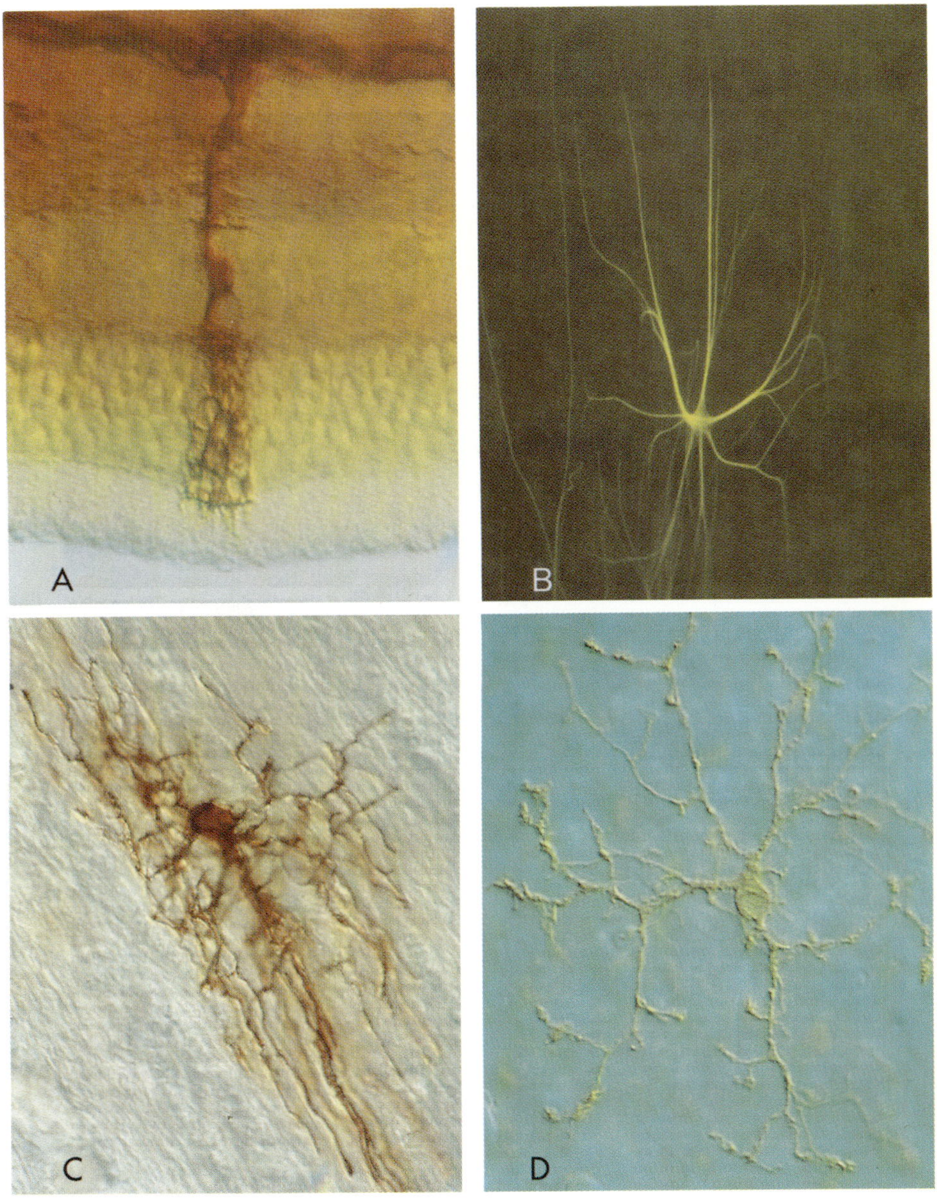

Figure 1.1. The four types of glial cells in the rabbit retina. A. Müller cell filled by intracellular injection with horseradish peroxidase. B. Astrocyte labeled with a monoclonal (MA1) antibody. C. An oligodendrocyte filled with biocytin extends processes that ensheathe the bundles of medullated nerve fibers that enter the optic nerve. D. A lectin-labeled microglial cell shows the multiple small branches that project from the cell body (Robinson, 1992). (Copyright 1992 *Today's Life Science*, reprinted with permission.)

Müller cells are ubiquitous, and save for the optic nerve head, are found in all retinal regions of every vertebrate studied, including the central fovea of primates (Yamada, 1969; Distler and Dreher, 1996). In species with avascular retinas (lizards, amphibians, birds, and some mammals), Müller cells are the only glial elements that can be detected in the neural retina (Pedler, 1963; Rasmussen, 1974; Stone and Dreher, 1987; Schnitzer, 1988a).

There are between 10^6 and 10^7 Müller cells in the mammalian retina (Robinson and Dreher, 1990; Dreher et al., 1992; Reichenbach and Robinson, 1995; Distler and Dreher, 1996), and they occupy between 6–10% of the total cytoplasmic volume (Rasmussen, 1975; Reichenbach and Wohlrab, 1986). The measurements of Jeon et al. (1998) on mouse, rabbit, and monkey retinas indicate that Müller cells account for 16–22% of the cell bodies in the inner nuclear layer (INL); these data are shown in Fig. 1.3B together with comparable values for some of the other retinal cell types. However, the population density and morphology of Müller cells vary in different parts of the retina. Cells located in regions close to the ora serrata are shorter, have stouter trunks and broader endfeet, and are present in lower density (Fig. 1.4) than in more central locations (Uga, 1974; Rasmussen, 1974, 1975; Dreher et al., 1988; Gaur et al., 1988; Reichenbach et al., 1989; Robinson and Dreher, 1990). In the far periphery of monkey retina, for example, cell density is about 6,000/mm^2, whereas in the parafoveal region, the Müller cells have long slender trunks and reach a peak density of more than 30,000/mm^2 (Distler and Dreher, 1996). The extreme variability in these morphometric parameters undoubtedly reflects the functional requirements of different retinal regions, and the special properties of the microenvironment in which the cells develop (Reichenbach et al., 1989).

Regional differences aside, Müller cells of all species have many common features. At the ultrastructural level (Fig. 1.5), the Müller cell cytoplasm appears more electron dense than neighboring neurons, and contains a well-developed endoplasmic reticulum and varying amounts of glycogen granules (Hogan et al., 1971; Uga, 1974). The cell nuclei are typically oval or polygonal and are generally located in the middle of the INL.

Mitochondria are located throughout the Müller cell cytoplasm, and in some retinas they may be found concentrated toward one end of the cell (cf. Uga and Smelser, 1973b; Rasmussen, 1974). It has been suggested that the location of the mitochondria within the Müller cell relates to the energy requirements of its neuronal neighbors (Rasmussen, 1973), but this seems unlikely. For example, mitochondrial density is often highest near the external limiting membrane in a region bordering the photoreceptor inner segments (Uga and Smelser, 1973a; Rasmussen, 1975); this area of the visual cell contains one of the highest concentrations of mitochondria of any body tissue. Recently, studies of mitochondrial migration and localization in

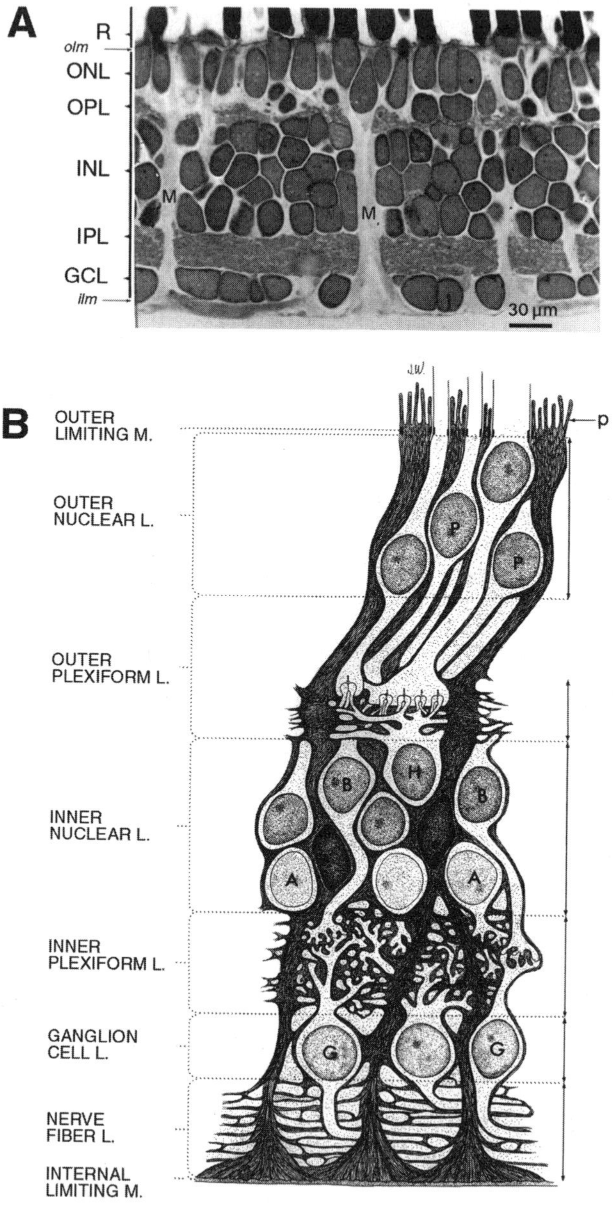
A
R
olm
ONL
OPL
INL
M
M
IPL
GCL
ilm
30 µm
B
OUTER
LIMITING M.
OUTER
NUCLEAR L.
OUTER
PLEXIFORM L.
INNER
NUCLEAR L.
INNER
PLEXIFORM L.
GANGLION
CELL L.
NERVE
FIBER L.
INTERNAL
LIMITING M.
p
P
P
B
H
B
A
A
G
G

intact and isolated mammalian Müller cells provided evidence that the distribution of these organelles is determined by the local cytoplasmic oxygen pressure. Clustering occurs in regions where the Po_2 levels exceed ~10–20 mmHg (Germer et al., 1998a, b), but the mechanism underlying mitochondrial migration, and the functional implications of this observation remain to be determined.

1.1.1. Relation to Neurons

As already mentioned, the intimate association between Müller cell processes and retinal neurons implies a close functional relationship between the two cell types. In this connection, an interesting concept put forward by Reichenbach et al. (1993) suggests that cooperativity exists among cells that arise from a common stem cell to form a columnar array (cf. Chapter 2, the developmental studies by Turner and Cepko, 1987 and Turner, Snyder and Cepko, 1990). The basic tenet of this hypothesis is that one Müller cell subserves many of the metabolic, ionic, and extracellular buffering requirements of those neurons with which it shares a common progenitor (Fig. 1.6).

The idea of a functional unit in which local interactions occur between a group of retinal neurons and their supportive glial cell is an attractive concept. On this view, Müller cells need not interact with each other to accommodate the far greater number of retinal neurons they subserve. By limiting its sphere of influence, each Müller cell may only have to meet the requirements of its immediate neuronal neighbors, e.g., to maintain a stable extracellular environment in the face of intense neural activity (cf. Chapters 4 and 5). Cooperativity and intercellular communication between Müller cells and retinal neurons will be considered further in Chapters 3 and 4.

Figure 1.2. A. Light micrograph of the mudpuppy retina shows two Müller cells extending from the internal limiting membrane (ilm) to the outer limiting membrane (olm), and their relationship to the nuclear and plexiform layers of the retina. R, receptors; ONL, outer nuclear layer; OPL, outerplexiform layer; INL, inner nuclear layer; IPL, inner plexiform layer; GCL, ganglion cell layer (Dowling, 1987). (Copyright 1987 Belknap Press, reprinted with permission. After Dowling [1970]. Copyright 1970 Association for Research in Vision and Ophthalmology, reprinted with permission.) B. Schematic drawing of mammalian Müller cell perikarya and their processes interdigitating with every class of retinal neuron; the figure was reconstructed from electron micrographs of the human retina. Shown are the photoreceptors (P) of the outer nuclear layer, the horizontal (H), bipolar (B), and amacrine (A) cells of the inner nuclear layer, and the ganglion cells (G) whose axons form the nerve fiber layer. Note the location of the oval-shaped, densely stained Müller cell nuclei near the innermost part of the INL. Intercellular junctions can be seen at the outer limiting membrane, distal to which villous processes (p) embrace the myoid region of the photoreceptor. Müller cells abut the internal limiting membrane at the vitreal border, where they terminate in pyramidal-shaped endfeet (Hogan et al., 1971). (Copyright 1971 W.B. Saunders Co., reprinted with permission.)

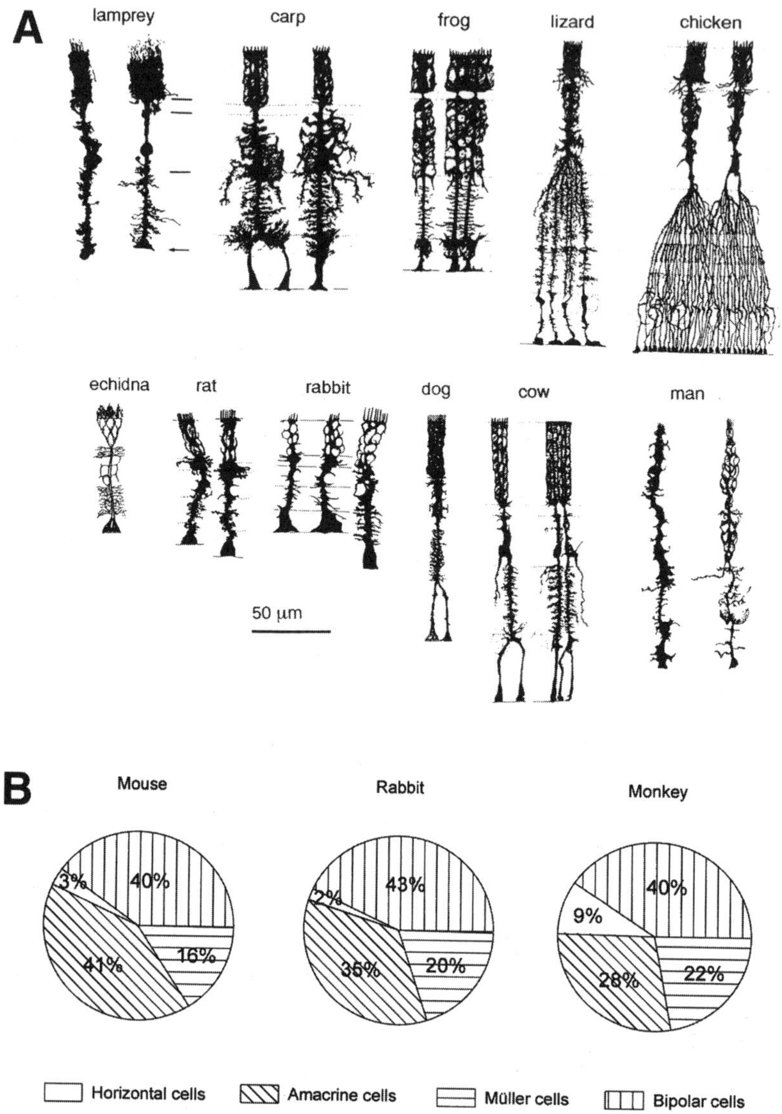

Figure 1.3. A. Camera-lucida drawings of silver-impregnated Müller cells in the retinas of various vertebrate species illustrates the remarkable morphological heterogeneity of the radial glia (Rubinson, 1990; Reichenbach and Robinson, 1995). (Lamprey copyright 1990 John Wiley & Sons, Inc., reprinted with permission. Echidna, rat, rabbit copyright 1995 Oxford University Press, reprinted with permission.) B. The relative distributions of the principal cells of the INL in the mouse, rabbit, and monkey retinas (Jeon et al., 1998). (Copyright 1998 the Society for Neuroscience, reprinted with permission.)

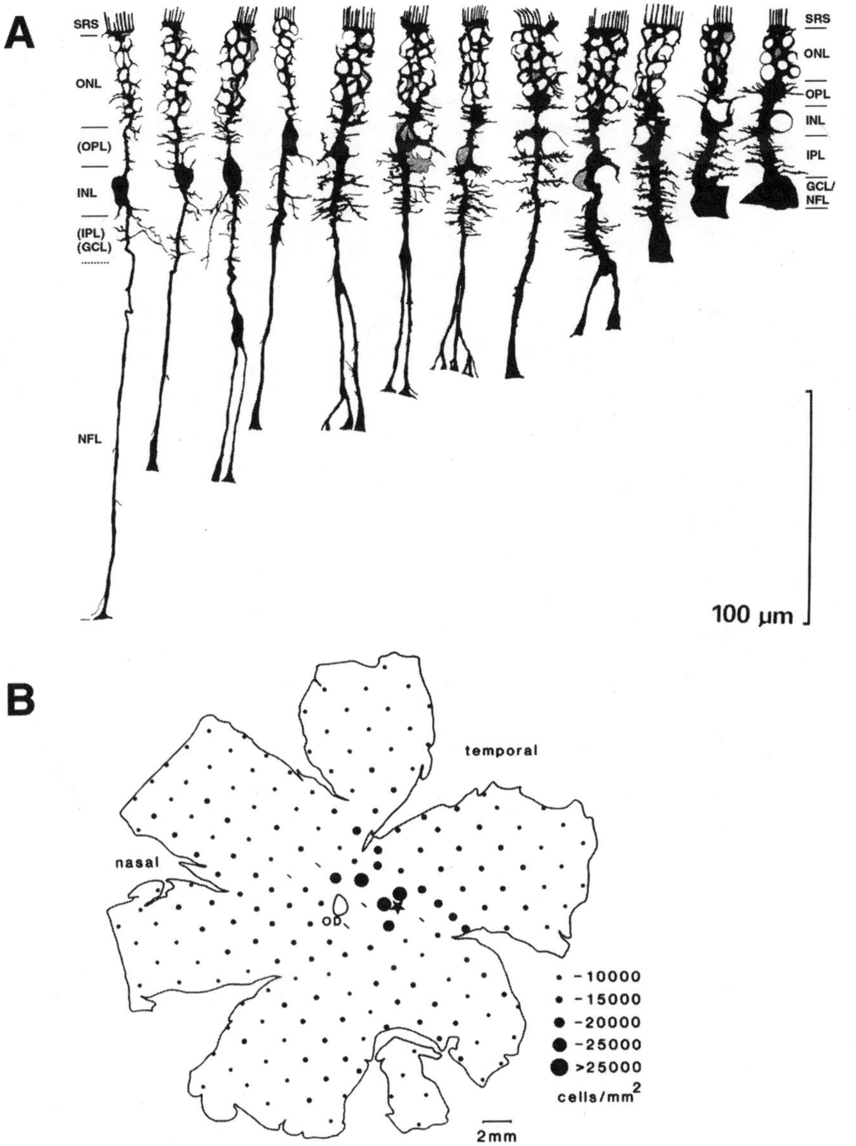

Figure 1.4. Müller cell structure and population density vary with retinal location. A. Camera-lucida drawings of Golgi-stained Müller cells across the rabbit retina extending from the densely populated medullary region (left) to the far periphery (right) (Reichenbach et al., 1989). (Copyright 1989 Springer-Verlag, reprinted with permission.) B. Density distribution of Müller cells in a monkey retina stained with a cell-specific monoclonal antibody (Distler and Dreher, 1996). (Copyright 1996 Elsevier Science, reprinted with permission.)

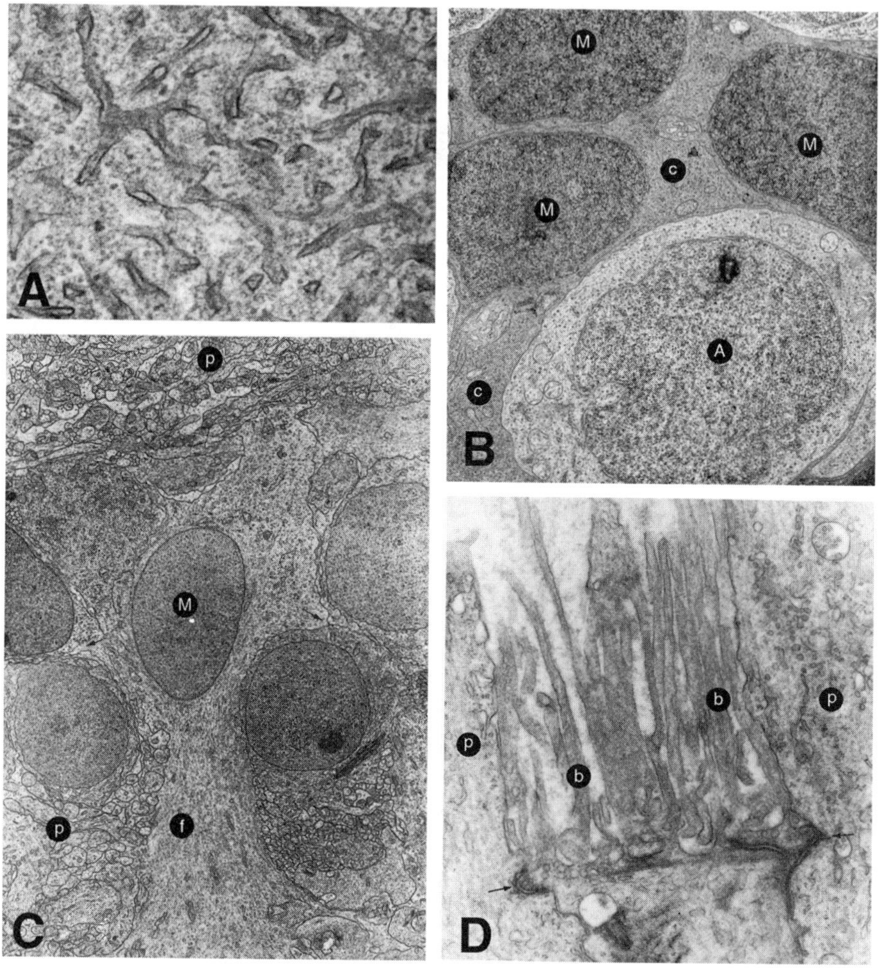

Figure 1.5. The ultrastructure of the Müller cell seen in electron micrographs of the human retina. A. The smooth-surfaced endoplasmic reticulum (arrows) is evenly dispersed throughout the cytoplasm of the Müller cell (magnified 16,500). B. The nuclei of three Müller cells (M) are located distal to the large nucleus of an amacrine cell (A). Note the more densely staining cytoplasm (c) of the Müller cells (magnified 8,000). C. Thick radial fibers (f) emerge from the nuclear region of the Müller cell (M), and extend lateral branches to form a meshwork within the plexiform layers (p) of the retina (magnified 4000). D. Villous processes extend beyond the external limiting membrane (arrows) to form baskets (b) that surround the inner segments of the photoreceptors (p) (magnified 21,600) (Hogan et al., 1971). (Copyright 1971 W.B. Saunders Co., reprinted with permission.)

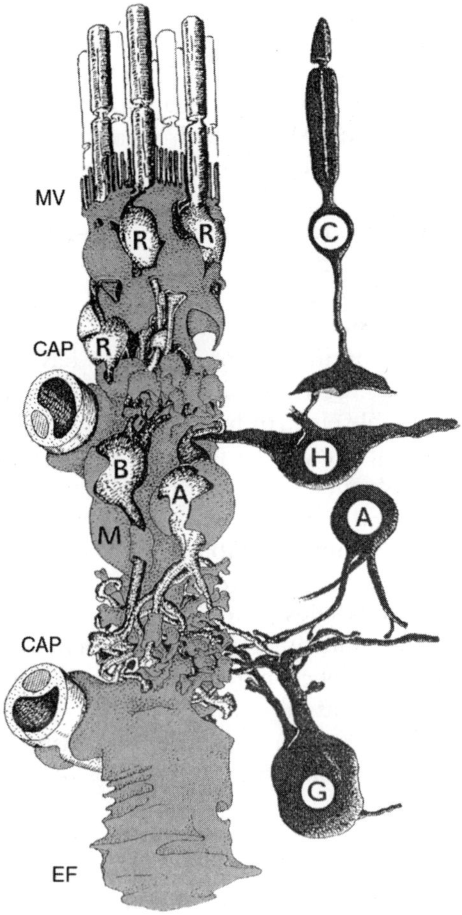

Figure 1.6. The columnar arrangement of stem cell progeny generated late in ontogenesis. The cell types include a Müller cell (M), rod photoreceptors (R), bipolar cells (B), and some classes of amacrine cell (A). Extracolumnar cells (cones, horizontal cells, ganglion cells, and other classes of amacrine cell) are born earlier within a thin immature neuroepithelium. Retinal capillaries (CAP), and the microvilli (MV) and endfoot (EF) of the Müller cell are also labeled (Newman and Reichenbach, 1996). (Copyright 1996 Elsevier Science, reprinted with permission.)

1.1.2. Intercellular Junctions

Glial cells do not make chemical synapses with other cells, but they often form gap junctions with neighboring glial cells to create an electrochemical syncytium. The aqueous pore of the gap junction channel, formed by hexameric assemblies of proteins (connexins) in the apposed membranes of adjoining cells (Fig. 1.7A), permit the passage of ions, metabolites, and other small molecules (<1 kDa). Although intercellular junctional complexes of various types, including gap junctions, are often found at various locations along the Müller cell, evidence of electrical coupling between neighboring Müller cells has been observed in some, but not all, of the species tested.

In many lower vertebrates, for example, cell–cell contacts bearing the morphological features of gap junctions have been identified ultrastructurally as well as in freeze-fracture replicas of Müller cells (Miller and Dowling, 1970; Uga and Smelser, 1973; Gold and Dowling, 1979; Tonus and Dickson, 1979). These communicating junctions have been seen primarily in the distal retina, near the external limiting membrane (Fig. 1.7B).

However, the available electrophysiological data suggest that intercellular coupling is not very strong, and an electric current injected into a single cell does not spread far from its origins. Recording simultaneously from two different Müller cells in the turtle retina, Conner et al. (1985) found that coupling between cells was undetectable when they were separated by more than 65 μm, and not all cell pairs separated by a lesser distance showed electrical coupling. Nevertheless, the fact that the input resistance of isolated Müller cells was significantly greater than when measured *in situ* is evidence that coupling exists between these cells, albeit with a greatly restricted length constant (~30 μm) as compared, for example, to the broad, well-coupled network (100–1000 μm) of horizontal cells in turtle (Lamb, 1976; Piccolino et al., 1984) and fish (Naka and Rushton, 1967; Qian and Ripps, 1992) retinas.

A study of Müller cell coupling in the axolotl retina (Mobbs et al., 1988) yielded results similar to those obtained by Conner et al. (1985) with respect to the spread of current (and dye) from the site of injection to neighboring cells, e.g., a 20 mV response to current injection in one cell gave rise to a response of only 0.45 mV in a cell 50 μm away (Fig. 1.8). Based on the foregoing results, electrical coupling between Müller cells does not appear to be a major factor in the transcellular spread of ionic current. However, relatively little is known of the factors that modulate the conductance of Müller cell gap junctions. Thus, it is possible that photically or pathologically induced changes in the chemistry of the extracellular environment could significantly enhance gap-junctional communication.

In the mammalian retina, cell–cell coupling appears to be more complex. There is general agreement that Müller cells do not form gap junctions with other Müller cells. Unlike the Bergmann (radial) glia of the cerebellum (Müller et al., 1996) or astrocytes in retina and central nervous system (CNS) (Robinson et al., 1993; Brightman and Reese, 1969; Hanani et

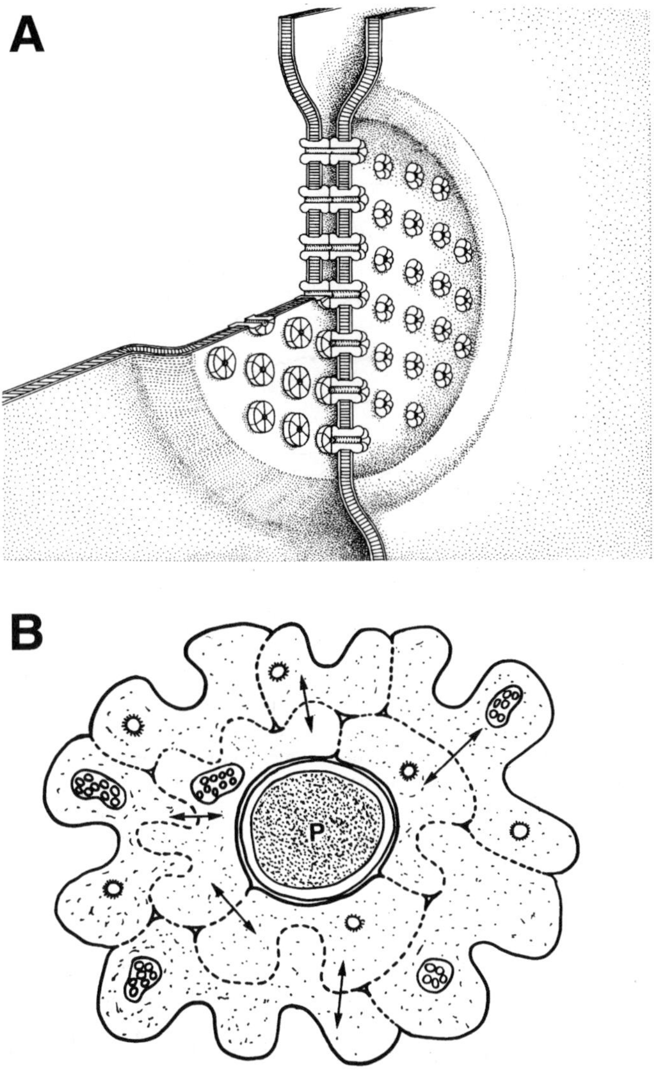

Figure 1.7. A. Schematic drawing of the membrane composition of gap junctions constructed from freeze–fracture images. Connexin proteins form the hexameric arrays (connexons) that span the lipid bilayers of adjacent cell membranes. The docking of pairs of connexons create the intercellular channel that allows passage of ions and small molecules (Staehelin and Hull, 1978). (Copyright 1978 Patricia J. Wynne, reprinted with permission.) B. Diagrammatic representation of a tangential section at the outer limiting membrane of the newt retina showing the communicating (gap) junctions between Müller cells (arrows) surrounding the distal part of the photoreceptor nucleus (P) (Tonus and Dickson, 1979). (Copyright 1979 Academic Press Inc., reprinted with permission.)

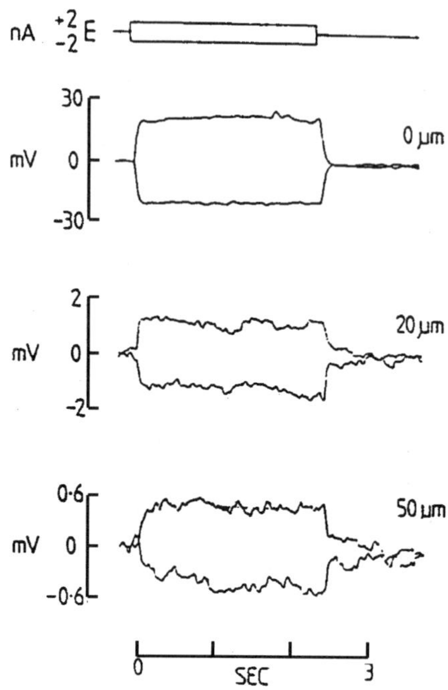

Figure 1.8. Voltage responses recorded from electrically coupled Müller cells. Current (± 2 nA) injected into a single cell (0 µm) produced hyperpolarizing and depolarizing responses of equal magnitude. Much smaller potentials of the same sign were elicited in cells located at distances of 20 µm and 50 µm from the cell in which current was injected. Note the change in scale for the lower traces, indicating the large response loss over a distance spanning only about three Müller cells (Mobbs et al., 1988). (Copyright 1988 Elsevier Science, reprinted with permission.)

al., 1989; Dermietzel et al., 1991; Batter et al., 1992; Lee et al., 1994; Zahs and Newman, 1997), in which intercellular coupling was demonstrated with various experimental methods, neither gap junctions (Uga and Smelser, 1973; Holländer et al., 1991) nor tracer coupling (Robinson et al., 1993; Zahs and Newman, 1997) has been observed between mammalian Müller cells. A surprising feature of the apparent lack of gap junction communication between mammalian Müller cells is that rabbit Müller cells in culture form gap junctions which they do not express in vivo, whereas other membrane proteins, e.g., orthogonal arrays of membrane particles (see Section 1.1.9), are typically seen in situ yet are not detected in cell explants (Wolburg et al., 1990).

Although mammalian Müller cells do not make gap junctions with each other, coupling between Müller cells and astrocytes seems to be widespread and can provide an effective pathway for signal propagation (Robinson et al., 1993; Zahs and Newman, 1997). Interestingly, coupling between the two cell types is highly unidirectional. Zahs and Newman (1997) found that when Müller cells were filled with lucifer yellow (LY) and neurobiotin, both tracers were confined to the injected cell. In contrast, when an astrocyte was filled with these tracers, the larger molecular weight LY spread to 2–10 neighboring astrocytes, whereas neurobiotin infiltrated clusters of 13–88 astrocytes as well as more than 100 Müller cells. Thus, there is an astrocyte–Müller cell network that may represent a signaling pathway by which the retinal glia influence neuronal activity (see Chapter 4).

1.1.3. Müller Cells as Insulators

In retinal sections, the processes of Müller cells are seen to enwrap the cell bodies and processes of retinal neurons, giving the impression that they provide electrical insulation, or act as a barrier to communication between neighboring neurons (Fig. 1.9). However, this is not their function. The plasma membrane preserves the electrical integrity of nerve cells, and there is no evidence that intercellular signaling via chemical or electrical synapses is impaired by the presence of Müller cell processes.

We have seen that gap junctions provide the morphological substrate for electrotonic spread of potentials, and almost every class of retinal neuron appears to communicate with its neighbors via this specialized type of contact (Vaney, 1994). Clearly, glial cells do not prevent these cell–cell

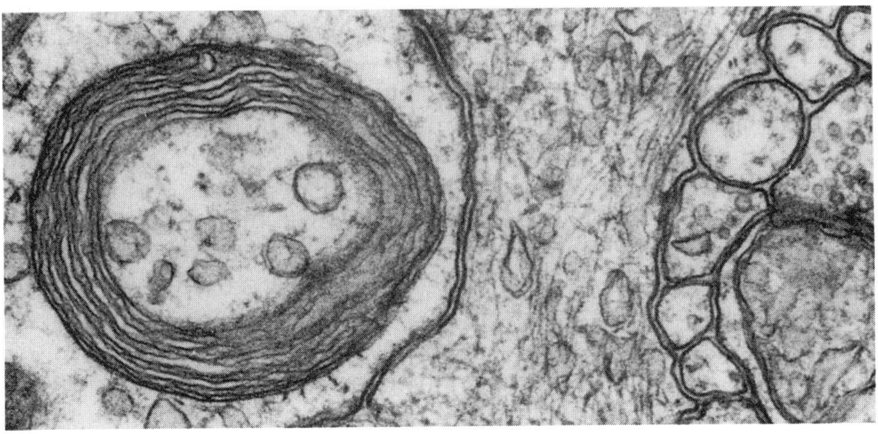

Figure 1.9. Müller cell processes elaborate membrane that wraps around a bipolar cell process (magnified 45,000) (Hogan et al., 1971). (Copyright 1971 W.B. Saunders Co., reprinted with permission.)

contacts. This is apparent even at the level of the photoreceptors, where Müller cell processes are interspersed between neighboring cells and form basket-like processes that ensheathe the receptor inner segments (Fig. 1.5 D). Despite the close apposition of the two cell types, there are distinct regions where photoreceptors are joined by gap junctions (Raviola and Gilula, 1973; Witkovsky et al., 1974; Fain et al., 1976; Cohen, 1989), and electrical signals can spread laterally from the source of excitation (Baylor et al., 1971; Lamb and Simon, 1976).

It is not immediately apparent why Müller cell processes often elaborate membranes that ensheathe ganglion cell somata, as well as some of the axonal, dendritic, and synaptic regions of retinal neurons, a pattern of ensheathment that is shared with astrocytic processes (Pedler, 1962; Holländer et al., 1991). Using Müller cell-specific fluorescent-tagged antibodies as markers, Robinson and Dreher (1990) were able to visualize the cooperativity of several adjacent Müller cells in contributing to the ensheathment of a single ganglion cell (Fig. 1.10). These observations, and complementary images seen in electron micrographs (Stone et al., 1995b), have been taken as a possible indication that the glial coat serves to control current flow from dendrite to axon, and to define the site of generation of action potentials. However, there has been no experimental evidence to support this interpretation. In contrast, based on results obtained in a number of recent studies (cf. Pfrieger and Barres, 1996), it seems more likely that the intimate association between the two cell types enables glial cells to promote the formation of synapses and helps maintain neuronal function by providing the nerve terminals with energy substrates and neurotransmitter precursors.

1.1.4. The Zonula Adherens of the Outer Limiting Membrane

Müller cell processes are an integral component of a diverse group of intercellular contacts that comprise the outer limiting membrane (OLM), a readily identifiable line of demarcation that extends from the margin of the optic nerve head to the ora serrata (cf. Fig. 1.2). The OLM is not a true "membrane," but rather is formed by a tangentially oriented series of structurally unique junctions, the zonulae adherentes (Cohen, 1965a; Spitznas, 1970). These areas of contact join the inner segments of rods and cones to the Müller cells, Müller cells to neighboring Müller cells, and occasionally photoreceptors to photoreceptors (Fig. 1.11). The junctions at these sites are generally assumed to function as support structures to help maintain the alignment and orientation of the photoreceptors. The composition of the electron-dense plaques that demarcate the junctions are consistent with this view.

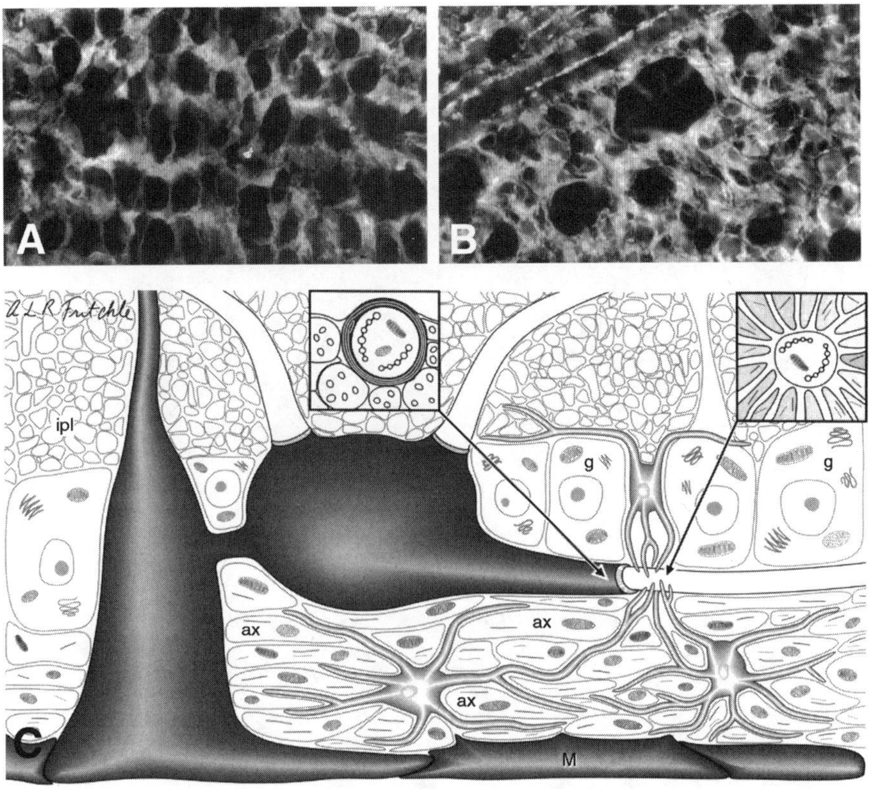

Figure 1.10. Müller cell processes ensheathe regions of the perikarya and processes of ganglion cells. A. The endfeet of Müller cells (stained with an antibody to vimentin) surround the densely-packed ganglion cells of the visual streak (medullary region) of the rabbit retina. B. In a more peripheral region, the ganglion cells are less densely packed, and the Müller cell processes can be seen enclosing fascicles of ganglion cell axons at the upper left of the figure (Robinson and Dreher, 1990). (Copyright 1990 Wiley-Liss, Inc., subsidiary of John Wiley & Sons, Inc., reprinted with permission.) C. Schematic diagram based on electron microscopic analysis of glial ensheathment of ganglion cell soma and axon in cat retina. An extension from the pyramidal endfoot region of the Müller cell surrounds the soma and axon hillock of the ganglion cell; more distally, the axon is contacted by astrocytic processes (Stone et al., 1995b). (Copyright 1995 Cambridge University Press, reprinted with permission.)

Light and electron microscopy immunolocalization studies have demonstrated the presence of circumferential bands of actin filaments at the zonula adherens junctions of the OLM, as well as the presence of the actin-filament-associated proteins myosin, α-actinin, and vinculin (Drenckhahn and Wagner, 1985; Williams et al., 1990).

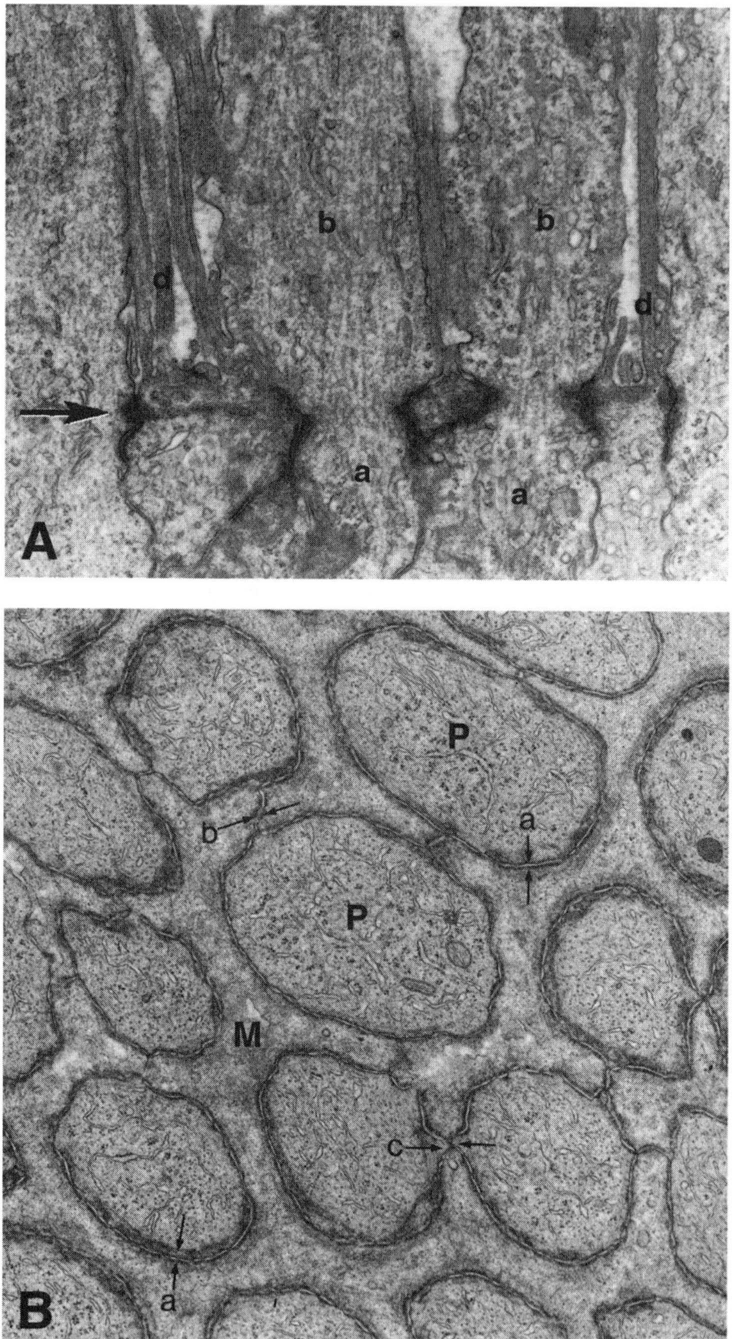
d
b
b
d
a
a
A
P
a
b
P
M
c
a
B

TABLE 1.1. The Pore Size of the Outer Limiting Membrane (OLM) of Rabbit Retina[a]

Biotinylated protein	Source	R_s (Å)	$M_r \times 10^{-3}$	Traverses OLM
Myoglobin	Horse muscle	21	18	Yes
Ovalbumin	Bovine milk	27	45	Yes
Peroxidase	Horseradish	30	44	Yes
Albumin	Bovine serum	36	67	No
Lactoperoxidase	Bovine milk	36	78	No
γ-globulin	Rabbit serum	53	160	No
IRBP (interphotoreceptor retinoid-binding protein)	Bovine retina	55	140	No

[a]Adapted and condensed from Bunt-Milam et al. (1985).

1.1.5. The Confines of the Subretinal Space

The junctional complexes at the OLM selectively influence the movement of macromolecules from the subretinal space (sometimes referred to as the interphotoreceptor matrix) to the inner retina. By applying biotinylated proteins of known size to the photoreceptor side of the isolated rabbit retina and determining the ability of the various probes to diffuse through the OLM, Bunt-Milam et al. (1985) inferred that the pore size of the zonulae adherentes forming the intercellular junctions of the OLM is between 30 and 36 Å (Table 1.1).

This barrier helps to confine molecules such as the interphotoreceptor retinoid-binding protein (IRBP), with a Stokes radius of 55 Å, to the extracellular matrix bordered distally by the tight junctions of the retinal pigment epithelium (RPE) and proximally by the zonulae adherentes of the OLM (Fig. 1.12). Thus, retinoids formed in the bleaching–regeneration cycle of the visual pigments can be transported between the photoreceptors and the RPE without loss through diffusion to the inner retina.

Figure 1.11. The ultrastructure of cells and processes at the outer limiting membrane of the human retina. A. Electron micrograph of the OLM (arrows) showing the relation between the photoreceptors and Müller cells at the junctional complexes. The distal processes (p) of Müller cells extend beyond the junctions. Magnification x 18,000. B. Tangential section through the region of the OLM showing the junctions between a Müller cell and a photoreceptor at (a), between two Müller cells at (b), and between two photoreceptors at (c). The cytoplasm of the Müller cell and photoreceptor are labeled M and P, respectively (Hogan et al., 1971). (Copyright 1971 W.B. Saunders Co., reprinted with permission.)

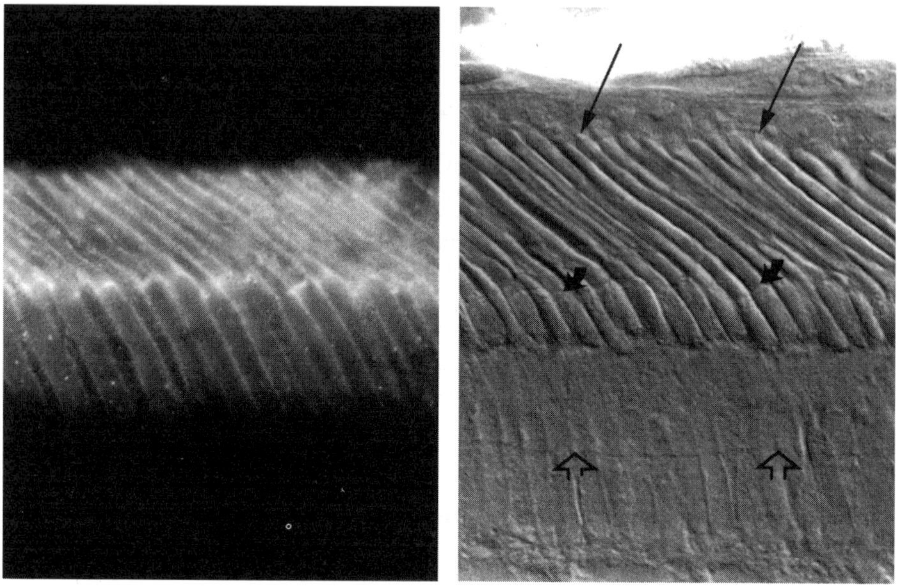

Figure 1.12. Skate retina reacted with a polyclonal fluorescein-tagged antibody to mammalian interphotoreceptor binding protein (IRBP). A comparison of the fluorescent image (left) and the Nomarski image of the same field (right) shows that the protein is confined to the photoreceptor layer, between the tight junctions of the RPE near the distal ends of the photoreceptors (long arrows) and the outer limiting membrane (open arrows). Heavy curved arrows show the junction between inner and outer segments of the photoreceptors (Duffy et al., 1993). (Copyright 1993 Academic Press, Inc., reprinted with permission.)

1.1.6. The Blood–Retina Barrier

In the distal retina, tight junctions (zonula occludens) between cells of the RPE restrict the entry of circulating macromolecules from the choriocapillaris to the neural retina (Cohen, 1965b; Ripps et al., 1989). In the inner retina, this barrier function is subserved primarily by the retinal endothelial cells, whose properties have been likened to the vascular endothelial cells forming the blood–brain barrier of the CNS (Ashton, 1965). The existence of these distinct cellular barriers at the two loci was demonstrated in tracer experiments conducted more than 30 years ago. One of the earliest reports (Rodriguez-Peralta, 1962) noted that after parenteral administration of diaminoacridines, the fluorescent dye could be seen permeating from the choroidal circulation to the level of the pigment epithelial cells, but not beyond, and to penetrate the ocular blood vessels, but not into the retina proper. These observations led to the suggestion that the blood–retina barriers were located at "that part of the membrane of the retinal vascular endothelium which faces the vascular lumen," and at "that part of the pigment epithelial cell membrane which faces the retina" (Rodriguez-Peralta, 1962).

Closer scrutiny of the endothelial cells of the retinal vessels revealed important similarities between their permeability characteristics and those of blood vessels in the CNS (Cunha-Vaz et al., 1966; Cunha-Vaz and Maurice, 1967). One innovative series of experiments involved the use of trypan blue or colloidal iron administered either intraventricularly or intravenously (Cunha-Vaz et al., 1966). After the tracers were allowed to circulate for more than 30 minutes, the vessels were perfused with saline, and the retinas examined both grossly and under the electron microscope. Neither trypan blue nor colloidal iron passed from the circulation into the neural retina, nor could they be detected adhering to the lumen of the vessels, a further confirmation of a barrier at the inner wall of the retinal vessels. In another study, the unique properties of the endothelial junctions of the retinal vasculature, and their similarity to those of cerebral vessels, was revealed by their failure to respond to histamine; this agent typically increases the permeability of blood vessels in other tissues (including conjunctiva and iris), but had no effect on the permeability characteristics of the retinal vessels (Ashton and Cunha-Vaz, 1965).

Although it has long been recognized that blood capillaries are ensheathed by Müller cell processes "with occasional contributions of accessory glia" (Ashton, 1965; Hogan and Feeney, 1963), little consideration was given until recently to the possibility that Müller cells might be involved in the structural organization of the blood–retina barrier. It had been assumed that Müller cells acted as a communicating system for metabolic exchange between vasculature and neurons, in much the same manner as postulated for brain astrocytes. However, in the case of cortical astrocytes, there is convincing evidence that they are capable of inducing the formation of "tight" endothelial junctions in cultures of brain capillaries (Arthur et al., 1987; Tao-Cheng et al., 1987), as well as in vascular tissue that normally does not exhibit barrier properties (Janzer and Raff, 1987).

In the case of the Müller cell, experimental evidence of a similar function is less compelling. Nevertheless, renewed interest in the morphology of the contacts between Müller cell processes and the walls of blood vessels, together with some of the results obtained in transplantation studies, have prompted a reassessment of this issue. At the electron microscope level, the perivascular space is <20 nm (Cunha-Vaz et al., 1966), and confocal laser-scanning micrographs of monkey retina (Distler and Dreher, 1996) have shown that Müller cell processes, which often appear as small endfeet at the outer wall of the vessel (Fig. 1.13), encircle completely the

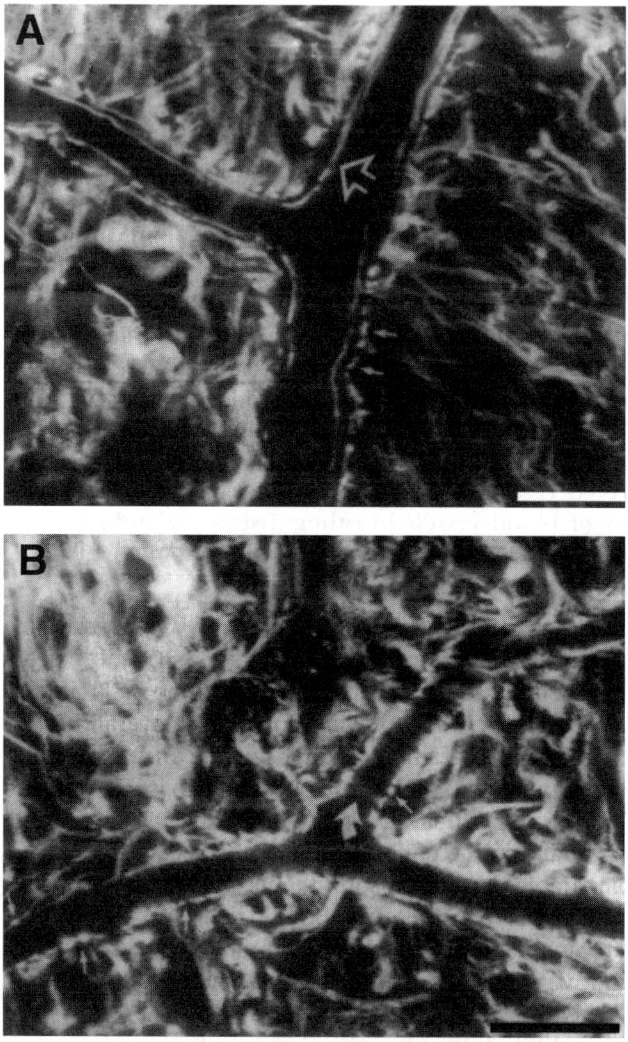

Figure 1.13. Indirect immunofluorescent images of a Müller-cell-specific monoclonal antibody in whole mounts of monkey retina. A. Tiny endfoot-like contacts of Müller cells at the walls of a blood vessel in the nerve fiber layer. Open arrow points to the inner lining of the vessel. B. Curved arrow points to sheath-like envelopment of a blood vessel by Müller cell processes. Scale bar represents 25 μm (Distler and Dreher, 1996). (Copyright 1996, Elsevier Science, reprinted with permission.)

blood vessels. The apposition of Müller cell and vasculature was evident not only at the large vessels of the nerve fiber and ganglion cell layers, but was seen also in relation to capillaries as far distally as the border between the INL and OPL (Distler and Dreher, 1996).

The close physical association between Müller cell processes and the retinal vessels still leaves open questions as to whether the Müller cell per se is an essential component of the blood–retina barrier, and whether it is capable of influencing its permeability characteristics. Attempts to resolve this issue through transplantation experiments modeled after those conducted by Janzer and Raff (1987) have produced conflicting results. In one such study (Tout et al., 1993), cultured Müller cells from the rabbit retina were implanted into the anterior chamber of the rat eye, where they became adherent primarily to the anterior surface of the iris and formed aggregates with its vascularized regions. Vessel permeability was then determined by electron microscopic examination of the tissue aggregates after perfusion with horseradish peroxidase (HRP). The results were compared with aggregates from implants of astrocytes and meningeal cells cultured from rat cerebral cortex. In the normal iris, or in the experimental control (meningeal cell implant), the vasculature was leaky to HRP, and its dark reaction product could be detected in the extracellular space outside the vessel wall. By contrast, the iris vessels from eyes implanted with astrocytes or Müller cells appeared impermeable to HRP (Tout et al., 1993).

Although these findings clearly point to a potential role for Müller cells in the formation of the barrier properties of vascular endothelium, a different result was obtained with purified cultures of Müller cells from neonatal guinea pig retina which had been injected into the anterior chamber of the adult guinea pig eye (Small et al., 1993). In this case, the cells aggregated at various sites within the ciliary body, largely within the connective tissue underlying the ciliary epithelium. The stromal blood vessels, which are normally permeable to HRP, remained so in the presence of Müller cells; i.e., leakage of HRP from the ciliary vessels was similar in saline-injected control eyes, and there was no sign of an endothelial barrier in ciliary vessels adjacent to the Müller cells. Unfortunately these results are difficult to interpret, because it was not possible to determine just how effective astrocytes would have been in inducing barrier properties in the fenestrated vasculature of the ciliary stroma. When cortical astrocytes, prepared from neonatal rat pups, were introduced into the anterior chamber, they adhered to the anterior surface of the iris, forming a broad monolayer over the tissue. Had the astrocytes formed vessel-related aggregates, a comparison with Müller-cell related activity still would not have been possible because the blood vessels of the normal guinea pig iris (unlike rat) are impermeant to HRP. Thus, the contribution of Müller cells to the barrier properties of the vascular endothelium at the blood–retina barrier is questionable; perhaps this experimental approach cannot offer a wholly satisfactory solution to the problem.

1.1.7. The Internal Limiting Membrane

Passing through the ganglion cell and nerve fiber layers of the retina, the Müller cell processes become intertwined with other glial elements (astrocytes, microglia) as they expand into pyramidal-shaped "endfeet" that terminate at the vitreal margin of the retina (Fig. 1.14).

The terminal ends of Müller cells are in intimate contact with, but separated from, a filamentous basement membrane that constitutes the inner limiting membrane (ILM) (Fig. 1.15) (Heegard et al., 1986; Fine, 1961; Hogan et al., 1971; Foos, 1972, 1974; Heegard, 1994, 1997). Whether the Müller cell contributes to the formation of this basal lamina is a question yet to be fully resolved. The ILM covers the entire inner surface of the retina,

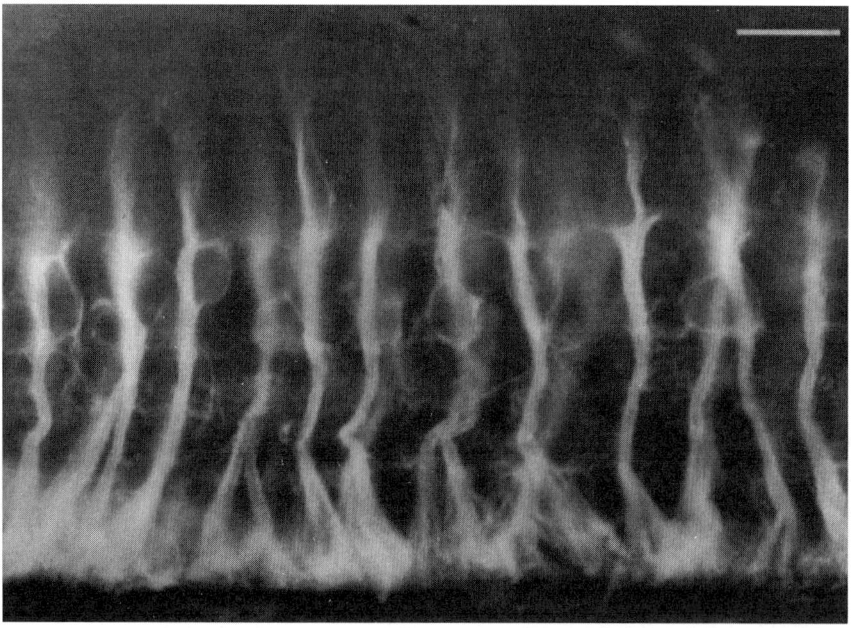

Figure 1.14. Photomicrograph of rabbit Müller cells at the upturned edge of a retinal whole mount labeled with antivimentin. There is a lack of continuity between the Müller cell endfeet at the vitreal margin of the retina. Scale bar = 20 μm (Robinson and Dreher, 1990). (Copyright 1990 Wiley-Liss, Inc., a division of John Wiley & Sons, Inc., reprinted with permission.)

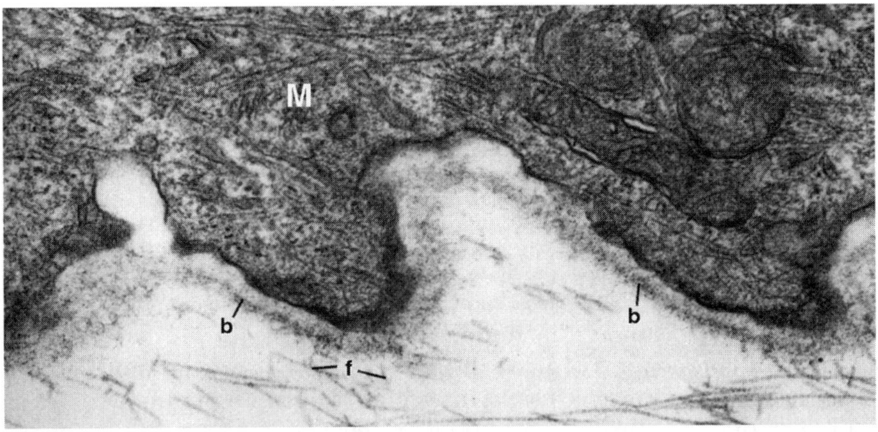

Figure 1.15. Electron micrograph of the inner limiting membrane of the human retina. A well-defined masement membrane (b) follows the contour of the Müller cell (M), and is closely related to the vitreous fibrils (f) (magnification 36,000) (Hogan et al., 1971). (Copyright 1971 W.B. Saunders, reprinted with permission.)

extending anteriorly beyond the ora serrata and posteriorly to the margin of the optic disc, but there appear to be large gaps between Müller cell end feet (cf. Dreher et al., 1988).

The thickness of the ILM varies considerably from the equator to the posterior retina (Heegard et al., 1986; Foos, 1972). It is highly attenuated at the margin of the optic nerve head, the fovea, and in regions where major retinal vessels are close to the inner retinal surface. Ultrastructural studies show that the ILM in primate retina consists of three distinct layers: the lamina rara externa which is contiguous with the vitreous cortex; the lamina densa which is made up of thin, unbranched filaments embedded in an electron dense (mucopolysaccharide) matrix; and the lamina rara interna which separates Müller cell end feet from the lamina densa. Vitreous fibers attach superficially but do not penetrate the lamina densa. Thin fibrils, more numerous in regions where Müller cell attachment plaques are found, are often seen traversing the sublaminar space (lamina rara interna).

Histochemical, immunocytochemical, and lectin-binding studies have all been used to define components of the ILM, and there is abundant evidence that the ILM consists of collagens, noncollagenous glycoproteins, and proteoglycans (Azuma et al., 1990; Russell et al., 1991; McLoon et al., 1988; Kohno et al., 1987; Sarthy et al., 1990; Sarthy, 1993; Perez and Halfter, 1993; Neugebauer et al., 1991; Newsome and Hewitt, 1985; Hagemann and Johnson, 1984; Halfter et al., 1988). In addition, many well known extracellular matrix (ECM) molecules such as laminin, type IV collagen, vitronectin, thrombospondin, and heparan sulfate are present at the ILM (Kohno et al., 1987, 1983; Sramek et al., 1987; Halfter et al., 1988; Jerdan et al., 1986). However, there is a good deal of interspecies variability in the expression of these molecules, and there is limited information as to their cellular origin, or the degree to which the composition of the ILM influences Müller cell function.

1.1.8. Cytoskeleton

Immunocytochemical techniques have been useful for detecting the cytoskeletal components of glial cells. The cytoplasmic organization of these proteins not only provides insight into cellular function but often serves as a marker or "signature" of a particular cell type that can be useful for routine identification or diagnostic histopathology. Ultrastructural studies show that the Müller cell cytoplasm contains microtubules, microfilaments, and intermediate filaments (Uga and Smelser, 1973). Perhaps the most widely studied cytoskeletal components of Müller cells are the intermediate filament proteins vimentin and glial fibrillary acidic protein (GFAP). Immunocytochemical experiments have demonstrated that vimentin is present in the Müller cells of all vertebrate species. Studies on rat retinas from animals ranging in age from embryonic day 14 to 1 year of age (Shaw and Weber, 1983) showed that vimentin is expressed in the radial fibers throughout development.

In the case of GFAP, however, there is appreciable variability in the results reported from different laboratories. GFAP is present in retinal astrocytes but not usually seen in Müller cells of the normal adult mammalian retina (Eisenfeld et al., 1984), although some investigators have observed GFAP immunostaining throughout the radial processes or in the endfoot region (Huxlin et al., 1995). On the other hand, the Müller cells of lower vertebrates are often GFAP-positive throughout their length (Bignami, 1984; Vaughan and Lasater, 1990; Semple-Rowland, 1991). These differences may be attributed to the antigen-specificity of different antibodies, epitope masking, the sensitivity of the detection method, species differences in GFAP expression, the retinal region studied, or some combination of these factors.

Although there is a paucity of GFAP in Müller cells of the mammalian retina, GFAP expression is strongly upregulated in response to optic nerve crush or penetrating wounds to the globe (Bignami and Dahl, 1979). Comparable changes occur in the dystrophic retinae of animals with various hereditary retinal degenerations (Eisenfeld et al., 1984; Ekstrom et al., 1988; Härtig et al., 1995), as well as after retinal detachment (Lewis et al., 1989). As shown in Fig. 1.16, a similar reaction can be produced experimentally with

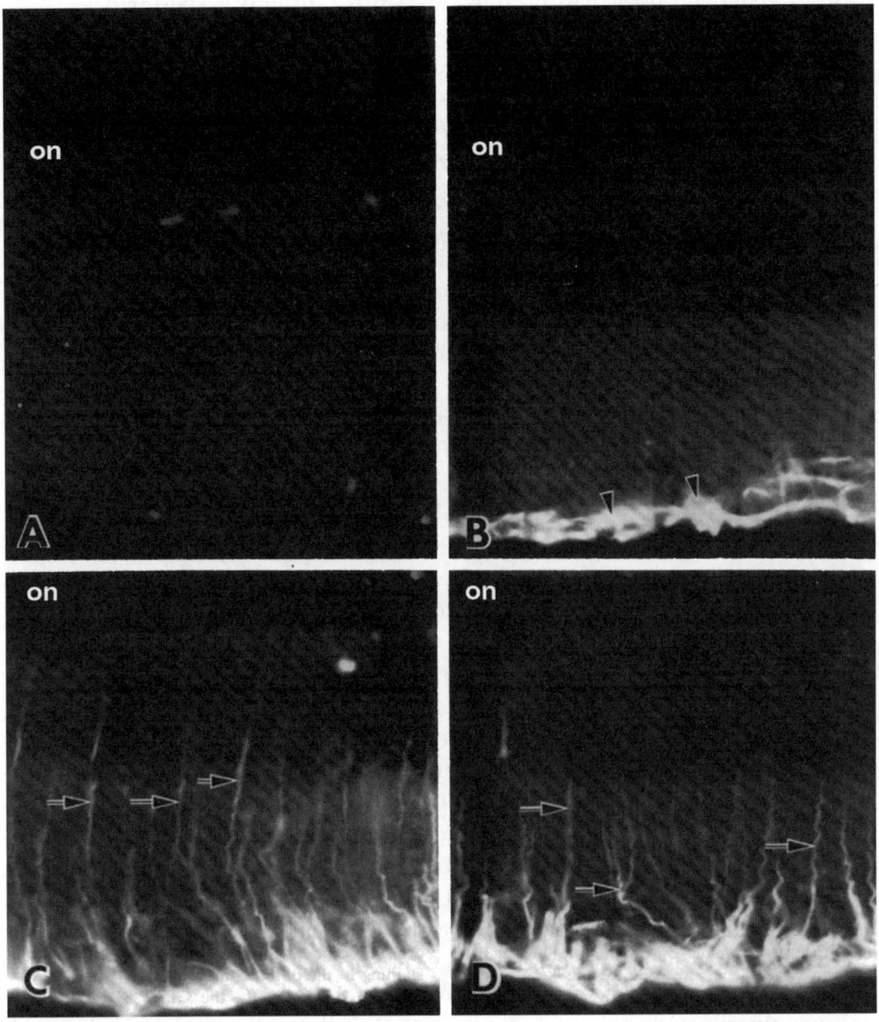

Figure 1.16. Rat retinas labeled with antibodies to GFAP before and after light-induced photoreceptor damage. A. Control section after a 3-day exposure to constant light, and treated with preimmune serum. B. Expression of GFAP in a normal retina shows labeling of astrocytes (arrows). After exposure to constant light for 3 days (C) and 2 weeks (D) GFAP immunoreactivity is seen in the terminals and radial fibers of the Müller cells (Eisenfeld et al., 1984). (Copyright 1984 Association for Research in Vision and Ophthalmology, reprinted with permission.)

light-induced photoreceptor damage (Eisenfeld et al., 1984). The regulation of GFAP expression in response to pathological conditions, and the issue of reactive gliosis will be described more fully in Chapter 6.

1.1.9. Orthogonal Arrays

Numerous aggregates of intramembranous orthogonal arrays of particles (OAPs) were detected initially in freeze–fracture images of astrocytic membranes adjacent to blood vessels and to the fluid surfaces of the brain (Dermietzel, 1973, 1974; Landis and Reese, 1974). The accumulation of OAPs in restricted domains, particularly when apposed to vascular structures, has been interpreted as evidence for a role in metabolic exchange. However, it has not been possible thus far to establish parallels between cellular activities and the distribution of astrocytic particle assemblies (Landis and Reese, 1981a).

The situation is no less uncertain in the case of Müller cells, where the unusual aggregation of particles forming small linear, rectangular, or square patterns (Fig. 1.17) is a prominent feature (cf. Wolburg and Berg, 1988; Gotow and Hashimoto, 1989; Berg-von der Emde and Wolburg, 1989; Richter, Reichenbach and Reichelt, 1990). In the well vascularized retinas of some species (e.g., mouse), Müller cells make extensive contact with blood vessels, and the presence of OAPs in relatively high density at these sites may permit interactions with the vascular system. However, in species with largely avascular retinae (e.g., rabbit, guinea pig), OAPs are found predominantly at the end feet of Müller cells, where they abut the vitreous body (Wolburg and Berg, 1987; Gotow and Hashimoto, 1989). While there is no opportunity for metabolic interaction with the retinal vasculature at the vitreous interface, it has been postulated that the OAPs form ionic channels for release (into the vitreous) of the potassium accumulated by Müller cells in the course of neuronal activity (see Chapter 5). Thus far there is no biochemical or physiological evidence to support the contention that OAPs constitute ion-selective channels. Although both OAPs and K^+ channels are seen predominantly in the endfoot region of some species, the association is too tenuous to link the two structures (Wolburg et al., 1992); potassium channels are distributed at many loci along the Müller cell membrane that lack OAPs.

The possibility that OAPs serve as transmembrane ion channels or represent enzymes of the Müller cell transport system cannot be ruled out, but as yet there is no convincing evidence to associate the arrays with any aspect of Müller cell function. From the appearance of deep-etch freeze–fracture images, Gotow and Hashimoto (1988) have suggested that OAPs provide physical stability against membrane bending at the Müller cell-vitreal interface, but it is not clear why this region is more subject to

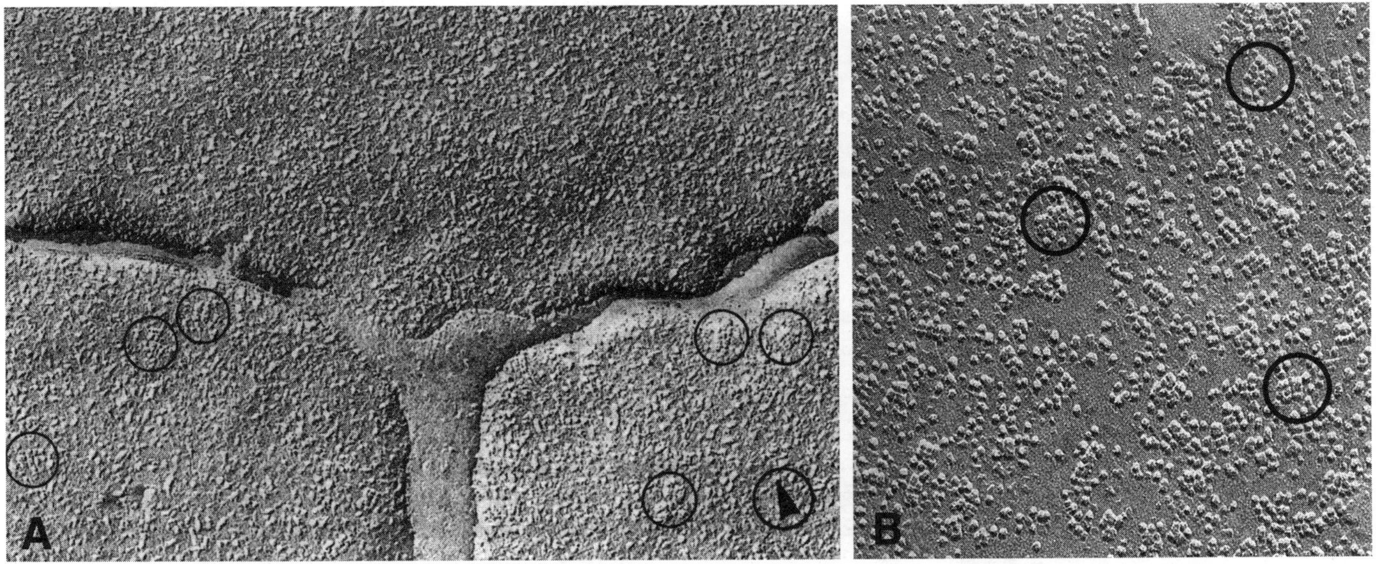

Figure 1.17. Freeze–fracture replicas showing orthogonal arrays of particles (OAP) among the multiple intramembranous particles at the vitreal surface of Müller cell membranes. The particle arrays (circled) are seen in P-face images of two Müller cell endfeet from a peripheral part of the rabbit retina (A), and a Müller cell endfoot in the goldfish retina (B) (Wolburg and Berg, 1987; Berg-von der Emde and Wolburg, 1989). ([A] Copyright 1987 Elsevier Science, reprinted with permission. [B] Copyright 1989 Wiley-Liss, a subsidiary of John Wiley & Sons, Inc., reprinted with permission.)

deformation than, for example, the distal processes that project to the subretinal space. A recent study of transgenic "knockout" mice deficient in AQP4 (aquaporin-4), a protein involved in the formation of water-permeable plasma membrane channels, demonstrated the absence of OAPs in kidney, brain, and muscle (Verbavatz et al., 1997). It has yet to be determined whether a similar loss occurs in Müller cell endfeet, but the similarities in OAP ultrastructure in various tissues suggests this may be the case.

The interest generated by the discovery of OAPs in glial membranes and the degree of speculation it has engendered may seem unwarranted. However, the association of the arrays with vasculature and vitreous, the fact that they are developmentally regulated (Richter et al., 1990), and evidence that their number and distribution can be influenced by circulatory arrest (Landis and Reese, 1981b) and the functional competency of the retina (Bolz and Wolburg, 1992), suggests that further characterization, particularly with regard to the molecular structure of the OAPs, may shed light on their functional significance. A study of the transgenic mice lacking aquaporin-4 would be an ideal place to start.

1.2. ASTROCYTES AND MICROGLIA

1.2.1. Astrocytes

Named for their stellate shape (Fig. 1.18A), astrocytes are confined almost exclusively to the innermost retinal layers. Using Golgi-impregnated whole mounts of retinas from human, baboon, and monkey, Ogden (1978) described two morphologically different forms of astrocyte, both with densely stained, compact, oval perikarya located in the nerve fiber layer. One type appeared elongated and had processes that parallelled the course of the nerve fibers without making specialized vascular contacts. The other type was the classic star-shaped variety, with shorter processes that traversed the nerve fiber bundles to make vascular attachments. In the electron microscope (Fig. 1.18B), these processes can be seen enveloping the blood vessels with apparently no perivascular space between the adventitia of the vessel wall and the glial cells (Hogan and Feeney, 1963; Hogan et al., 1971).

More definitive studies, particularly with respect to the development, organization, and distribution of retinal astrocytes, have since been performed with immunocytochemistry using antibodies against GFAP, a major constituent of astrocytic intermediate filaments (Bignami et al., 1972). Under normal circumstances neither microglia nor mammalian Müller cells contain significant amounts of GFAP. Thus, GFAP immunoreactivity provides a reliable marker for the identification and localization of astrocytes in

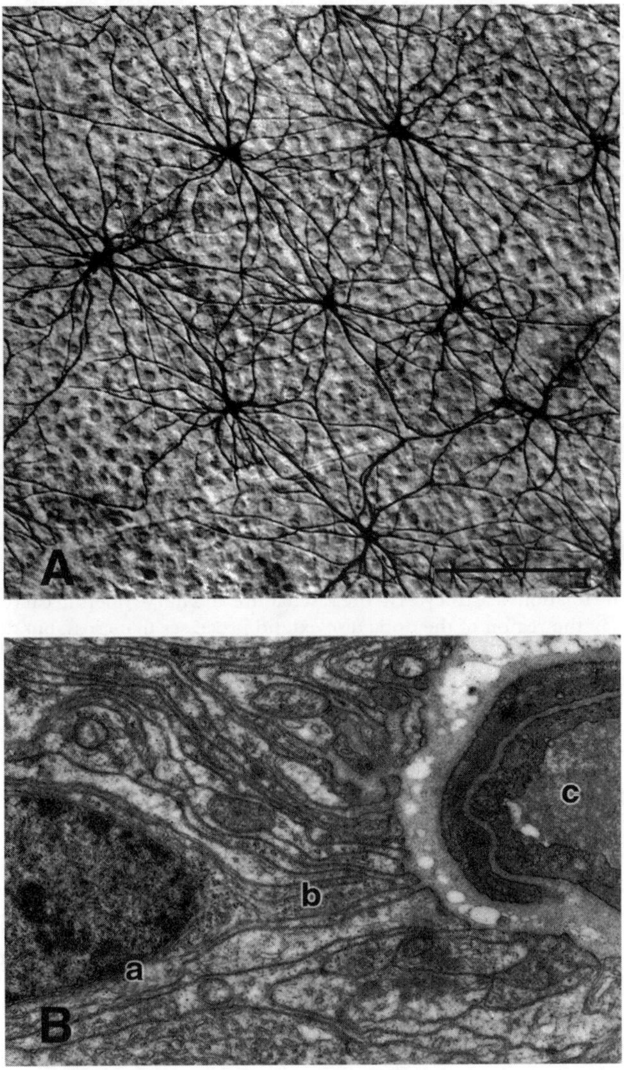

Figure 1.18. A. Light micrograph of GFAP-positive astrocytes in a whole-mount preparation from the peripheral retina of cat. Bar = 100 μm (Karschin et al., 1986). (Copyright 1986 Wiley-Liss, Inc., a division of John Wiley & Sons, Inc., reprinted with permission.) B. Electron micrograph showing the processes of an astrocyte (a) contacting the wall of a capillary (c) (magnification 12,000) (Hogan et al., 1971). (Copyright 1971 W.B. Saunders, reprinted with permission.)

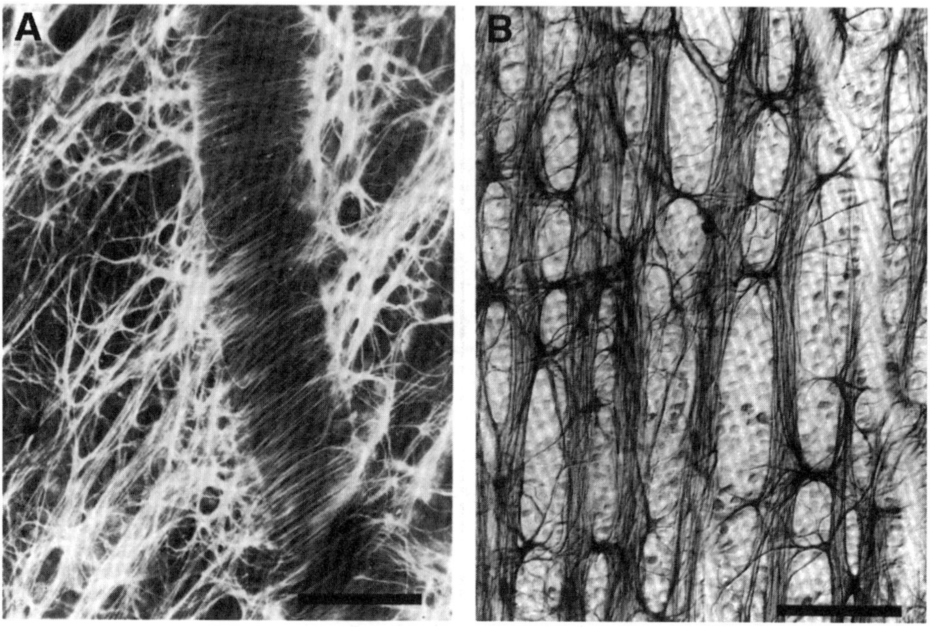

Figure 1.19. The relation of astrocytes to the vasculature and nerve fibers in cat retina. A. Astrocytic processes in the region of the optic disc extend processes that cross, but do not seem to contact, blood vessels. B. Processes of GFAP-labeled astrocytes form thick bundles that are aligned in parallel with the ganglion cell axons of the nerve fiber layer (Karschin et al., 1986). (Copyright 1986 Wiley-Liss, Inc., a division of John Wiley & Sons, Inc., reprinted with permission.)

mammalian retina. Using this method it has been possible to trace the development of retinal astrocytes, and to demonstrate their close association with the ganglion cell axons and vasculature (Fig. 1.19) of the nerve fiber layer of the retina (Karschin et al., 1986; Stone and Dreher, 1987; Schnitzer 1987b).

In the partially vascularized rabbit retina, for example, astrocytes are located in the band of medullary rays, a horizontally elongated group of myelinated axons of ganglion cells that also defines the extent of the vascular area (Stone and Dreher, 1987, Schnitzer, 1987a; Tout et al., 1988). In the retinas of horse (Schnitzer, 1987b; Schnitzer, 1988a) and opossum (Stone and Dreher, 1987), which are almost completely devoid of retinal vessels, the vasculature is confined to the region of the optic disc and the immediately adjacent surrounding area; this is the only region containing astrocytes. Since in these species, as well as in rabbit, ganglion cell axons devoid of astrocytes are present throughout the retina, the findings tend to support the contention that retinal astrocytes relate closely to the presence of intraretinal vessels (Schnitzer, 1987b, 1988b). Evidence that retinal astrocytes differentiate in the optic nerve close to the eye and migrate into the retina later in development (Ling and Stone, 1988; Watanabe and Raff, 1988; Ling et al., 1989; Chan-Ling and Stone, 1991; Sarthy and Fu, 1990), suggests that they may in fact enter the retina with its vasculature (Stone and Dreher, 1987).

Their numbers and distribution, on the other hand, appear to be influenced by the density of nerve fibers, which they invest. In the fully vascularized retina of the cat, for example, astrocytes are distributed throughout the inner retina in a pattern that mirrors closely that of the axon bundles (Karschin et al., 1986; Stone and Dreher, 1987). Moreover, reducing the number of viable ganglion cell axons by photocoagulation results in a reduction in the number of astrocytes within the lesioned area (Karschin et al., 1986). But despite this intimate relationship (Büssow, 1980), the structural development of explanted murine retina is apparently independent of the presence of astrocytes, i.e., explants that lack astrocytes grow, differentiate, and acquire the same neuronal structure as do retina well populated with astrocytes (Huxlin et al., 1992). Relatively little is known of the functional relations between astroglia and the nerve fibers and blood vessels with which they are associated. Because astrocytes are known to produce vascular endothelial growth factor (VEGF), they may be involved in the development of the retinal vasculature (Stone et al., 1995a).

Recently, confocal microscopy and computer-assisted image reconstruction of astrocytes in the vascularized retina of pigs, rats, and cats have provided striking three-dimensional views of astroglial ensheathment of the retinal vessels and their association with ganglion cell axons (Rungger-Brändle et al., 1993). Differences were evident in the disposition of astrocytes among the various species, but a number of general features emerged. Reconstructed images from retinas double-stained for GFAP (astrocytes) and for α-smooth muscle actin or collagen IV (blood vessels) enabled visualization of the asymmetric astrocytic ensheathment of blood vessels. Most notable was the preponderance of GFAP-positive fibers clustering on the vitreal and lateral sides of the blood vessels in close apposition to the vitreal surface of the retina. In addition, this type of confocal imagery enabled the authors to detect individual astrocytes that extended fibers to insert into axonal bundles, while other of its fibers simultaneously contacted the blood vessel wall; in some instances, astrocytic processes extended to the vitreoretinal surface. The authors hypothesized that these elements function as communication links between ganglion cell axons, the retinal vasculature, and the vitreous body (Rungger-Brändle et al., 1993). Although the means

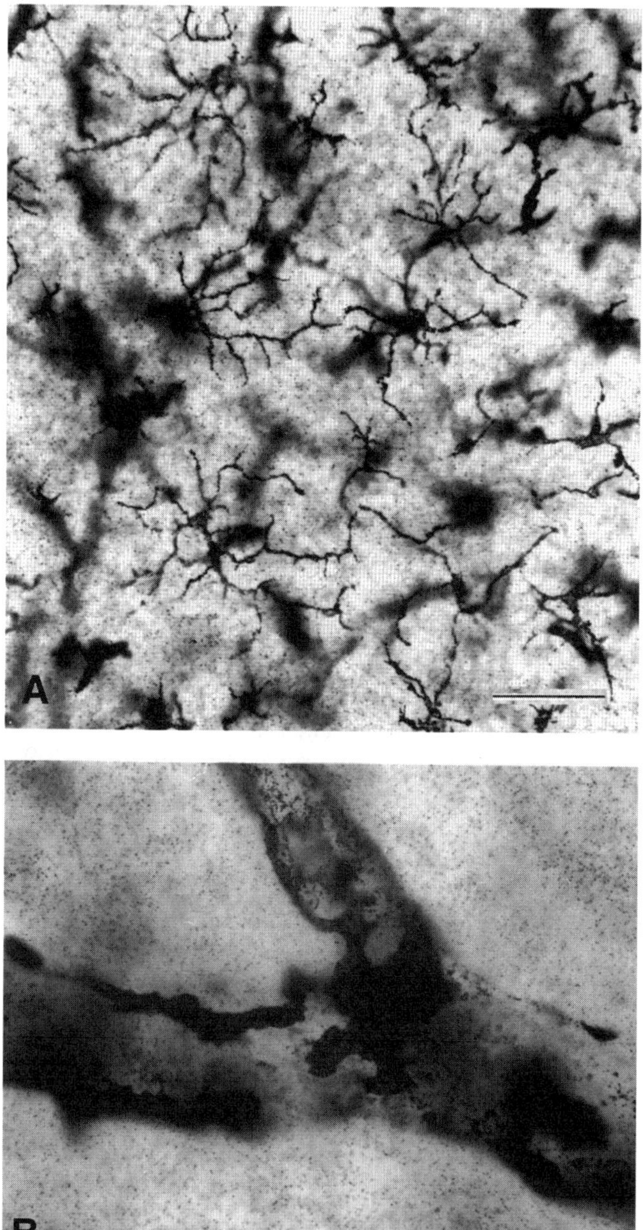
A
B

of communication and the nature of this interaction have yet to be firmly established, the suggestion that this arrangement may provide a system for the regulation of extracellular potassium (spatial buffering) in the inner retina is not without merit. The channel properties of retinal astrocytes (Newman, 1986; Clark and Mobbs, 1992), the communicating (gap) junctions they make with other astrocytes (Marc et al., 1988; Burns and Tyler, 1990; Robinson et al., 1993; Ramirez et al., 1996), and evidence that both the vitreous and blood vessels can serve as a "sink" for K^+ (Newman et al., 1984; Newman, 1986) are consistent with this view.

1.2.2. Microglia

Unlike Müller cells and astrocytes, which are derived from the neuroectoderm, the microglia have a mesenchymal origin similar to endothelial cells and pericytes (Ling, 1981). Microglia are typically small cells with a thin cytoplasmic rim surrounding the nucleus, and short branching processes that often encircle retinal capillaries (Wolter, 1959; Vrabec, 1970); some microglial cells extend fine processes to, or are located within the plexiform layers (Fig. 1.20; cf. Provis et al., 1995; Boycott and Hopkins, 1981; Hume et al., 1983). Their dense cytoplasm contains Golgi complexes and prominent wide cisternae of rough endoplasmic reticulum (Hogan et al., 1971) but few cytoskeletal elements. Although expression of vimentin in CNS microglia has been reported following injury (Graeber et al., 1988), this feature has not yet been studied in the vertebrate retina. Differences in antigenic properties indicate that the retinal microglia consist of a heterogeneous population of cells (Provis et al., 1995), but it is not known whether the various subtypes differ functionally.

The distribution of microglia appears to be age dependent (Ling, 1982), and it has been suggested that changes in their distribution may reflect the pattern of cell death at various stages of retinal development (Hume et al., 1983). In rat retina, microglia are detected as early as embryonic day 12, precursor cells having entered the retina from the blood stream probably via the hyaloid circulation (cf. Ling, 1981; Ashwell et al., 1989).

←

Figure 1.20. Flat mounts of normal human retina showing microglia labeled with antibodies directed against MHC-II antigens and the leukocyte common antigen, CD45. A. CD45 immunoreactive microglia within the inner plexiform layer of the retina illustrates their typical ramified morphology. Scale bar = 100 μm. B. Perivascular microglial cells immunoreactive to MHC-II antibody are closely apposed to a medium caliber vessel of the inner retina. Scale bar = 10 μm (Provis et al., 1995). (Copyright 1995 Wiley-Liss, a subsidiary of John Wiley & Sons, Inc., reprinted with permission.)

However, the notion that the initial entry of microglia into the retina is triggered by the developmentally determined onset of neuronal death (Hume et al., 1983) has not been borne out in studies showing that their appearance in embryonic retina precedes by at least five days the wave of ganglion cell death (Ashwell et al., 1989).

The microglia are usually seen in association with the retinal vasculature (Fig. 1.20). In the monkey (*Macaca mulatta*) retina they are found in all layers from the margin of the inner retina to the outer plexiform layer, where they are closely apposed to the outermost retinal capillaries (Vrabec, 1970). A similar distribution was also seen in the rabbit retina (Vrabec, 1970), but their presence in the plexiform layers of the rabbit cannot be linked to vascular sites because capillaries do not enter the neural retina of this species. These observations have been confirmed and extended by means of light- and electron-microscopic studies of Golgi-impregnated retinas from other mammalian species, e.g., cat, squirrel monkey, and rabbit (Boycott and Hopkins, 1981). However, the distribution of microglia in the mature rat retina appears to be somewhat different. Through the use of peroxidase-conjugated lectins that label selectively the microglia and endothelial cells, it was possible to trace the development of microglia from embryonic to early postnatal stages (Ashwell et al., 1989). In embryonic retina, the microglial cells were indeed seen throughout the thickness of the retina, but as the retina differentiated and a laminar structure began to form, they were progressively confined to the inner half of the retina. It is not yet known whether differences in the retinal distribution of microglia in rat and that reported for rabbit and primates are indicative of species differences, or are due to the different methods and criteria for identifying these cells.

There is general agreement that microglia are analogous to the histiocytes of the CNS, and that they exhibit similar phagocytic properties in response to injury. In the retina, resident microglia proliferate and display "ameboid" motion to engulf and phagocytose the debris of dying cells during the period of neuronal death that accompanies normal retinal development (Potts et al., 1982; Ling, 1982; Hume et al., 1983; Ashwell et al., 1989). Presumably, these wandering scavengers then deliver the lipids of the destroyed neurons to the vascular system for disposal (Wolter, 1959; Jacobiec, 1982). If retinal astrocytes behave similarly to microglia in other parts of the nervous system, it is likely that their phagocytic activity will be stimulated also in response to exogenously induced forms of neuronal injury (cf. Graeber et al., 1988; Streit et al., 1988). The ability of microglia to produce growth factors, e.g., the cytokine interleukin-1 (Giulian et al., 1986), has been taken to suggest that these cells may play a role in the inflammatory response to nerve damage (Brenneman et al., 1992).

Role in Retinal Development

2

Cell–cell interactions play a crucial role in the genesis, determination, and differentiation of neurons and glial cells in the developing nervous system. The interactions are mediated by cell adhesion molecules, growth and neurotrophic factors, and extracellular matrix components. Many of these substances are expressed by glial cells and interact in turn with their receptors on developing neurons or neuronal precursors. Conversely, macromolecular signals arising from neurons appear to influence the mitotic activity and differentiation of glial cells.

The retina has proven to be a useful model for studying questions about cell lineage and mechanisms of cell fate determination in the developing CNS (Adler, 1993). Despite many attempts to identify the role of Müller cells in the development, differentiation, or migration of retinal neurons, little is known about how Müller cells contribute to these processes. Because of their radial orientation, Müller cells have been implicated in cell migration and columnar organization of retinal neurons, and consistent with this idea, Müller cells express molecules that are potentially important in development.

A consideration of the developmental aspects of Müller cells embraces two related lines of inquiry: (1) the origin, birth, and differentiation of Müller cells and the environmental factors that regulate these events; and (2) the degree to which Müller cells influence the development and maturation of retinal neurons. In this context, it is important to recognize that most Müller cells are born relatively late in retinal development and are not likely to participate in the differentiation and migration of retinal neurons that develop at earlier stages. However, Müller cells may play a role in later events such as rod and bipolar cell maturation and synapse formation. It should be mentioned at the outset that although there is a general consensus that the Müller cell is an important player in retinal development, its precise function in the developmental process is largely speculative. Nevertheless, the rapid progress in this field can be expected to help define specific details of neuron–Müller cell interactions in developing retina. This chapter will describe the lineage relationship of Müller cells to retinal neurons, current ideas about the determination of Müller cell fate, and the potential role of Müller cells in development of the retina.

2.1. LINEAGE, BIRTHDATE, AND DEVELOPMENT OF MÜLLER CELLS

In recent years, there has been tremendous interest in understanding lineage relationships in the vertebrate nervous system. The cell lineage can tell us which cells are sisters or cousins. It can also point to the developmental decisions made by a precursor cell, and it can provide clues to the mechanisms that control a specific developmental pathway.

In developing organisms, cell determination has been found to occur by two different schemes: a cell lineage-dependent mechanism and a lineage-independent process. In the lineage-dependent mechanism, cell fate is largely determined by intrinsic instructions provided by the progenitor to the progeny. In this case, the developmental program is either cell-autonomous or occurs by cell–cell interactions that are highly reproducible from one animal to the next. On the other hand, the lineage-independent process involves multipotential precursors in which the developmental fate of the precursor is controlled not by the cell's ancestry, but by environmental or positional cues provided by neighboring cells (Anderson, 1992). These mechanisms are not mutually exclusive. Indeed, the "history" of a precursor may limit its developmental choices.

2.1.1. Lineage Relationship between Müller Cells and Retinal Neurons

Do neurons and glial cells share a common lineage? This question has been long-debated in neurobiology. Three different hypotheses have been proposed to explain cell lineage (Fig. 2.1): (1) Neurons and glial cells derive from different populations of precursors; (2) a precursor population first produces neurons and later gives rise to glial cells; and (3) neurons and glial cells are generated from a common precursor (cf. Purves and Lichtman, 1985). It is only recently that this topic has been examined carefully, and in the case of the developing retina, experimental evidence favors the last hypothesis.

It is worth noting that the lineage relationship between neurons and glial cells was first clearly demonstrated in the mammalian retina. In these studies, the relationship between Müller cells and retinal neurons was examined by following the developmental fate of single precursor cells that were marked with lineage tracers by two different methods. The first method involved injecting a tracer into a precursor cell and subsequently identifying

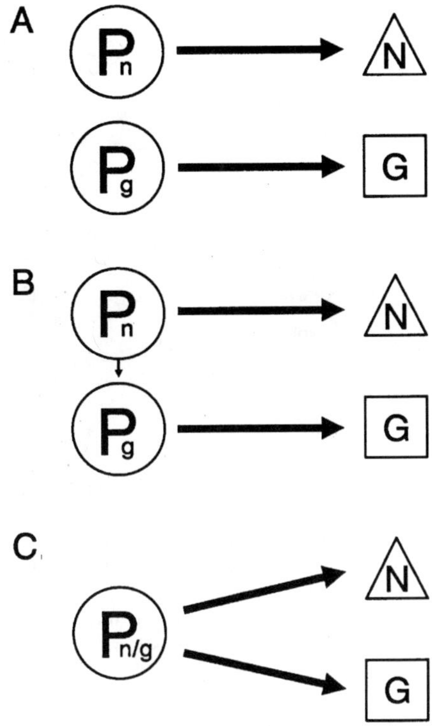

Figure 2.1. Schemes for generation of neurons (N) and glial cells (G). The three major schemes represent distinct pathways: (A) Neurons and glial cells are derived from distinct precursors, neurons from Pn and glia from Pg; (B) Neurons are first generated from a neuronal precursor (Pn) which later switches to a glial precursor (Pg) that gives rise to glia; and (C) Neurons and glial cells are derived from a common precursor (Pn/g). In the mammalian retina, neurons and Müller (glia) cells are derived from a common precursor (Scheme C).

the progeny that retained the tracer in the mature retina (Holt et al., 1988; Wetts and Fraser, 1988; Fraser et al., 1989; Wetts et al., 1989). Intracellular injections of single neuroepithelial cells in the optic vesicle with either HRP or fluorescent-dextran blue showed that an individual progenitor cell can give rise to a clone of radially oriented cells which are located in all layers of the *Xenopus* retina (Holt et al., 1988; Wetts and Fraser, 1988). More important, all the major classes of retinal neurons and the Müller cell were represented in the clone, indicating a common lineage.

The second method for lineage analysis is based on the ability of replication-incompetent retroviruses to infect and integrate into the genome of dividing cells, thereby serving as a permanent genetic marker in the progeny (Cepko, 1988). Studies with a retrovirus vector carrying the

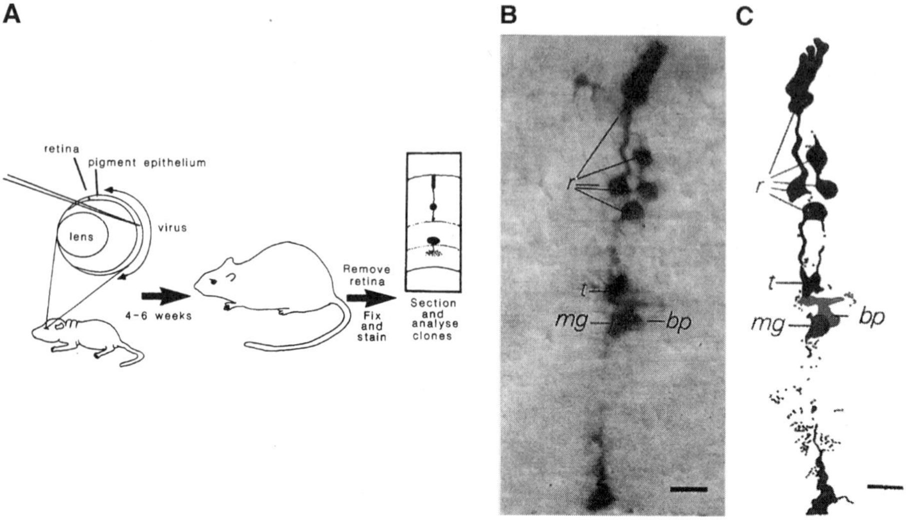

Figure 2.2. Müller cells and retinal neurons are derived from a common precursor. A. Outline of a lineage tracing experiment in which a rat eye is injected with a BAG retroviral vector; B. X-gal-stained retinal section showing a clone containing rods (r), a bipolar (bp), and a Müller cell (mg); C. Camera lucida drawing of the clone (Turner and Cepko, 1987). (Copyright 1987 Macmillan Magazines Limited, reprinted with permission.)

β-galactosidase (*lacZ*) reporter gene showed that all major classes of neurons and Müller cells were present in some clones (Fig. 2.2). Clearly, an individual progenitor can give rise to all retinal cell types (Turner and Cepko, 1987; Turner et al., 1990). Moreover, clones containing Müller cells alone were not observed. Thus, both experimental approaches demonstrate that a single precursor can give rise to all the major classes of retinal neurons as well as the Müller cell.

Although a multipotential precursor cell can give rise to both neurons and Müller cells in the retina, there is reason to believe that neurons and glial cells in the cerebral cortex are derived from distinct precursors fairly early in development (Luskin et al., 1993). It seems that generation of neurons and glial cells does not follow a common plan for the entire CNS, and that the specific scheme used to produce neurons and glial cells might differ in different regions of the CNS.

2.1.2. Lineage Relationship with Astrocytes and Microglia

In addition to Müller cells, the vertebrate retina is populated by other types of glial cells: the astrocytes and microglial cells (see Chapter 1).

Because Müller cells are functionally related to these glial types, one might ask whether the Müller cell shares a lineage relationship with other retinal glia, as it does with retinal neurons.

On the basis of morphological studies in the developing brain, Rakic (1978) suggested that cortical astrocytes could be derived by morphological differentiation of radial glial cells. Similarly, retinal astrocytes have been postulated to be derived from Müller cells or some other precursors in the retina (Ohira et al., 1984; Reichenbach and Wohlrab, 1986). However, several experimental observations suggest retinal astrocytes are not generated in the retina per se. In cell cultures prepared from prenatal retinas, astrocytes could not be detected, although they were readily found in postnatal retinal cultures (Watanabe and Raff, 1988). Moreover, immunocytochemical studies in many species suggest astrocytes migrate into the retina from the optic nerve (Dixon and Eng, 1981; Shaw and Weber, 1983; Kondo et al., 1984; Ling and Stone, 1988; Ling et al., 1989; Sarthy et al., 1991; Huxlin et al., 1992; Verderber et al., 1995). Perhaps the best experimental evidence comes from lineage tracing experiments which show that although clones containing various types of retinal neurons and Müller cells are frequently found, clones containing astrocytes are seldom observed (Turner and Cepko, 1987; Turner et al., 1990).

In the case of microglia, histochemical studies with chimeric animals indicate retinal microglia are of non-neural origin and are likely derived from vascular monocytic cells (Linden et al., 1986; Hickey and Kimura, 1988; Hume et al., 1988). Thus, the three glial types in the retina—Müller cells, astrocytes, and microglia—do not share a common lineage, but are derived from distinct precursors (Fig. 2.3). Müller cells arise from a precursor that can give rise also to retinal neurons; astrocytes are derived from an astrocyte precursor present in the optic nerve; and microglia originate from monocytes.

2.1.3. The Birth of Müller Cells

Although Müller cells share lineage with retinal neurons, their time of birth appears to be different. The birth dates of retinal neurons and Müller cells were determined in mammals, birds, amphibia, and fish by following the fate of ^{3}H-thymidine-labeled cells (Sidman, 1961). In this technique, a brief pulse of ^{3}H-thymidine is administered to the developing retina. Cells in the S phase incorporate the labeled nucleotide into DNA. Cells undergoing a final round of mitosis at the time the label is administered remain heavily labeled, whereas cells undergoing further rounds of mitosis dilute out the label (Sidman, 1961). Therefore, by administering ^{3}H-thymidine at different times during development, it is possible to establish a time span during which a specific cell type becomes postmitotic.

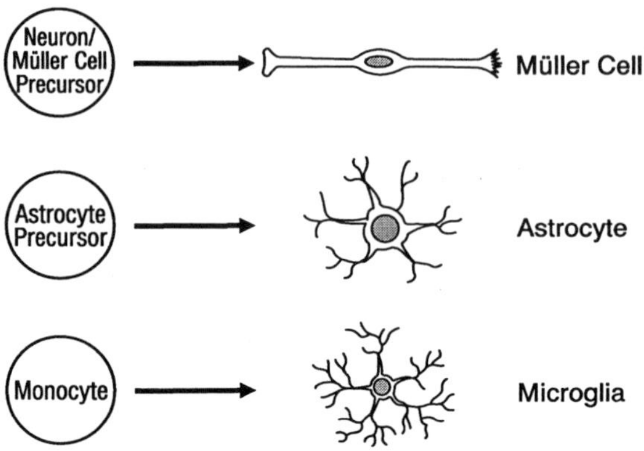

Figure 2.3. Lineages of retinal glia. The three classes of retinal glia are derived from distinct precursors.

The ^{3}H-thymidine labeling studies showed retinal cell types are generated in an orderly fashion (cf. Robinson, 1991; Cepko, 1993). In all retinas examined so far, cells are born first in the central retina and later in more peripheral regions. Moreover, apart from minor differences, the sequence in which specific retinal cell types are generated is largely conserved among vertebrates: Ganglion cells are the first cells to be born followed closely by horizontal cells, cones, and displaced amacrines. At later stages, amacrine cells, bipolars, rods, and Müller cells are generated (Fig. 2.4). The general consensus is that Müller cells are the last cells to become postmitotic.

The ^{3}H-thymidine birth-dating data clearly show that the majority of Müller cells are born late in development. Because cell differentiation usually occurs immediately following cessation of mitosis, it was assumed that Müller cells are also the last cells to differentiate in the developing retina. However, results obtained with ultrastructural and immunocytochemical methods suggest that a population of Müller cells are formed rather early in some species (Bhattacharjee and Sanyal, 1975; Weidman, 1975; Meller and Tetzlaff, 1976; Vrabec, 1983; Prada et al., 1989a, b; Sheffield and Li, 1987; Tout et al., 1989; Lemmon and Reiser, 1983; Willbold and Layer, 1992a, b). For example, developmental studies using immunohistochemistry for the Müller cell-specific marker, vimentin, suggest Müller cells are present at embryonic stages long before the major wave of cytogenesis occurs in the chick retina (Lemmon and Reiser, 1983; Willbold and Layer, 1992a, b).

How does one reconcile the discrepancy between the thymidine-labeling and immunocytochemical data? Several explanations have been offered. One idea is that two waves of Müller cell formation occur, one at

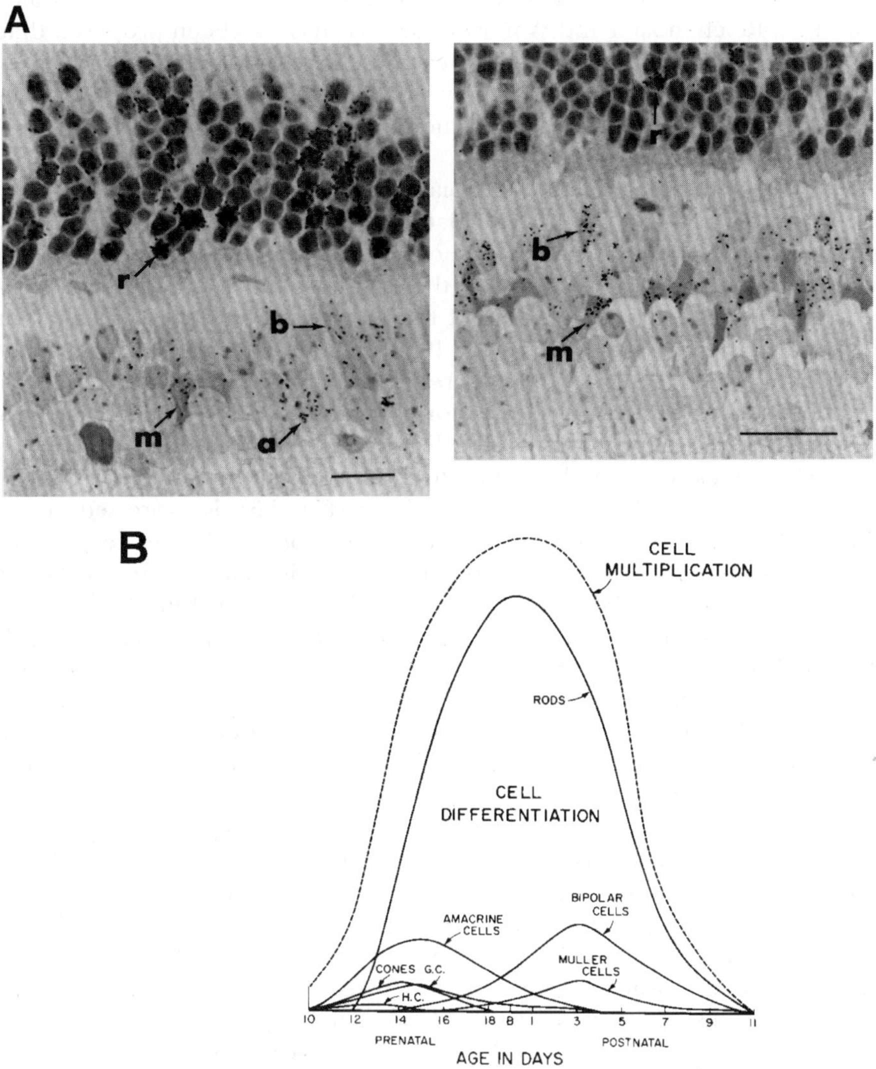

Figure 2.4. Birthdate of Müller cells in the mouse retina. A. Sections showing a ^{3}H-thymidine-labeled rods (r), bipolar cells (b), amacrine cell (a), and Müller cells (m). B. Time course of Müller cell generation relative to that of other cell types in the mouse retina. Müller cells appear to be one of the last cells to become postmitotic in the developing retina (Young, 1985). (Copyright 1985 John Wiley & Sons, Inc., reprinted with permission.)

early stage and a second much later in retinal development. Another explanation is that Müller cells are produced throughout development, but those born early either die or are transformed into other kinds of glia (Ohira et al., 1984; Reichenbach and Wohlrab, 1986). It has also been proposed that Müller cells are born early, but re-enter the cell cycle and are detected in ^{3}H-thymidine labeling experiments only at later developmental stages. At present, there is no unequivocal experimental evidence to support these possibilities (cf. Robinson, 1991).

Perhaps the most plausible explanation for the observed discrepancy between the ^{3}H-thymidine labeling data and immunocytochemical results is that the antigenic markers used are not reliable identifiers of Müller cells in the developing retina. It is possible that retinal (neuron–Müller cell) precursor cells indeed express some of the antigens (Sheffield and Li, 1987; Tout et al., 1999; Seiler and Turner, 1989; Lemmon and Reiser, 1983; Willbold and Layer, 1992a, b), but as the retinal cells differentiate, the antigen is lost in neurons whereas its expression is maintained in Müller cells. This would lead to exclusive antigen localization in mature Müller cells.

There is experimental evidence to support this argument. For example the Müller cell marker, carbonic anhydrase (Fig. 2.5), is expressed by virtually all cells in embryonic tissue, but is found only in Müller cells in the adult chick retina (Linser and Moscona, 1981a). Similarly, a glucocorticoid receptor essential for glutamine synthetase induction, is found on all embryonic retinal cells but is expressed only by Müller cells in the mature chick retina (Gorovits et al., 1994). These examples illustrate the point that some well-accepted Müller cell-specific markers are probably expressed by all cells in the embryonic retina, but their expression becomes restricted to Müller cells in the adult. Therefore, some commonly used "Müller cell markers" may not be useful indicators of Müller cell differentiation during development. It appears that negative regulatory elements acting in conjunction with neuron-specific transcription factors are responsible for turning off the expression of carbonic anydrase and glucocorticoid receptor genes in retinal neurons during or after their differentiation, whereas the genes remain active in Müller cells (cf. Schoenherr and Anderson, 1995). Moreover, cellular retinaldehyde binding protein (CRALBP), the most reliable marker for Müller cells, is not expressed until postnatal stages in the retina, when Müller cells are born. It is likely, therefore, that the ^{3}H-thymidine-birth data are more reliable, and this helps support the idea that Müller cells are generated late in retinal development.

2.1.4. Müller Cell Determination

As previously described, cell determination in developing retina occurs through a lineage-independent mechanism involving multipotential pre-

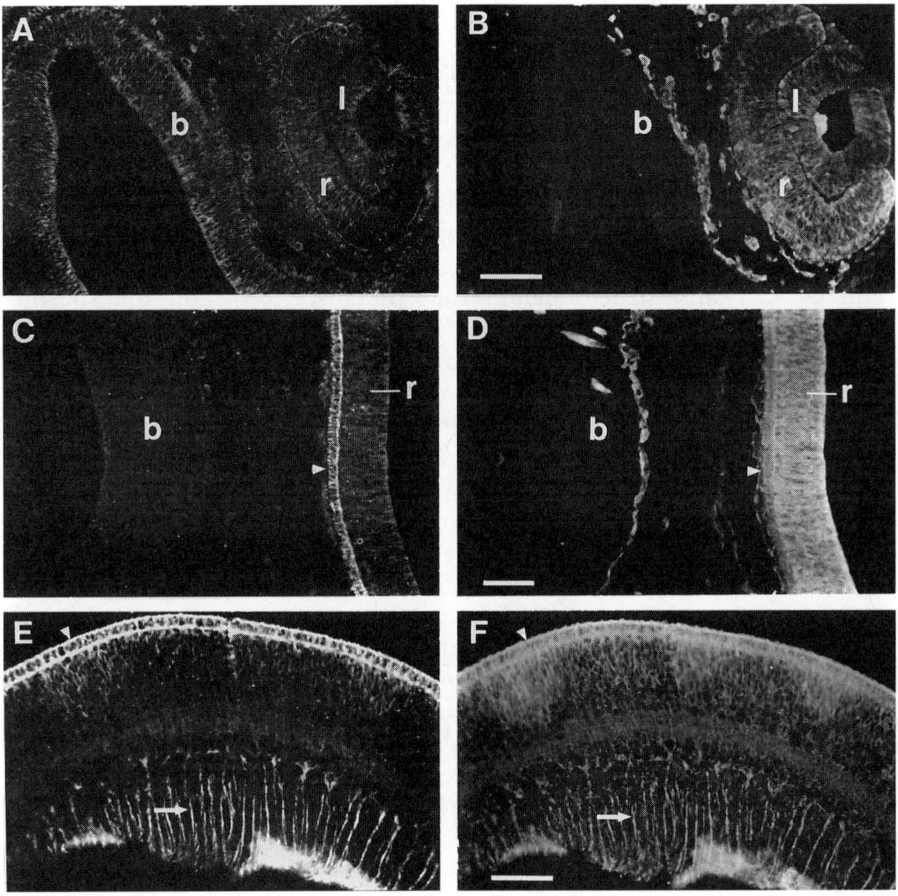

Figure 2.5. Developmental pattern of 5A11 antigen (A,C,E) and carbonic anhydrase (B,D,F) in chick retina. Whereas the two antigens are present in many retinal cells at day 3 (A,B) and day 5 (C,D), by day 11 (E,F), their immunoreactivity is confined to the radial Müller cells (arrows). B, brain; l, lens; and r, retina. Arrows point to Müller cells and arrowheads to the RPE (Linser, 1988). (Copyright 1988 Springer-Verlag New York, Inc., reprinted with permission.)

cursors. Therefore, cell determination may be mediated either by a stochastic process in which a precursor gives rise to progeny with a given probability or by an instructive mechanism in which cell–cell interactions direct a precursor to generate a particular progeny (Cepko et al., 1996). Although retinal cells are generated in a reproducible order, it is not known how individual precursors choose specific cell fates. On the available evidence, it appears that both the intrinsic character of the precursor and extracellular signals in the environment are involved in influencing the developmental choice (Harris, 1997). Moreover, both of these features appear to change as the retina develops.

In their classic study, Reh and Tully (1987) observed that when dopaminergic amacrine cells were selectively destroyed in the frog retina, there was an overproduction of the dopaminergic cells in the regenerated retina by a process involving recruitment of uncommitted precursors (Reh, 1987). This observation and the results of related studies (Cepko, 1993), support the idea that cells decide their fate by interacting with more differentiated neighboring cells. It is likely that a signaling molecule secreted by a differentiated cell can instruct a neighboring precursor to commit to a specific cell fate.

It appears that both cell fate and cell number are largely regulated by instructive cell–cell interactions between differentiated cell types and precursors in the developing retina. Neither birthdate nor lineage appears to be a deciding factor. Indeed, more than one retinal cell type can be generated at any time (Cepko et al., 1996; Harris, 1997).

We have yet to learn when Müller cells become "determined" or what signals drive retinal precursors to differentiate into Müller cells in the developing retina. Indeed, Müller cell determination may depend on extracellular signals, or it might occur in their absence. Some experimental observations indicate neuronal signals can influence Müller cell differentiation. In chick retinal cultures, contact with neurons is required for induction of glutamine synthetase, a cell-specific marker for Müller cells (Linser and Moscona, 1979). Moreover, in retinal cell cultures, glial cell differentiation has been reported to occur in clumps containing neurons and glia, but not in isolated cells with no neuronal contact (Adler et al., 1982).

There is even more compelling evidence that extracellular signals such as growth and neurotrophic factors as well as cytokines, determine whether a precursor differentiates into a Müller cell or into another cell type (Lillien, 1995; Kirsch et al., 1996; Ezzeddine et al., 1997; McFarlane et al., 1998). In retinal cell cultures, high concentrations of TGF-α have been found to inhibit differentiation of retinal precursors into rods; since there is a concomitant increase in the number of Müller cells, it appears that TGF-α levels may decide the differentiation of precursors into Müller cells (Lillien, 1995). In accord with this data, overexpression of the EGF receptor increases the proportion of clones containing Müller cells in the rat retina (Fig. 2.6). These findings suggest that the levels of EGF receptor or its ligand, TGF-α, is naturally limiting and could serve as a potential determinant of Müller cell fate in the mammalian retina.

Growth and neurotrophic factors generally activate signaling pathways that lead to activation or expression of specific transcription factors (Segal and Greenberg, 1996). These molecules in turn direct the expression of cell-

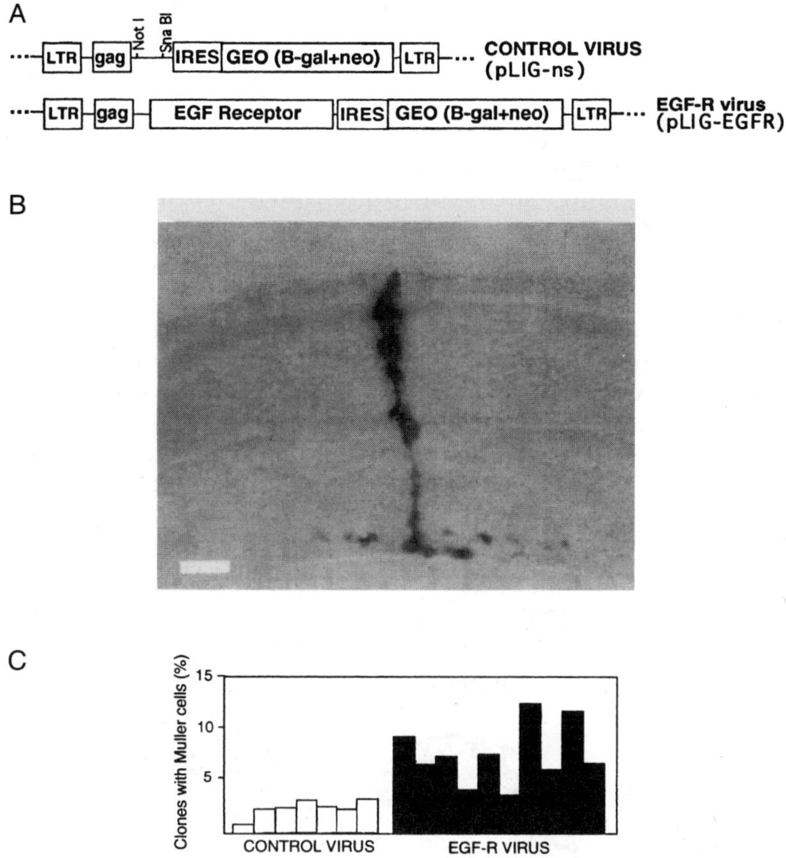

Figure 2.6. Effect of EGF receptor expression on Müller cell development. A. Constructs used to infect cells contained the EGF-R gene and the β-gal marker gene. B. A EGF-R-infected Müller cell. EGFR overexpression increased the number of Müller cells. C. Histogram shows proportion of Müller cells in rat retinas infected at birth. Each bar represents a retina (Lillien, 1995). (Copyright Macmillan Magazine Limited, reprinted with permission.)

specific proteins as well as restrict the choice of cell fate. The transcription factor MASH-1, for example, appears to be restricted to cells that are potential precursors for bipolar cells and Müller cells (Jasoni and Reh, 1996).

At present, we do not know the identity or even the origin of neuronal signals for Müller cell determination. It is not clear whether instructions for Müller cell determination come from multiple neuronal types or are derived from a late-developing neuron, such as the rod or the bipolar. Alternatively, the signal for Müller cell determination could be derived from the

RPE. When perinatal, mouse retinal cultures are treated with Sonic Hedgehog (Shh) protein, which is known to be expressed by RPE *in vivo,* there is an increase in the number of rods and Müller cells suggesting that Shh stimulates proliferation of late retinal precursor cells (Levine et al., 1997; Jensen and Wallace, 1997).

An alternative suggestion is that Müller cell differentiation occurs by default rather than through an instructive pathway, i.e., if a retinal precursor does not differentiate into a neuron its only fate would be to become a Müller cell. The best experimental evidence for Müller cell genesis by a default pathway comes from developmental studies of *Xotch* (Dorsky et al., 1995), the *Xenopus* homolog of *Notch,* which encodes a well-known neurogenic protein found in dividing neuronal precursors (Artavanis-Tsakonas et al., 1999). Dorsky and associates examined *Xotch* expression in developing *Xenopus* retinas in which cells had been prelabeled with BrdU to mark their time of birth. They observed that while most of the cells lost *Xotch* expression, a small number of *Xotch*-expressing, mitotic cells persisted in the inner nuclear layer of postembryonic central retina, even after the majority of cells had become postmitotic (Fig. 2.7). Immunostaining with a glia-specific marker suggested that the *Xotch*-expressing cells were indeed Müller cells. These data indicate that Müller cells are among the last cells to differenti-

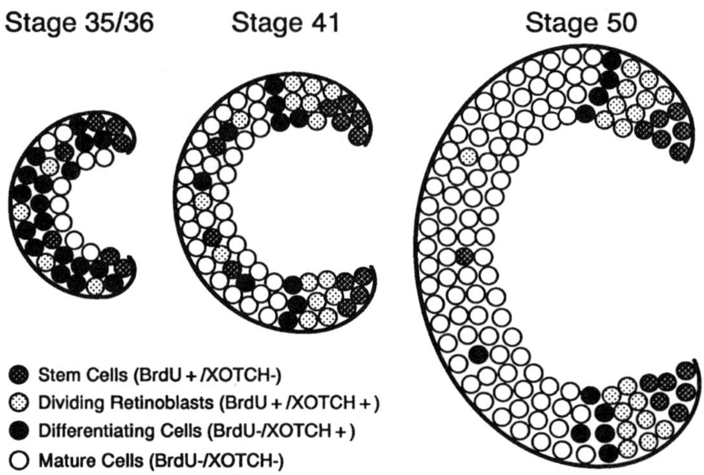

Figure 2.7. Relationship between *Xotch* expression and cell differentiation in the *Xenopus* retina. The proportion of stem cells ($BrdU^+/Xotch^-$), dividing retinoblasts ($BrdU^+/Xotch^+$), differentiating cells ($BrdU^-/Xotch^+$), and mature cells ($BrdU^-/Xotch^-$) varies with the developmental stage. The few cells (located in the inner nuclear layer) that are mitotic and continue to express Xotch ($BrdU^+/Xotch^+$) in stage 50 retinas are likely to be Müller cells. The data support the idea that Müller cells may arise by a default pathway (Dorsky et al., 1995). (Copyright 1995 Cell Press, reprinted with permission.)

ate, and therefore might arise by a default pathway in *Xenopus* retina. These findings are consistent with the general observation that among all retinal cells, Müller cells alone retain the capacity to become mitotic in the adult mammalian retina (see Chapter 6).

Although Müller cells may arise by a "default" pathway in *Xenopus* retina, it is not evident that they are also generated by default in other vertebrate retinas. There are known exceptions. In the goldfish, for example, where rod photoreceptors are the last cells to be generated (Johns, 1982), Müller cell differentiation must be determined by other mechanisms.

From a mechanistic point of view, one can argue whether Müller cell genesis can be called a "default" event. The results of cell culture experiments suggest that a set of genes that is active in progenitors drives them to become neurons of one type or another, when the progenitors are isolated. Therefore, to generate a Müller cell, the neurogenic pathway has to be inhibited. This can be achieved, for example, by continued expression of *Xotch* which activates *Hairy* and enhancer of split homologues and inhibits activity of *Xash* and other neurogenic genes. Based on this mechanism, the Müller cell can be considered to have been generated through the "default" pathway because it was formed in the absence of a neurogenic signal. However, it could also be argued that the formation of neurons occurs by "default" because the progenitors were primed to make neurons, unless *Xotch* was kept activated in them. Nonetheless, it is clear that the formation of Müller cells in the retina is not simplistic and must involve some kind of cell–cell interaction with neurons.

In summary, it is clear that Müller cells share lineage relationship with retinal neurons but not with retinal astrocytes or microglia. The majority of Müller cells withdraw from mitotic activity rather late in retinal development and may be the last cells to differentiate in the retina. It remains to be determined whether Müller cells arise by a default pathway in all vertebrate retinas or arise by an instructive mechanism in some cases (Fig. 2.8). At issue are questions such as which extrinsic factors influence Müller cell fate, what transcription factors determine Müller cell fate, and which retinal cell interactions are crucial. Finally, the signal involved in suppressing Müller cell mitotic activity in the adult retina needs to be identified.

2.2. ROLES IN RETINAL DEVELOPMENT

So far, the discussion has focused on the developmental aspects of the Müller cell. Do Müller cells play an influential role in development of neurons? The role of the Müller cell in retinal development is poorly understood. It is generally assumed that since Müller cells make and secrete many

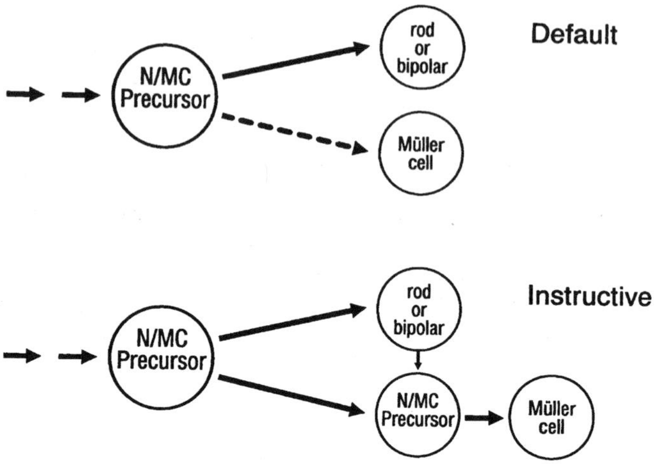

Figure 2.8. Schemes for generation of Müller cells in the vertebrate retina. Müller cell determination could occur by a default pathway, or it could be mediated by a instructive signal coming from a neighboring rod or bipolar cell.

developmentally important molecules, they must have a role in this process. This remains to be established. The following section will discuss potential roles of Müller cells in neuronal migration, columnar development, and retinal histogenesis and will also consider the role of Müller cell-derived extrinsic factors in retinal development.

2.2.1. Neuronal Migration and Retinal Histogenesis

Based on their morphological similarity to cerebellar Bergmann glia and the radial glial cells of the developing cortex, Müller cells have been implicated in the migration of neurons from the ventricular zone into the vitread regions of the developing retina (Meller and Tetzloff, 1976; Silver and Robb, 1979). As discussed at the beginning of the chapter, the majority of Müller cells are formed late in retinal development and their involvement in neuronal migration seems unlikely, at least during early stages of retinal development (Hinds and Hinds, 1978). However, there are special situations in which Müller cells might participate in retinal cell migration.

In the developing retina of the goldfish growth occurs primarily by the addition of cells in a circumferential zone at the margin of the expanding retina. However, rod photoreceptors are continuously added throughout the central retina from a scattered population of precursor cells located in the distal part of the inner nuclear layer. Thus, the rod density does not

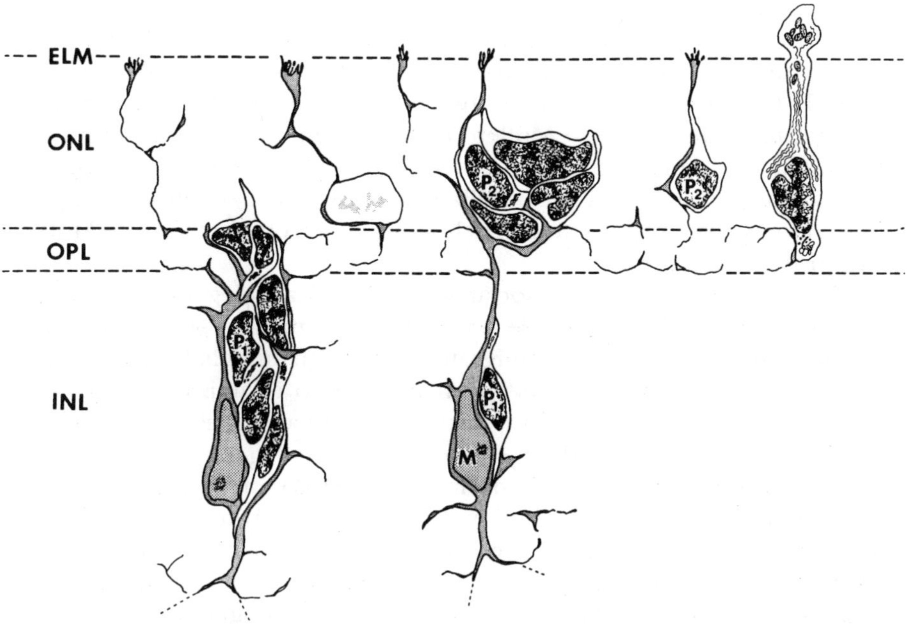

Figure 2.9. A schematic diagram showing ontogenesis of rods in the goldfish retina. The figure shows a migrating rod precursors (P_1) in contact with a Müller cell (M) in developing retina (Raymond and Rivland, 1987). (Copyright 1987 Academic Press, Inc., reprinted with permission.)

change significantly as the retina increases in size throughout the animal's lifetime. Since rods generated from the precursors end up in the outer nuclear layer, the question arises as to how they migrate through the OPL to reach their destination. In an ultrastructural study of developing goldfish retina, Raymond and Rivland (1987) observed that migrating ^{3}H-thymidine labeled rod photoreceptors were in close apposition to Müller cells suggesting a potential guidance of newly formed rods in fish retina (Fig. 2.9).

A role for Müller cells in migration and developmental organization of photoreceptors is also indicated by results from experiments that examined the effects of gliotoxins on developing retina. Rich and associates (1995) looked at the effects of intravitreally injected DL-α-aminoadipic acid (α-AAA), a well known gliotoxin, on photoreceptor development. They found that many regions of the retina were unaltered after treatment. In some regions, however, the photoreceptor layer was disorganized, and clumps of photoreceptor nuclei were present next to the RPE. To explain this observation, the authors suggested that the outer limiting membrane formed by Müller cell–photoreceptor junctions (see Chapter 1) might normally serve as a barrier to prevent aberrant migration of photoreceptors, and disruption of this barrier would result in photoreceptor cell bodies accumulating near the RPE. One problem with these studies is that the toxic effects of α-AAA are probably not limited to Müller cells; the drug may also affect photoreceptors, RPE, and constituents of the interphotoreceptor matrix. An alternate approach would be to utilize an "early" Müller cell-specific promoter to express a cellular toxin such as diphtheria toxin to ablate Müller cells in developing retina.

Mary Hatten and her associates developed an elegant cell culture system for the developing mouse cerebellum that made it possible to study the mechanism of granule cell migration on Bergmann glial cells (Hatten, 1990). Unfortunately, it has not been possible to establish a similar cell culture system in the retina. When dissociated retinal glial cells are plated in culture, they rapidly lose their elongated morphology and transform to "flat" cells. A cell culture method that preserves the radial morphology of retinal Müller cells would be highly useful in testing whether Müller cells have a role in cell migration. Moreover, because there are few Müller cells during early developmental stages of the retina, it is likely that the cell migrations occur on radially oriented processes of progenitor cells.

In addition to their potential role in rod photoreceptor migration, Müller cells have also been implicated in the columnar organization of developing retina. Lineage tracing studies show that retinal cells belonging to a single clone are generally limited to a column in the retina (Holt et al., 1988; Wetts and Fraser, 1988; Turner and Cepko, 1987; Turner et al., 1990). Moreover, studies on cell dispersion patterns in a transgenic mouse strain with the *lacZ* gene integrated into the X chromosome suggest that cells born early in development, such as cones and horizontal cells, can disperse laterally, whereas cells generated later (e.g., bipolars, rods, and Müller cells) tend to stay in a column (Reese et al., 1995). The mechanism that keeps these cells from migrating laterally in the retina is not known, but it is conceivable that the radial processes of Müller cells help to spatially organize the newly formed cells to stack up into columns (see Chapter 1). Moreover, since photoreceptors depend on Müller cells for their energy supply (see Chapter 3), this need may also limit the lateral spread of newly formed rods.

Müller cells have also been implicated in the formation of the nuclear and plexiform layers (Glees and Meller, 1978; Rhodes, 1984; Sheffield and Li, 1984; Robinson and Dreher, 1990), the regeneration of photoreceptor neurites (Fig. 2.10) (Akagawa et al., 1987; Kljavin et al., 1994; Milam et al., 1995), and possibly in the formation of the primate fovea (Robinson, 1991). Recent studies show that glial cells can influence the synaptic activity and number of connections in retinal ganglion cell cultures (Pfrieger and

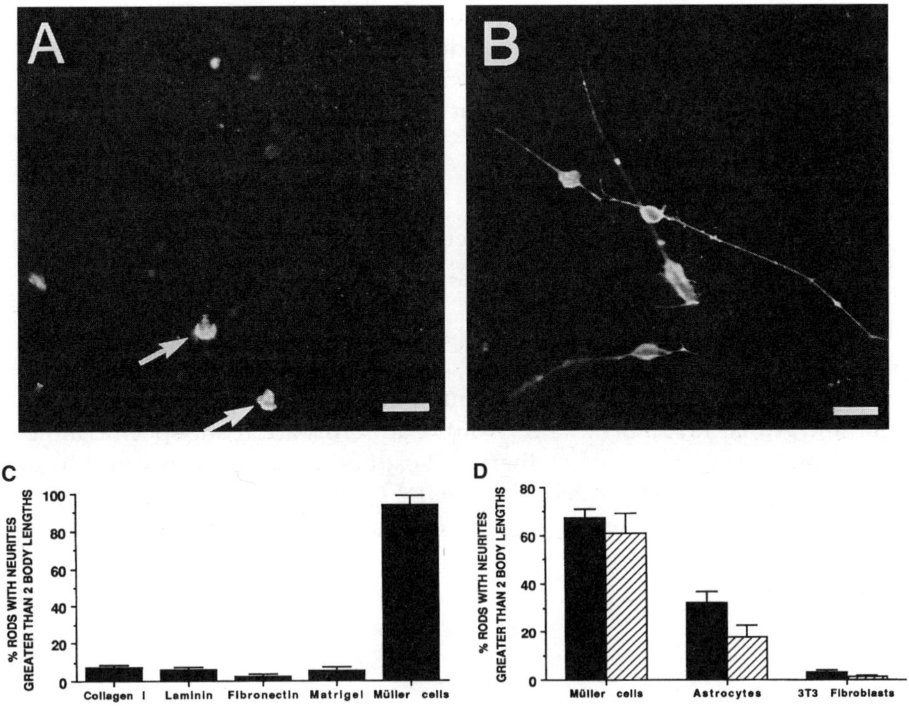

Figure 2.10. Neurite extension by rods on various substrates. Process extension by rods on Matrigel (A) and Müller cell cultures (B). Histogram in (C) shows percentage of rods with neurites on various substrates. (D) Comparison of neurite outgrowth on Müller cells, astrocytes, and 3T3 fibroblasts after 24 hr (solid bar) and 6 days (hatched bar). The results suggest that Müller cells are a preferred substrate for neurite outgrowth (Kljavin and Reh, 1991). (Copyright 1991 *The Journal of Neuroscience*, reprinted with permission.)

Barres, 1996, 1997). Whether Müller cells play a similar role in shaping retinal synapse formation is an exciting possibility that remains to be explored.

2.2.2. Extrinsic Molecules in Retinal Development

It is well recognized that neuronal development is governed by a diverse group of macromolecules which include cell adhesion receptors, growth and neurotrophic factors, and extracellular matrix (ECM) constituents, acting in a concerted fashion. These molecules are expressed at specific stages in the developing retina and are intimately involved in processes such as cell determination, differentiation, and migration of retinal neurons (Harris, 1997). *In situ* hybridization and cell culture studies show that Müller cells synthesize many of these developmentally important molecules, but we know very little about the specific roles of Müller cell-derived molecules in shaping retinal development.

2.2.3. Cell Adhesion Receptors

The interaction of cells with other cells or with the extracellular matrix is generally mediated by a diverse group of cell adhesion receptors that belong to at least four distinct classes: cadherins, Ig family members, selectins, and integrins. Fig. 2.11 presents salient features of these molecules.

The cadherins: These cell adhesion receptors are integral membrane proteins that mediate Ca^{2+}-dependent, homophilic cell–cell binding (Geiger and Ayola, 1992; Grunwald, 1996). Cadherins are tissue-specific molecules, and a retina-specific cadherin (R-cadherin) has been isolated from the chicken retina (Inuzuka et al., 1991). Cadherins are concentrated at adherens junctions where they are connected to the actin cytoskeleton

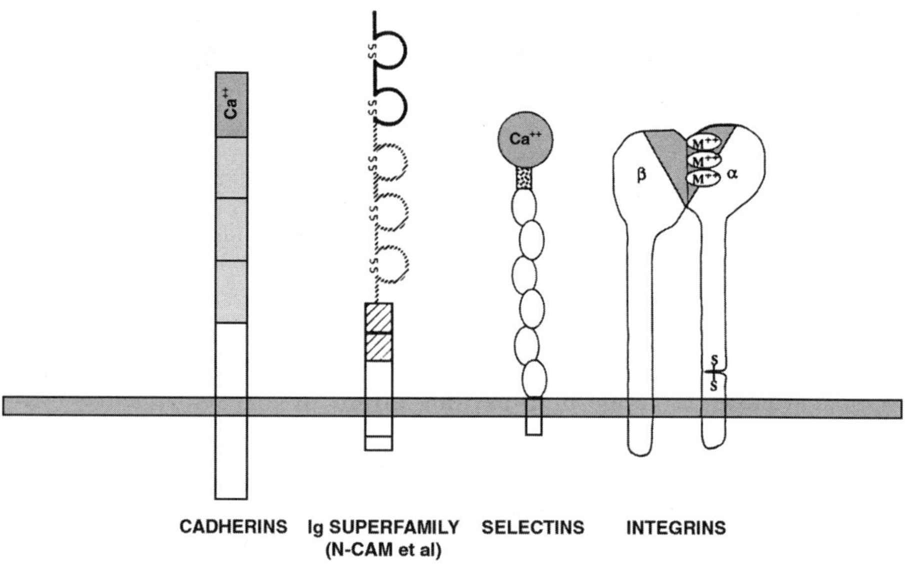

Figure 2.11. Major classes of cell adhesion receptors. Cadherins are Ca^{2+}-dependent, homophilic cell adhesion molecules. The Ig superfamily members contain immunoglobulin domains (loops) and frequently fibronectin repeats (cross-hatched boxes). The selectins contain a Ca^{2+}-dependent C-type lectin domain, and the intergins are heterodimeric receptors made up of α and β chains (Hynes and Lander, 1992). (Copyright 1992 Cell Press, reprinted with permission.)

(Gumbiner, 1993). At these junctions, cadherins interact with cytoskeletal associated proteins such as catenins, α-actinin, and p120cas. This association is required for their cell binding activity. Both cadherins and Ig family members are expressed by neurons and glial cells.

The immunoglobulin (Ig) superfamily: A second class of cell adhesion molecules contain immunoglobulin domains and frequently fibronectin III repeats; and they promote heterophilic and homophilic cell–cell interactions in the absence of Ca^{2+} (Edelman and Crossin, 1991). Many well-known adhesion proteins (such as neural cell adhesion molecule [NCAM], Thy-1, and L1) belong to the Ig superfamily.

Selectins and Integrins: The third type of adhesion receptors, known as selectins, mediate Ca^{2+}-dependent, lectin-like binding among endothelial cells and blood cells (Varki, 1994); their role in the nervous system has not been well studied. The fourth class is comprised of heterodimeric receptors called integrins, which are primarily involved in binding of cells to basement membranes and extracellular matrices (Reichardt and Tomaselli, 1991; Powell and Kleinman, 1997).

The roles of cell adhesion receptors in nervous system development have been extensively studied using genetic as well as cell biological methods. The results indicate that, in general, cadherins and integrins appear to promote neurite outgrowth, growth cone motility, and neuronal migration, whereas the Ig family members are likely to be involved in axonal fasciculation and interactions with other neurons or glial cells (Tessier-Lavingne and Goodman, 1996). Surprisingly, many of the cell adhesion molecules are also present in the adult nervous system, including the retina, which suggests that the molecules might also be required to maintain the structural integrity of adult tissues.

Cell adhesion molecules are found in abundance in the developing retina (Table 2.1), and their importance is supported by several lines of evidence (Buskirk et al., 1980; Rutishauser et al., 1985; Hofmann et al., 1986; Kljavin et al., 1994). Perhaps the best evidence for the involvement of Ig superfamily members and cadherins in retinal development comes from antibody-perturbation experiments. When NCAM and neuron-glia CAM (NgCAM) antibodies are added to reaggregating cultures of embryonic chick retina, retinal histogenesis is disrupted (Buskirk et al., 1980; Rutishauser et al., 1985; Hofmann et al., 1986). Similarly, cadherin-specific antibodies also disrupt histotypic aggregation of dissociated retinal cells (Matsunaga et al., 1988), and recent experiments with dominant negative mutants suggest that cadherin inhibition prevents ganglion cell axon outgrowth in *Xenopus* retina (Riehl et al., 1996).

Cell culture experiments have also examined the effect of NCAM antibodies on neurite outgrowth on Müller cell cultures. Results from these

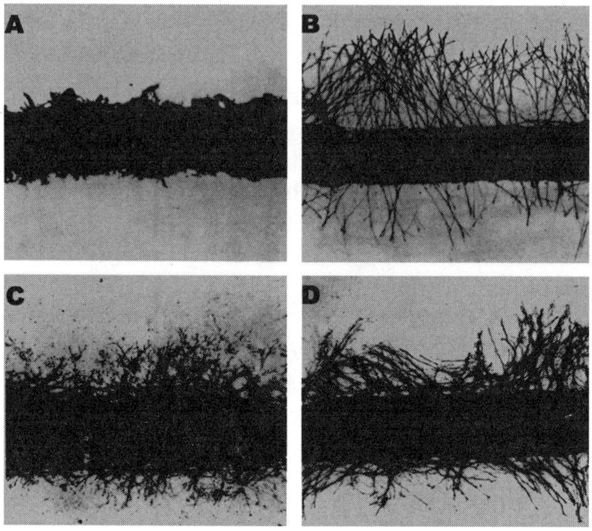

Influence of Antibodies on Retinal Ganglion Cell Axon Outgrowth of Various Substrates

Substrate	Antibody	N	Average inhibition (%) Neurite length	Average inhibition (%) Neurite density
Poly-L-lysine	0.1 mg/ml JG22	3	3	11
Laminin	0.1 mg/ml JG22	3	100**	100**
Müller glia	0.1 mg/ml JG22	6	4	2
Müller glia	1 mg/ml JG22	6	17	13
Fibroblasts	1 mg/ml JG22	6	+7	+16
Laminin	RaC L1	9	21	17
Müller glia	RaC L1	6	30*	44*
Fibroblasts	RaC L1	6	12	23
Laminin	Ra NCAM	3	31	20
Müller glia	Ra NCAM	3	36*	49*
Laminin	Ma N-cadherin	4	27	12
Müller glia	Ma N-cadherin	4	37**	52**

Note. N = number of experiments. *$P < 0.02$, **$P < 0.002$ for ≥ two-thirds of experiments using a paired Student's t test. Data values lacking * had $P > 0.05$. This table summarizes data from over 1000 measurements of both neurite lengths and densities (approximately 18 measurements per experiment). RaC L1, rabbit anti-chick L1; Ra NCAM, rabbit anti-NCAM; Ma N-cadherin, mouse anti-N-cadherin.

Table 2.1. Presence of Retinal Cell Adhesion Molecules in Müller Cells

Cell adhesion molecule	Müller cell	Reference
Bravo/Nr-CAM	? (R)	Kayyem et al., 1992
CD44	+ (R/C)	Chaitin et al., 1994, 1998
F11	+ (R/C)	Willbold et al., 1997; Dalessandri et al., 1995
Gicerin	? (R)	Tsukamoto et al., 1997
N-cadherin	+ (R/C)	Matsunaga et al., 1988
NCALCAM	? (R/C)	Crittenden et al., 1987
NCAM	+ (R/C)	Bartsch et al., 1990; Hoffmann et al., 1986
Neurothelin	+ (R/C)	Schlosshauser, 1991
NgCAM, L1	+ (R/C)	Bartsch et al., 1990; Drazba and Lemon, 1991
R-cadherin	+ (R/C)	Inuzuki et al., 1991
R-cognin	− (C)	Hausmann et al., 1993

R, results with retina; C, data from retinal cell cultures.

studies suggest that NCAM and other Ig family members expressed by Müller cells can influence neurite outgrowth (Fig. 2.12) (Neugebauer et al., 1988; Drazba and Lemmon, 1991). However, retinal abnormalities have not been reported in NCAM- and NgCAM-knockout mice (Cremer et al., 1997; Demyanenko et al., 1999).

Although cell adhesion molecules derived from Müller cells are unlikely to have significant influence on early retinal development, these molecules may be involved in later stages of retinal histogenesis, including synaptogenesis, and in maintaining the structural integrity of the adult retina. Moreover, cell adhesion receptors expressed by Müller cells may play a role in retinal remodeling following injury or degeneration (Dabin and Barnstable, 1995).

In addition to CAMs and cadherins, other proteins have been reported to mediate aggregation of dissociated retinal cells. Their role in retinal-Müller cell interactions is, however, poorly understood. R-cognin, a well-studied retinal protein, is found exclusively on neurons, and it has been suggested that Müller cells express a cognin receptor that mediates Müller

←

Figure 2.12. Neurite outgrowth on various substrates and the effects of antibodies on retinal axonal growth on Müller cells. The figures show growth of neurites from chick retinal explants on polylysine (A), laminin (B), Müller cells (C), and fibroblasts (D). The table shows the influence of antibodies on neurite outgrowth on various substrates. Many of the antibodies suppress neurite extension on Müller cells suggesting the involvement of adhesion molecules (Drazba and Lemmon, 1991). (Copyright 1991 Academic Press, Inc., reprinted with permission.)

cell–neuron interactions (Hausmann et al., 1993). Another Müller cell-protein, the 5A11 antigen (neurothelin), has also been implicated in neuron–Müller cell interactions because a monoclonal antibody against 5A11 has been reported to suppress glutamine synthetase induction in Müller cells, a phenomenon that is dependent on normal neuron–Müller cell interactions (Schlosshauser and Herzog, 1990; Fadool and Linser, 1993; Linser and Moscona, 1979).

2.2.4. Extracellular Matrix Molecules and Integrins

CAMs and cadherins mediate cell–cell interactions, whereas the integrin family of cell adhesion receptors promotes interaction of cells with the extracellular matrix (Fig. 2.11). The ECM includes secreted molecules that are immobilized outside cells, and the major constituents of the ECM are the collagens, noncollagenous glycoproteins, and proteoglycans. Each of these subclasses shows great diversity. Laminin, for example, a well-known member of the ECM glycoproteins, turns out to be a member of a large family of laminin-like molecules (Aumailly and Rousselle, 1999). The func-

Table 2.2. Localization of Some Retinal Extracellular Matrix Molecules

Extracellular matrix molecule	Müller cell	Reference
Agrin	± (R)	Kroger et al., 1995
Claustrin	+ (R)	McCabe, 1992
Collagen I	+ (R/C)	Burke and Kower, 1980
Collagen II	± (R)	Von der Mark et al., 1977
Collagen IV	− (R)	Sarthy, 1993
EAP-300	+ (R)	McCabe, 1992; Kelly et al., 1995
Fibronectin	± (R)	Kohno et al., 1987
Laminin	− (R)	Sarthy et al., 1991
Merosin	± (R)	Morissette and Carbonetto, 1995
S-laminin	± (R)	Hunter et al., 1992
Tenascin	± (R)	Barsch et al., 1992; Perez and Halfter, 1993
Thrombospondin	± (R)	Neugebauer et al., 1991
Vitronectin	± (R)	Neugebauer et al., 1991
Chondroitin sulfate	± (R)	Morris et al., 1987; Snow et al., 1991; Brittis et al., 1992; McAdams and McLoon, 1995
Heparan sulfate	± (R)	Halfter and Schurer, 1994; Chai and Holt, 1997
Keratan sulfate	± (R)	McAdams and McLoon, 1995; Holt, 1997
Decorin	± (R)	Inatani et al., 1999

R, results with retina; C, data from retinal cell cultures; ±, these substances are found in the internal limiting membrane. It is not known whether they are synthesized by Müller cells.

tional diversity of ECM molecules is further broadened by their interactions with other kinds of regulatory molecules. Many cell adhesion molecules like NCAM interact with the ECM. Growth factors such as b-FGF bind to the ECM, and their activity and stability appears to be regulated by interactions with the ECM (Hall and Schachner, 1998).

In the developing nervous system, ECM molecules play a vital role in neuronal migration and axonal growth (Reichardt and Tomaselli, 1991; Letourneau et al., 1994). Many ECM molecules are found in both developing and adult retina (Table 2.2). However, neither the function nor the cellular origin of these molecules is well studied. The ECM molecules are generally prominent at the ILM and could be derived from ganglion cells, astrocytes, Müller cells, or the retinal vasculature, all of which abut or are in close proximity to the ILM. Müller cells appear to be a likely source for some of these molecules. However, *in situ* hybridization studies demonstrate Müller cells do not synthesize either laminin B1 mRNA or collagen IV mRNA during development or in the adult mouse retina (Fig. 2.13; Sarthy et al., 1991; Sarthy, 1993). Whether other ECM molecules such as merosin, vitronectin, and thrombospondin present at the ILM are derived from Müller cells remains to be established. On the other hand, two inhibitory ECM molecules, EAP-300 (embryonic avian polypeptide of 300 kDa) and clausterin (a 320 kDa keratan sulfate proteoglycan), are expressed by Müller cells in the developing retina (McCabe and Cole, 1992). EAP-300 and clausterin may be involved in retinal stratification (McCabe and Cole, 1992).

As described previously, cells interact with the extracellular matrix via the integrin family of cell surface receptors. Integrins are heterodimers made up of α and β subunits each of which belongs to a large subfamily (Reichardt and Tomiselli, 1991; Powell and Kleinman, 1997). Müller cells have been reported to express some integrins (Elner and Elner, 1997). Although we know very little about integrin function in Müller cells, we might speculate that integrins mediate adhesion of the Müller cell end feet to the vitreous collagen (see Chapter 1), and may also be involved in Müller cell migration into epiretinal membranes (Elner and Elner, 1997).

2.2.5. Growth and Neurotrophic Factors

It is widely recognized that growth and neurotrophic factors serve as important signaling molecules in development of the nervous system. Many of the factors present in the developing CNS are also found in the developing retina (Tanihara et al., 1997) (Table 2.3). Some factors such as TGF-α regulate cell proliferation while others appear to promote neuronal differentiation and survival (Harris, 1997). Indeed, there is growing evidence that

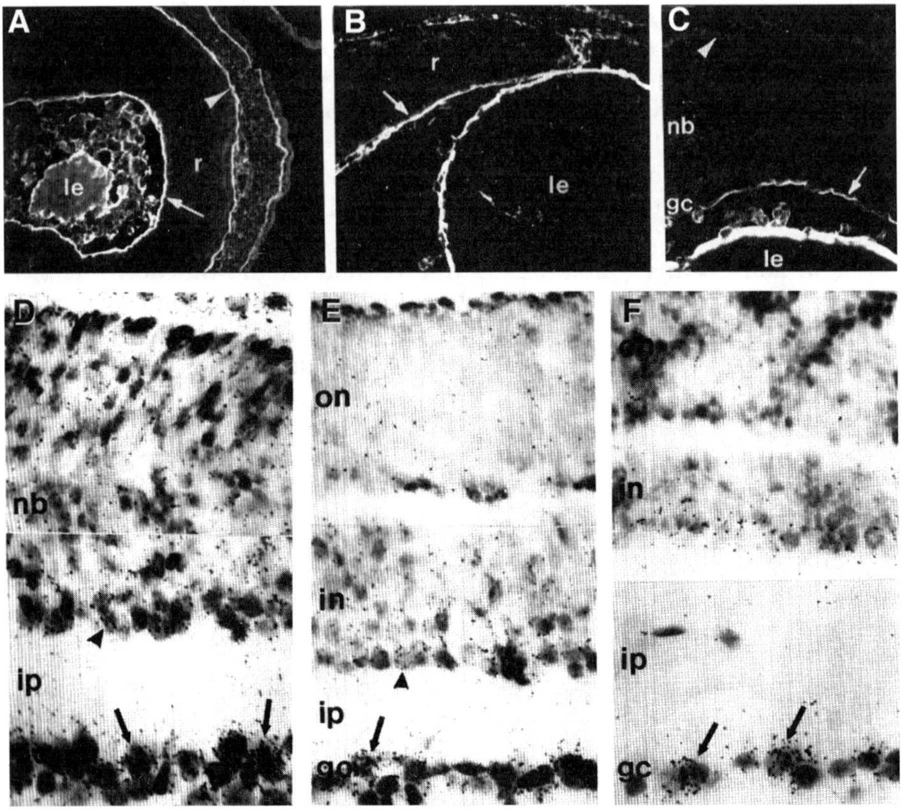

Figure 2.13. Immunolocalization and *in situ* hybridization data for laminin B1 expression in developing mouse retina. Immunostaining shows that laminin is present at the internal limiting membrane in embryonic retina at E12 (A), E-15 (B), and E-20 (C). It is also present, albeit in a smaller quantity, in the adult retina. In situ hybridization data show that laminin B1 mRNA is not present in Müller cells at P3 (D), P-7 (E), or adult retina (F). The data show that laminin is highly expressed at the INL but the site of laminin synthesis is not the Müller cell. Arrows in A, B, C show internal limiting membrane. Arrowhead, RPE. Nb, neuroblast layer; le, lens; and r, retina. Arrows in D, E, F show ganglion cell bodies (Sarthy and Fu, 1990). (Copyright 1990 The Rockefeller University Press, reprinted with permission.)

Müller cells secrete extrinsic factors that can influence neuronal differentiation and survival.

One of the best examples comes from the studies of Neophytou and associates (1997) who observed that serum had a dramatic effect on rod cell differentiation in cultures of neonatal mouse retina. In medium containing 10% fetal calf serum (FCS), very few rods were observed whereas in cultures

Table 2.3. Cytokines, Growth Factors, and Their Receptors in Retinal Müller Cells

	Müller cell	Reference
Cytokines/Growth factors		
bFGF	+ (R/C)	Morimoto et al., 1993; Hageman et al., 1991; Raymond et al., 1992; Gao and Hollyfield, 1992
CNTF	+ (R)	Wen et al., 1995; Cao et al., 1997
γ-interferon	+ (C)	Roberge et al., 1991
Glucagon	+ (R)	Das et al., 1985
IGF I and II	− (R)	Chakrabarti et al., 1990
Insulin	+ (C)	Das et al., 1987
LIF	+ (C)	Neophytou et al., 1997
NGF	+ (R)	Chakrabarti et al., 1990
TGFβ2	+ (R)	Anderson et al., 1995
Receptors		
aFGFR	+ (C)	Mascarelli et al., 1991
bFGFR	+ (C)	Mascarelli et al., 1991
EGF-R	+ (C)	Roque et al., 1992; Lillien, 1995
IGFBP2	+ (C)	Lee et al., 1992
LNGFR	+ (R)	Carmignoto et al., 1991; Takahashi et al., 1993
PDGF-α-R	+ (R)	Mudhar et al., 1993
trk B	− (R)	Perez and Caminos, 1995
Protooncogenes		
fyn	+ (C)	Biscardi et al., 1993
Rek	+ (R)	Fiordalisi and Maness, 1999
yes	+ (C)	Biscardi et al., 1993

R, results with retina; C, data from retinal cell cultures.

with little serum, a large number of rods were found. Moreover, it appeared that when serum was present in the medium, rod development was somehow arrested. Further studies showed that the serum effect was indirect. It was found that serum acted by stimulating Müller cells in the culture to proliferate and release the cytokine, leukemia inhibitory factor (LIF), which in turn arrested rod development (Fig. 2.14). Similar experiments in rat retinal cultures indicate that ciliary neurotrophic factor (CNTF) and LIF inhibit rod development by redirecting rod-precursors to become bipolar cells (Kirsch et al., 1996; Ezzedine et al., 1997). Although these findings point to the exciting possibility that Müller cells regulate rod differentiation, it remains to be shown that CNTF/LIF released by Müller cells has an inhibitory effect *in vivo*.

There are other situations where Müller cell-derived extrinsic factors appear to promote neuronal survival. Müller cell-conditioned medium has

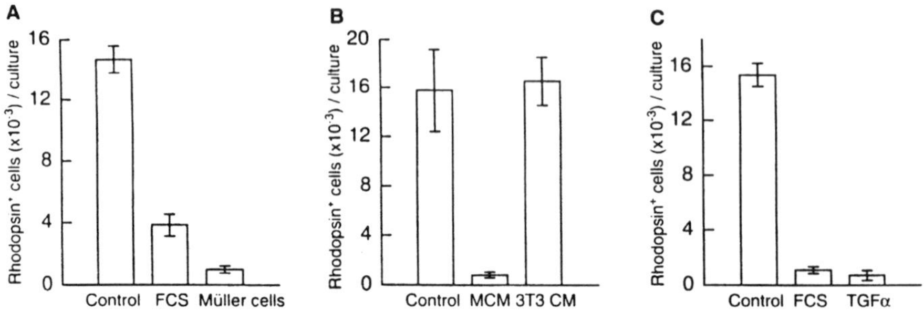

Figure 2.14. The inhibitory effect of Müller cells on rod differentiation. (A) The histograms show the number of rhodopsin$^+$ cells (rods) is decreased in retinal cultures that contain fetal calf serum (FCS) or Müller cells. (B) There is a dramatic decrease in the number of rods in Müller cell-conditioned medium (MCM) but not in 3T3 conditioned medium (3T3CM). (C) the suppression of rod generation in FCS can be mimicked by adding transforming growth factor-α (TGFα) which is a known Müller cell mitogen. It appears that inhibition of rod generation by FCS is an indirect effect, and arises from proliferation of Müller cells in FCS-containing culture medium. Also, the conditioned medium effect could be blocked by antibodies to leukemia inhibitory factor (data not shown) which suggests that LIF released by Müller cells inhibits rod generation (Neophytou et al., 1997). (Copyright 1997 Company of Biologists Ltd., reprinted with permission.)

been reported to support the survival of ganglion cells (Sarthy et al., 1985; Armson et al., 1987), and recent studies show that a combination of growth factors and neuronal stimulation can mimic the effects of the Müller cell-conditioned medium on ganglion cell survival (Meyer-Frank et al., 1995). This exciting observation provides good evidence that Müller cells may produce growth factors that are important for long-term neuronal survival in the retina.

2.2.6. Retinoic Acid

Retinoic acid (RA) has long been recognized as a morphogenetic factor in the developing retina (McCaffrey et al., 1991; Hyatt et al., 1996; Hyatt and Dowling, 1997). Treatment of zebrafish with RA stimulates precocious rod differentiation, and conversely, inhibition of RA synthesis in the eye leads to retarded rod development (Hyatt et al., 1996). In dissociated rat retinal cultures, RA application results in an increase in the number of progenitors that develop as rods (Stenkamp et al., 1993; Kelly et al., 1994). In agreement with these observations, the retinas of double null mice lacking the retinoic acid receptors β2 and γ2 are thinner, and show limited photoreceptor differentiation (Grondona et al., 1996).

The cellular sources of retinoic acid in the developing retina have yet to be identified. Recent studies suggest that Müller cells may be a major source

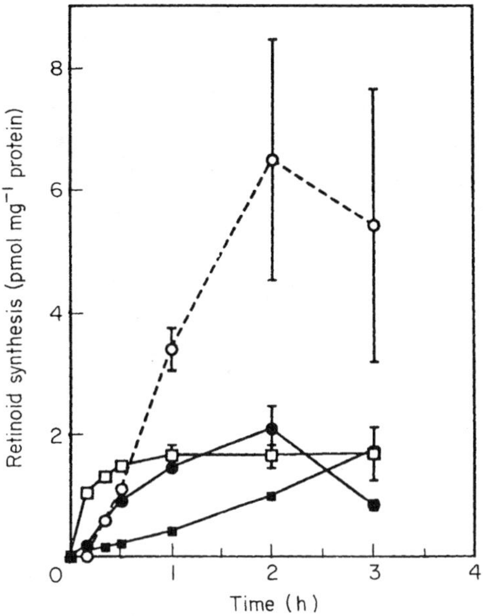

Figure 2.15. Müller cells can synthesize and secrete retinoic acid. Müller cell cultures from rabbit retina were initially incubated with [11,12-^{3}H]all-*trans*-retinol, and the retinoid content of the media and the cells were subsequently analyzed by HPLC. The figure shows time course of accumulation of retinoic acid in medium (○), and retinaldehyde (□), retinyl esters (■), and retinoic acid (●) in cells. The data show that significant amounts of retinoic acid are secreted into the medium (Edwards et al., 1992). (Copyright 1992 Academic Press, Inc., reprinted with permission.)

of this retinoid. Immunocytochemical experiments, for example, show that an aldehyde dehydrogenase (ADH-2) which catalyzes the conversion of retinaldehyde to retinoic acid is present in Müller cells (McCaffrey et al., 1991). Direct evidence that Müller cells can synthesize and secrete retinoic acid was demonstrated in a recent biosynthetic study (Edwards et al., 1992). In that study, Müller cell cultures obtained from adult rabbit retina were incubated with radioactive retinol, and the cells and incubation medium were analyzed by high performance liquid chromatography (HPLC). The results showed that Müller cells synthesized both retinoic acid and retinaldehyde. Although the retinaldehyde was retained within the Müller cells, most of the retinoic acid was rapidly released into the medium (Fig. 2.15).

If retinoic acid is synthesized and secreted by Müller cells in the postnatal retina, Müller cells could exert a significant influence on the development of the late-generated retinal neurons, e.g. rods and bipolars. The observation that *in vitro* rod differentiation is influenced by retinoic acid (Kelley et al., 1994; Stenkamp et al., 1993) is consistent with this idea.

2.3. DEVELOPMENTAL REGULATION OF GENE EXPRESSION

The discussion so far has focused on the potential roles of Müller cells in retinal development, and it is clear that Müller cells produce extrinsic agents that affect neuronal differentiation and survival. Do retinal neurons influence Müller cell differentiation? There is some evidence that the biochemical differentiation of Müller cells depends on contact interactions with retinal neurons. This idea is best illustrated by the pioneering studies of Moscona and his associates who examined the role of neuron–Müller cell interactions in the induction of glutamine synthetase, a Müller cell-specific protein, in the chick retina (cf. Moscona, 1987; Linser and Moscona, 1979, 1983; Linser, 1987; Vardimon et al., 1988, 1983; Gorovits et al., 1994). In the developing chick retina, the levels of glutamine synthetase are fairly low until days 15–16, but the enzyme can be induced precociously by administering glucocorticoids to the embryo (Fig. 2.16). Thus glucocorticoid receptors

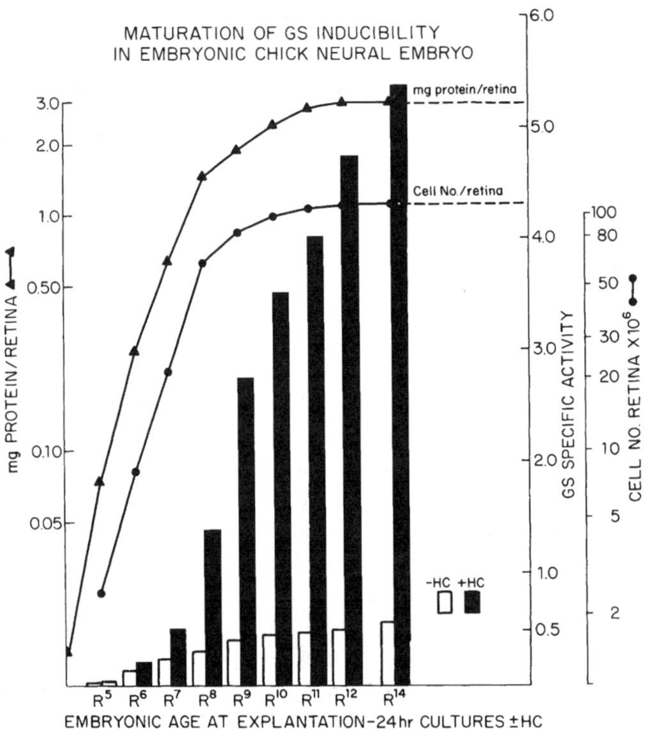

Figure 2.16. Glutamine synthetase induction in developing chick retina. Development of inducibility for glutamine synthetase in chick retina between 5 and 14 days of incubation; relation to embryonic age and to changes in cell number and in total protein per retina. Retinas dissected from embryos were cultured for 24 hr in medium with cortisol (+H) or without it (−HC). The black bars show the levels of GS activity induced in retina of different embryonic ages. The white bars show the levels of GS in the absence of the steroid inducer (Moscona and Moscona, 1979). (Copyright 1979 Springer-Verlag New York, Inc., reprinted with permission.)

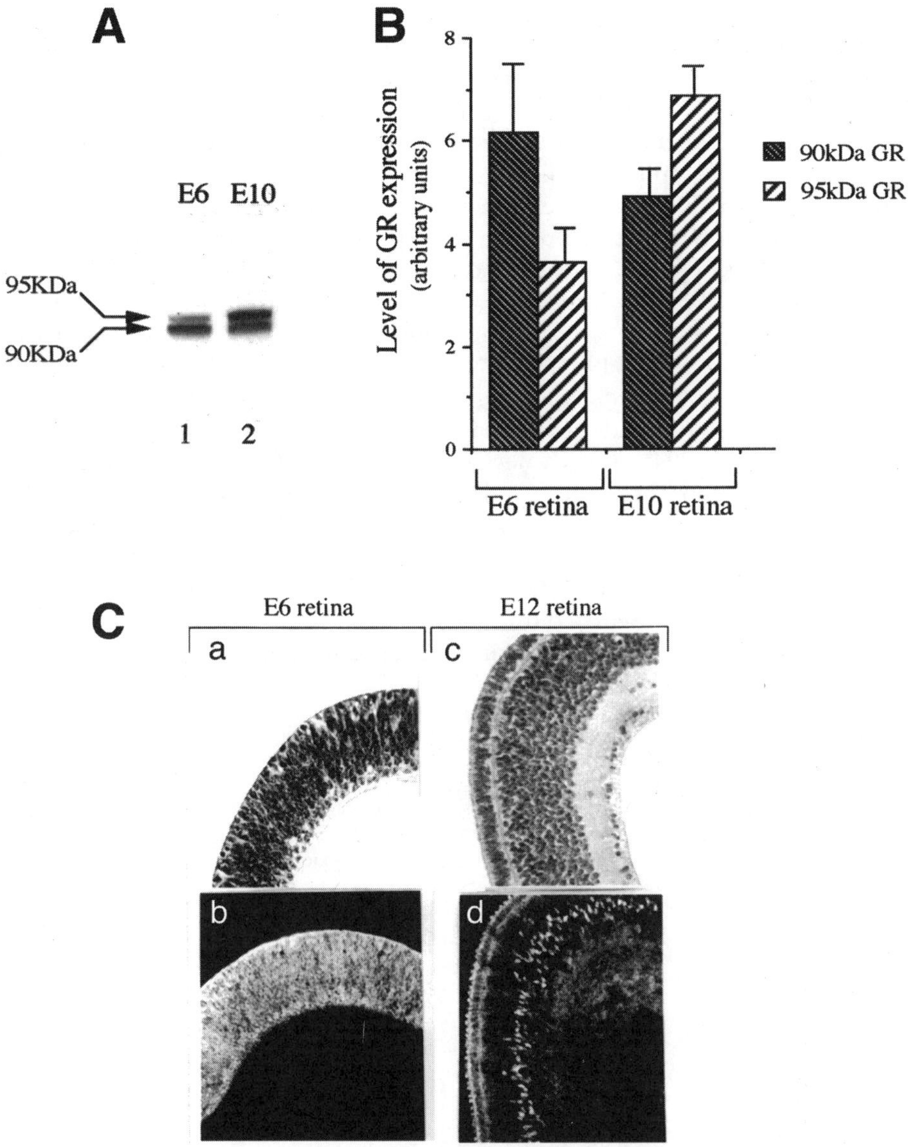

Figure 2.17. Glucocorticoid receptor expression changes in developing retina. A. Immunoblotting studies with GR in E6 and E10 retinas. The blots show nuclear and cytoplasmic distribution of GR isoforms in early and middevelopmental ages. B. The histogram shows differential levels of the GR isoforms in E6 and E10 retinas. C. Immunocytochemical localization of GR in developing chick retina. E6 (A,C) and E12 (C,D). A and C are paraffin sections, and B and D are cryostat sections stained with glucocorticoid receptor antibody. The immunostaining is present in all cells initially but becomes restricted to a small group of cells (and their nuclei) as the retina matures (Gorovits et al., 1994). (Copyright 1994 National Academy of Sciences, U.S.A., reprinted with permission.)

are present in the retina before day 15, but the lack of circulating glucocorticoids prevents induction of glutamine synthetase. Moreover, glutamine synthetase induction in this system appears to accompany Müller cell differentiation.

Although the molecular mechanisms underlying glutamine synthetase induction are complex and not completely understood, experimental evidence suggests that Müller cell-specific expression is achieved through the concerted involvement of positive and negative *cis* elements, a glucocorticoid response element (GRE) and a neural-restrictive silencer (Zhang et al., 1993; Li et al., 1996; Avisar et al., 1999). In addition, it is quite clear that glucocorticoid receptors play an important role in this process (Zhang et al., 1993; Grossman et al., 1994; Gorovits et al., 1994). Results from a recent developmental study indicate that glutamine synthetase expression in chick retina is strongly dependent on the level of a 95 kDa glucocorticoid receptor (Gorovits et al., 1994). At early developmental stages, all retinal cells express the receptor, and consequently all cells express glutamine synthetase. At later developmental stages, however, receptor expression appears to be lost from all retinal cells except Müller cells. As a consequence, glutamine synthetase can be induced only in Müller cells (Fig. 2.17). Interestingly, transcription of the glucocorticoid receptor gene is strongly repressed by the c-Jun protein, which is expressed in proliferating neuroblasts (Vardimon et al., 1999).

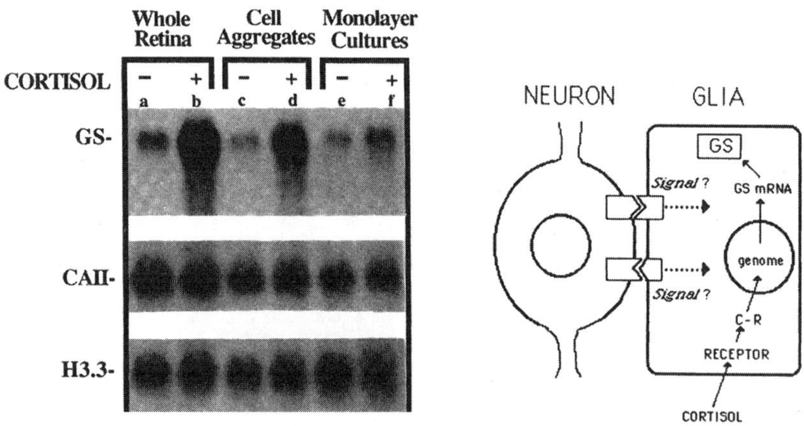

Figure 2.18. Glutamate synthetase induction is dependent on cell contact. Northern blots show that GS mRNA induction is higher in cell aggregates and retina than in monolayer cultures of Müller cells. Note that CAII and H3.3 levels do not change appreciably. The figure to the right shows a proposed model for glutamine synthetase induction in Müller cells (GLIA) (Moscona and Vardimon, 1988). (Copyright 1988 Wiley-Liss, Inc., a subsidiary of John Wiley & Sons, Inc., reprinted with permission.)

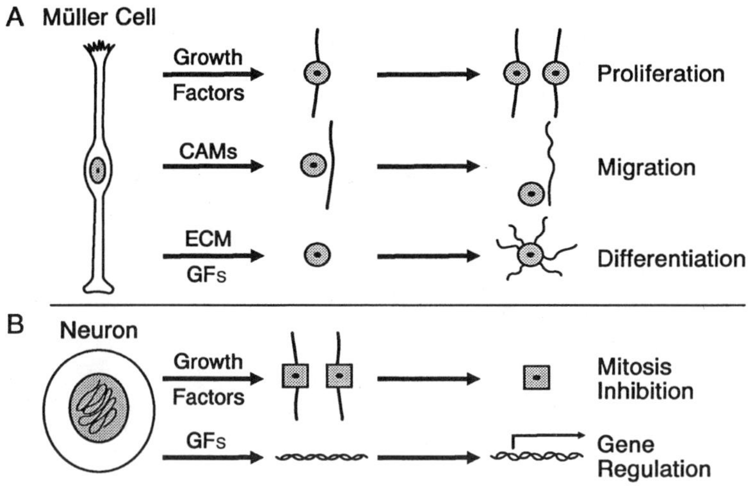

Figure 2.19. Potential Müller cell–neuron interactions in developing retina. A. Influence of Müller cells on the proliferation, migration, and differentiation of neurons. These processes are mediated by growth factors, cell adhesion molecules, and extracellular matrix molecules; B. Neuronal influence on Müller cell activity. Neurons may regulate the mitosis and gene activity in glial cells.

Perhaps the most remarkable feature of the system is that contact interaction with neurons appears to be essential for glutamine synthetase induction in Müller cells. When retinal cultures containing both neurons and Müller cells are treated with glucocorticoids (cortisol), glutamine synthetase is strongly induced in Müller cells (Moscona, 1983). However, if Müller cells are grown in monolayer cultures in the absence of neurons, hormone treatment fails to induce glutamine synthetase (Fig. 2.18). Although the mechanism underlying this phenomenon is still not established, it appears that disruption of neuron–glia contact leads to activation of the c-Jun signaling pathway which in turn inhibits transcription of glucocorticoid receptor gene (Reisfeld and Vardimon, Vardimon et al., 1999). In the absence of the receptor, glutamine synthetase induction becomes nonresponsive to glucocorticoids.

Neuronal interaction is apparently not a universal requirement for gene expression in Müller cells (Fig. 2.19). Although glutamine synthetase induction is dependent on neuronal contact, synthesis of CA II, filamin, or CRALBP is not affected by the absence of neurons (Linser and Moscona, 1981; Hicks and Courtois, 1986; Lewis et al., 1988; Lemmon, 1986; Sarthy et al., 1998; Roque et al., 1997). Clearly, the expression of many Müller cell-specific proteins does not require contact with neurons.

Metabolic Interactions with Neurons 3

Glial cells have long been considered as providers of glucose and other nutrients to neurons (Golgi, 1886). This idea stems from the observation that, in the brain, glycogen granules are normally found in glial cells. Furthermore, because glial processes are interspersed among brain capillaries and neurons, it was thought that they served as a conduit for passage of glucose from capillaries to neurons. Early experimental evidence for metabolic coupling between neurons and glial cells came from the studies of Hydén and Lange (1960, 1962) who used manually isolated cells to demonstrate that neuronal stimulation caused changes in the activities of metabolic enzymes in neighboring glial cells. Subsequent work revealed that in addition to energy metabolism, glial cells participate in many biochemical activities in the nervous system and display complex metabolic interactions with neurons. Indeed, normal glial functioning is essential for neuronal health and survival, and glial cell dysfunction can lead to severe neurological conditions.

Because the Müller cell is the main support cell in the retina, it is not surprising that it plays an active role in a number of metabolic processes vital to normal retinal function. In this regard, its metabolic activities are analogous to those of astrocytes in the brain. This chapter deals with the role of Müller cells in retinal energy metabolism, neurotransmitter degradation, pH regulation, and retinoid metabolism.

3.1. ENERGY COUPLING

The retina is a metabolically active tissue that utilizes a great deal of energy to support processes such as phototransduction, maintenance of ionic gradients, and synaptic activity (Graymore, 1970; Sickel, 1972; Winkler, 1981; Ames et al., 1992). Its energy demands are normally met through the uptake of glucose and oxygen derived from the choroidal or retinal circula-

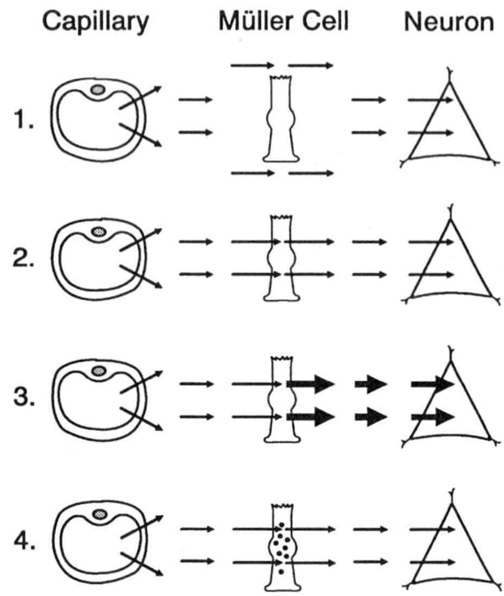

Figure 3.1. Pathways by which glucose (arrows) released from capillaries can end up in neurons. The first pathway does not involve the Müller cell. In the second pathway, glucose is taken up by the Müller cell and released for neuronal uptake. In the third pathway, glucose enters the Müller cell where it is converted to a metabolite, such as lactate (heavy arrows), and is released for use by neurons. In the final pathway, glucose is taken up by the Müller cell and converted to glycogen (dots) for storage; subsequently, the glycogen is broken down to glucose and released for uptake by neurons.

tion. How does glucose released from capillaries end up as carbon fuel in retinal neurons? Do Müller cells serve as an intermediary in this process?

There are several routes by which glucose released from capillaries can end up in neurons (Fig. 3.1; cf. Coles, 1996). One route would be direct transfer of glucose from capillaries to neurons without any Müller cell involvement; other routes can involve active participation of glial cells either as a conduit for glucose transfer or as a site for glucose storage and metabolism before release to neurons. At least four pathways can be invoked to describe glucose transfer (Coles, 1996; Fig. 3.1).

In pathways one and two, glucose does not undergo any change, whereas in pathways three and four glucose is converted to a metabolite in Müller cells. In the first pathway, glucose can diffuse away from capillaries and be taken up directly by neurons without intervention of Müller cells. In the second pathway, Müller cells can take up glucose from blood and redistribute it to neurons. In the third pathway, glucose taken up by Müller cells can be converted to a metabolite, which is released for use by neurons.

And in the fourth pathway, glucose can be transformed to glycogen in Müller cells and later broken down to provide glucose for neurons. These pathways are not mutually exclusive. Although none of the pathways can be ruled out completely, the available experimental evidence indicates that in the CNS, glucose is taken up by glial cells, converted to a metabolite, and subsequently released for neuronal use (Magistretti et al., 1999).

3.1.1. Glucose Uptake and Neuronal Activity

If Müller cells provide neurons with energy metabolites, they must first take up glucose in quantities sufficient for their own needs and for the retinal neurons they serve. The cellular sites of glucose uptake have been determined in a number of vertebrate retinas by autoradiographic studies using ^{3}H-2-deoxyglucose (^{3}H-2-DG), a nonmetabolizable analog of glucose. Early studies showed that photoreceptors are able to accumulate 2-DG (Basinger et al., 1979; Witkovsky and Yang, 1982; Sperling et al., 1982) irrespective of whether ^{3}H-2-DG was supplied in an incubation medium or was injected intravitreally. Based on these data, it appeared glucose could be directly taken up by photoreceptors in some species.

In contrast, Poitry-Yamate and Tsacopoulos (1991; 1992) found that when isolated guinea pig retina was incubated with ^{3}H-2-DG, labeling was confined to Müller cells; photoreceptors and other retinal neurons were not labeled. The authors further demonstrated that freshly dissociated Müller cells could take up and phosphorylate 2-DG (Fig. 3.2). This result, and evidence that the glucose transporters, GLUT1 and GLUT2, are localized to the apical processes of Müller cells (Mantych et al., 1993; Watanabe et al., 1994), are a good indication that Müller cells are capable of avidly taking up exogenous glucose.

Further support for the importance of glucose uptake comes from experiments showing that neuronal activity can stimulate glucose uptake by glial cells (cf. Tsacopoulos and Magistretti, 1996). How does the glial cell sense the energy needs of the neurons? There is growing experimental evidence that synaptically released glutamate serves as the signal that links neuronal activation and glial uptake of glucose. Pellerin and Magistretti (1994) have shown directly that the addition of 0.2 mM glutamate strongly stimulates glucose uptake by astrocyte cultures (Fig. 3.3).

Surprisingly, the action of glutamate in promoting glucose uptake does not involve glutamate receptors but is mediated by the glial glutamate transporters (Pellerin and Magistretti, 1994; Takahashi et al., 1995). Glutamate-induced glucose uptake appears not to be affected by glutamate receptor antagonists, but is inhibited by preincubating astrocyte cultures with the glutamate transport inhibitor, DL-threo-β-hydroxyaspartate (THA) (Pel-

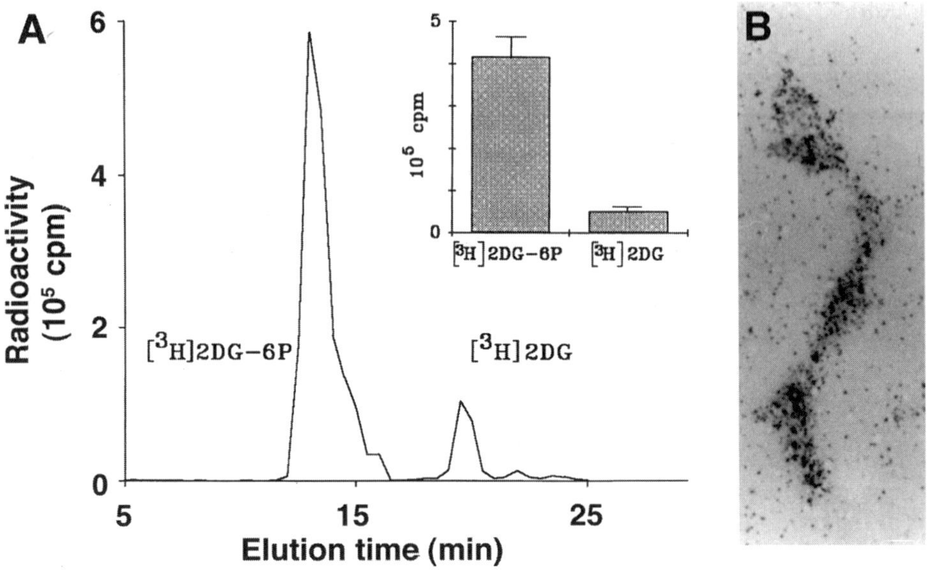

Figure 3.2. Glucose uptake by Müller cells. A. Chromatographic profile showing intracellular [^{3}H] 2DG and [^{3}H] 2DG-6P in Müller cells incubated with [^{3}H] 2-deoxyglucose (2DG). Note that >80% of ^{3}H-radioactivity is associated with 2DG-6P. B. Autoradiographic localization of ^{3}H-radioactivity in a dissociated Müller cell (Poitry-Yamate and Tsacopoulos, 1992). (Copyright 1992 John Wiley & Sons, Inc., reprinted with permission.)

lerin and Magistretti, 1994). Uptake is also blocked by the Na^+/K^+ ATPase inhibitor, ouabain (Takahashi et al., 1995). Moreover, treatment with monensin, a Na^+-ionophore, was shown to stimulate 2-DG uptake into astrocytes (Yarowsky et al., 1986).

The following sequence of events has been proposed to explain how glucose uptake and utilization might be triggered by the influx of glutamate. Glutamate transport, which is coupled to the inward movement of sodium ions, results in a significant elevation in intracellular [Na^+] (Kimelberg et al., 1993). The increase in $[Na^+]_i$ leads in turn to activation of Na^+/K^+ ATPase, resulting finally in stimulation of glucose uptake and glycolysis (Fig. 3.3). Since glutamate is the neurotransmitter released by photoreceptors (Massey, 1990), it is likely that glucose uptake into Müller cells is linked to photoreceptor synaptic activity. Is the glucose taken up by glial cells transferred to neurons or is it first converted to other substrates before transfer? The fate of glucose taken up by glial cells will be considered in the following sections.

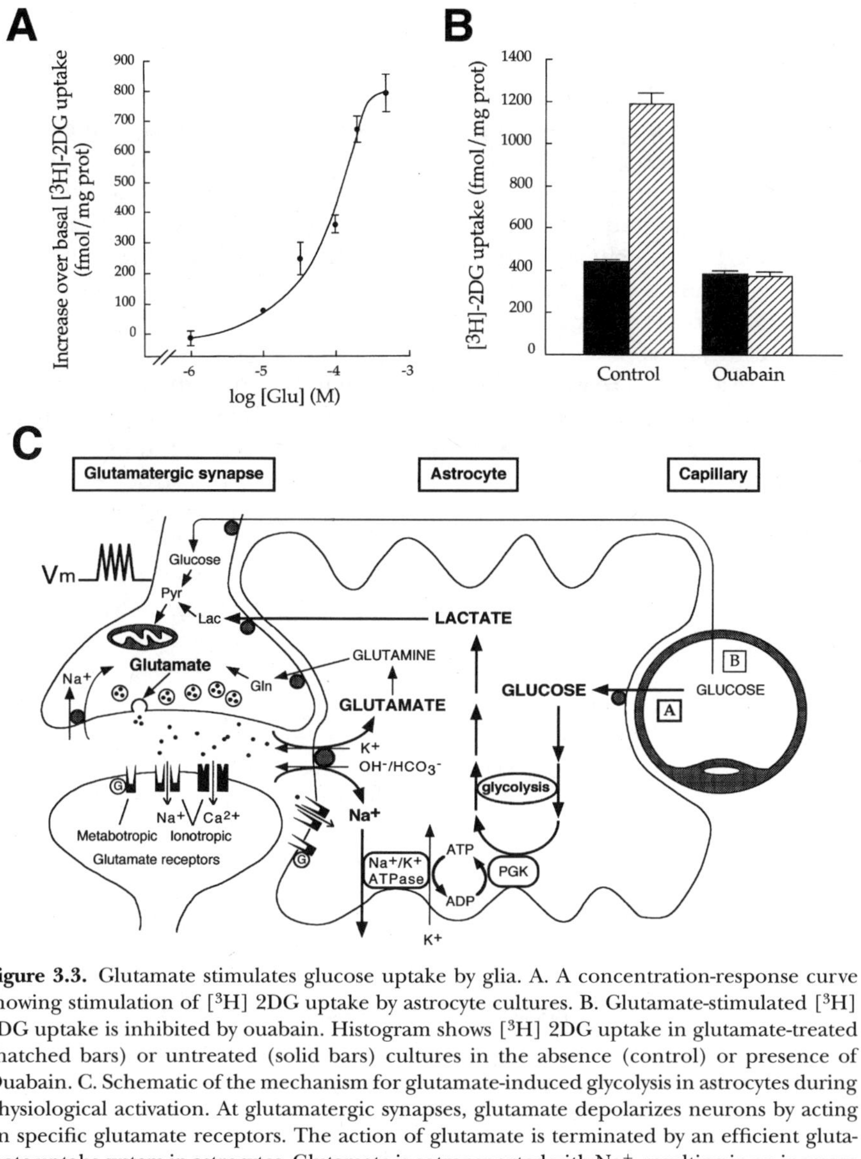

Figure 3.3. Glutamate stimulates glucose uptake by glia. A. A concentration-response curve showing stimulation of [^{3}H] 2DG uptake by astrocyte cultures. B. Glutamate-stimulated [^{3}H] 2DG uptake is inhibited by ouabain. Histogram shows [^{3}H] 2DG uptake in glutamate-treated (hatched bars) or untreated (solid bars) cultures in the absence (control) or presence of Ouabain. C. Schematic of the mechanism for glutamate-induced glycolysis in astrocytes during physiological activation. At glutamatergic synapses, glutamate depolarizes neurons by acting on specific glutamate receptors. The action of glutamate is terminated by an efficient glutamate uptake system in astrocytes. Glutamate is cotransported with Na^+, resulting in an increase in $[Na^+]_i$, leading to activation of Na^+–K^+-ATPase. The pump fueled by ATP provided by membrane-bound glycolytic enzymes, activates glycolysis, i.e., glucose utilization and lactate production in astrocytes. Lactate, once released, can be taken up by neurons to serve as an adequate energy substrate. Direct glucose uptake into neurons under basal conditions is also shown (labeled B) (Pellerin and Magistretti, 1994). (Copyright 1994 National Academy of Sciences, U.S.A., reprinted with permission.)

3.1.2. Lactate in Metabolic Exchange

We cannot rule out the possibility that the Müller cell serves as a conduit for transferring glucose from blood vessels to neurons (Niemeyer, 1997). However, it appears that glucose may not be the principal metabolic substrate for transfer to neurons (Fig. 3.4). The monocarboxylates, pyruvate and lactate, appear to be the preferred candidates for metabolic exchange between neurons and glial cells throughout the nervous system (Larrabee, 1983, 1992; Hamprecht and Dringen, 1996). Lactate is an excellent substrate for energy metabolism because after conversion to pyruvate through a lactate dehydrogenase-mediated reaction (Fig. 3.4), one mole of lactate can generate 18 moles of ATP via oxidative phosphorylation.

There are good reasons for supposing that lactate might be the metabolite transferred between neurons and glial cells: Lactate and pyruvate are excellent energy substrates for brain slices *in vitro* (cf. Izumi et al., 1994). First, lactate is the major energy metabolite released by astrocyte cultures. Second, glycogenolysis, even in the presence of exogenous glucose, results in lactate formation in astrocyte cultures (Dringen et al., 1993a, b). And last, lactate compartmentation and lactate transfer from astrocytes to neurons has been observed in cell cultures (Dringen et al., 1993a, b; McKenna et al.,

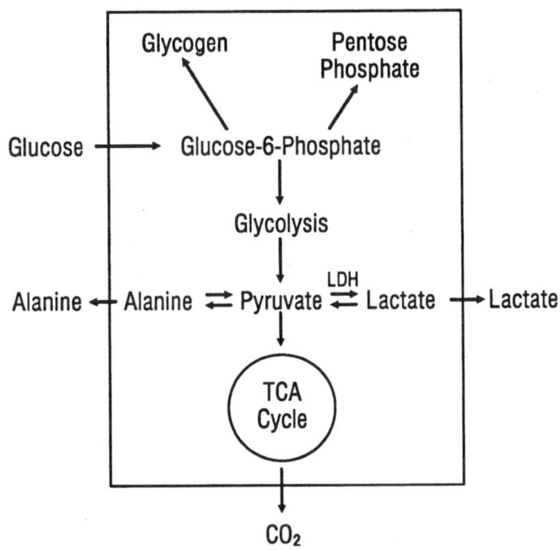

Figure 3.4. Pathways of glucose metabolism through glucose-6-phosphate. Glucose can end up as glycogen, enter the pentose pathway, or can be channeled through glycolysis to produce lactate or enter the TCA cycle.

1993). Perhaps the strongest evidence for lactate involvement in metabolic exchange comes from experiments which show that the energy needs of glutamate-activated neurons is met by glial oxidative glucose consumption which significantly increases lactate production (Schurr et al., 1999).

Lactate is a major energy metabolite in the retina. Isolated mammalian retina produces and secretes large amounts of lactate under both aerobic and anaerobic conditions (Krebs, 1972), and exogenous lactate alone can sustain retinal oxidative metabolism and photoreceptor function for long periods of time *in vitro* (Winkler, 1981). The idea that lactate is the main substrate transferred from Müller cells to photoreceptors is supported by recent studies in which the fate of ^{14}C-glucose was examined in dissociated retinal preparations consisting of either Müller cells or Müller cell–photoreceptor aggregates (Fig. 3.5; cf. Poitry-Yamate et al., 1995). The results

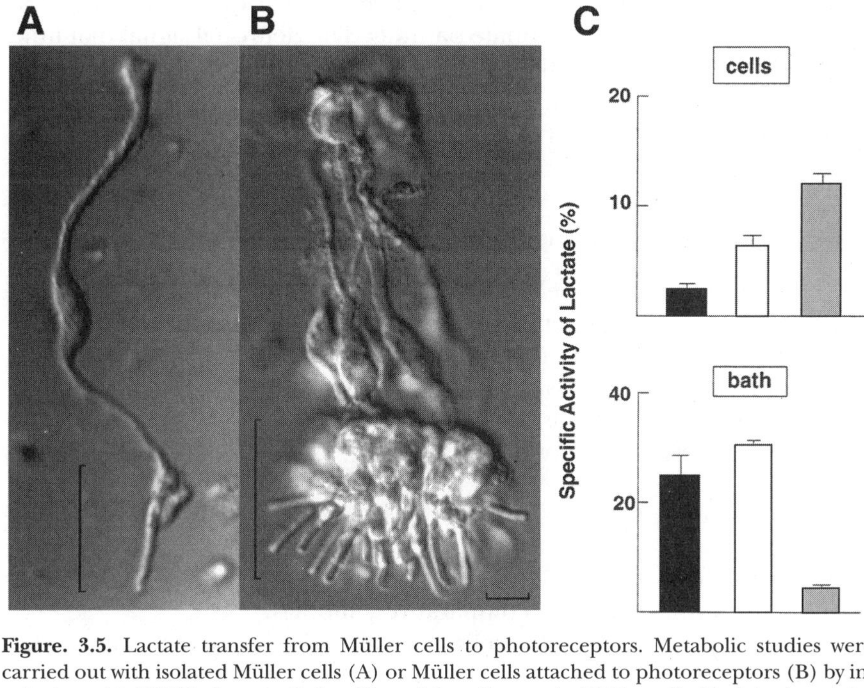

Figure. 3.5. Lactate transfer from Müller cells to photoreceptors. Metabolic studies were carried out with isolated Müller cells (A) or Müller cells attached to photoreceptors (B) by incubating with $^{14}C(U)$-glucose. C. Specific activity of lactate in Müller cells (dark bar) and the rod–Müller cell complex in light (open bar) and in dark (gray bar), and in the bath (Poitry-Yamate et al., 1995). Note the high lactate level in the bath showing that newly made lactate is mostly released (Poitry-Yamate et al., 1995). (Copyright 1995 *The Journal of Neuroscience*, reprinted with permission.)

show that Müller cells synthesize and release large amounts of lactate and much smaller amounts of amino acids and Krebs cycle metabolites. Moreover, based on CO_2 production, lactate was found to be a better substrate than glucose for oxidative metabolism in photoreceptors. Finally, metabolism of lactate, glutamate and glutamine is modulated by light in Müller cell–photoreceptor aggregates, indicating an association between photoreceptor activity and lactate production (Fig. 3.5). It appears, therefore, that lactate produced in the Müller cell results from the high level of glycolysis, and that the release of lactate fuels oxidative metabolism in photoreceptors.

If lactate were the substrate transferred to neurons, we would expect that neurons should have a lactate uptake system, as well as the enzymes necessary for entry of lactate into the TCA cycle. Although these requirements have been fulfilled in other neuronal tissues (Dringen et al., 1993a, b; Tholey et al., 1981; Bittar et al., 1995), they remain to be demonstrated in the retina. As previously discussed, neurons also take up and phosphorylate glucose, and much of the glucose taken up by neurons is channeled to glycolysis or the pentose phosphate pathway. The neuronal signal that links photoreceptor metabolic activity to lactate release from Müller cells has yet to be identified, but a likely candidate is glutamate, the neurotransmitter released at the photoreceptor synapse (Fig. 3.6; cf. Poitry-Yamate et al., 1995).

3.1.3. Glycogen Storage and Mobilization

The idea that Müller cells utilize glucose to synthesize glycogen as an energy store received considerable support from the early studies of Kuwabara and Cogan (1959) who used selective histochemical stains to show that enzymes involved in glycolysis and glycogen synthesis were highly concentrated in retinal Müller cells. From a survey of the glycogen content and distribution in several vertebrate retinas, they concluded that the observed variations in glycogen content in vertebrate retinas were related to the blood supply and glial cell content of the retina (Kuwabara and Cogan, 1961). For example, in the avascular rabbit and guinea pig retinas, glycogen was abundant, whereas in the highly vascularized human retina, the glycogen content was found to be comparatively modest.

It is now clear that Müller cells can synthesize, store, and metabolize glycogen rapidly (Kuwabara and Cogan, 1959a, b, 1960, 1961; Kuwabara et al., 1959; Hutchinson and Kuwabara, 1962; Lessell and Kuwabara, 1964; Kuwabara, 1965; Schabadasch and Schabadasch, 1972; Goldman, 1988, 1990; Babel and Stangos, 1973; Poitry-Yamate and Tsacopulos, 1991, 1992; Uga and Smelser, 1973b; Magalhaes and Coimbra, 1979, 1972; Pfeiffer et al., 1994;

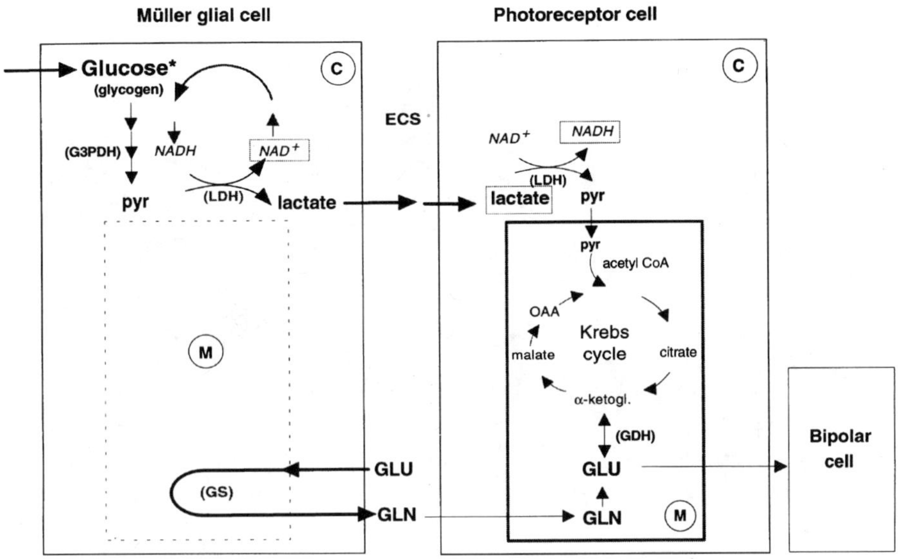

Figure 3.6. A scheme for lactate and amino acid trafficking between photoreceptors and Müller cells. In the presence of exogenous glucose and oxygen, lactate formed from glucose or glycogen in the Müller cell is released into the extracellular space (ECS) and fuels the photoreceptor. Also shown are the glutamate to glutamine conversion in the Müller cells and glutamate's downstream target, the bipolar cell. C, cytoplasm; and M, mitochondrion (Poitry-Yamate et al., 1995). (Copyright 1995 *The Journal of Neuroscience*, reprinted with permission.)

Niemeyer et al., 1997; Runger-Braddle et al., 1996). In electron micrographs, glycogen granules appear as 25–30 nm, electron-dense, isodiametric β particles scattered throughout the Müller cell cytoplasm (Fig. 3.7). In the rabbit retina, ischemia leads to a rapid and progressive loss of available glycogen (a decline of ~50% in 15 min) (Wassilewa et al., 1976; Johansson, 1977). Similarly, in rats with streptozotocin-induced diabetes, the glycogen content in Müller cell processes increased 2–4-fold when the blood glucose was elevated 5-fold (Sosula et al., 1974). Although many studies show that the glycogen level *in situ* appears to vary with changes in retinal function, there are few estimates of the glycogen content in the retina. The glycogen content of dissociated Müller cells from juvenile guinea pig retina has been estimated to be at least 175 nmol glycosyl units/mg protein (Poitry-Yamate and Tsacopoulos, 1992).

The external signal for glycogen breakdown has not been identified. Glycogen mobilization in Müller cells is likely to be regulated primarily by neuroactive substances released by retinal neurons (Table 3.1). For exam-

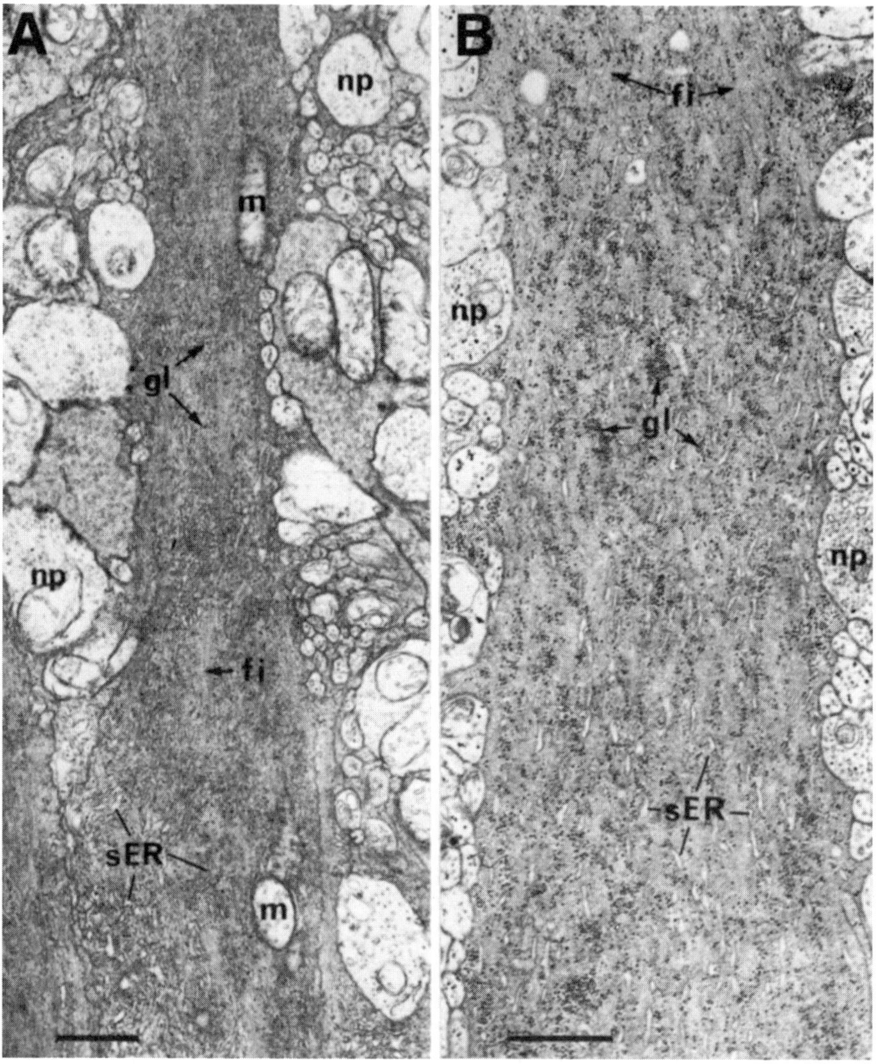

Figure 3.7. Glycogen granules in retinal Müller cells. The panels show electron micrographs from the inner plexiform layer in the vascular retina (opossum; A) and the avascular retina (Necturus; B). Many glycogen granules (gl) can be seen in the Müller cell cytoplasm. sER, smooth endoplasmic reticulum; fi, fine filaments; m, mitochondrion; and np, neural processes (Uga and Smelser, 1973). (Copyright 1973 Association for Research in Vision and Ophthalmology, reprinted with permission.)

Table 3.1. Modulation of Glycogen in Brain Slices and Astroglial Cultures

Substance/treatment	Effect[a]	Reference
Adenosine	–	Magistretti et al., 1986
Adenosine-triphosphate	–	Sorg et al., 1995
Ammonium ions	–	Swanson et al., 1989a; Dombro et al., 1993
Arachidonic acid	–	Sorg et al., 1995
Histamine	–	Quach et al., 1980; Arbones et al., 1990
Noradrenaline	–	Quach et al., 1978; Pearce et al., 1985; Rosenberg and Dichter, 1987; Cambray-Dekin et al., 1988a; Subbarao and Hertz, 1990
Secretin	–	Martin et al., 1992
Serotonin	–	Quach et al., 1982
VIP	–	Magistretti et al., 1981; Magistretti, 1988
Glucose deprivation	–	Rosenberg and Dichter, 1985; Dringen and Hamprecht, 1992
Elevated cytosolic Ca^{2+}	–	Ververken et al., 1982; Hof et al., 1988; Hamprecht et al., 1993
Elevated extracellular K^+	–	Cambray-Dekin et al., 1988b; Hof et al., 1988; Subbarao et al., 1995; Cummins et al., 1983
Glucose	+	Cummins et al., 1983; Swanson et al., 1983b; Dringen and Hamprecht, 1992
IGF I	+	Dringen and Hamprecht, 1990, 1992
IGF II	+	Dringen and Hamprecht, 1990
Insulin	+	Dringen and Hamprecht, 1990, 1992
Glutamate	+	Swanson et al., 1990
Aspartate	+	Swanson et al., 1990
K252a	+	Dringen, 1992
Methionine sulfoxamine	+	Swanson et al., 1989b; Hevor and Delorme, 1991

[a]–, decrease, +, increase. The references listed are cited in Wiesinger et al. (1997).

ple, an increase in intracellular cAMP brought about by the action of biogenic amines and neuropeptides can lead to glycogen breakdown in Müller cells (Memo et al., 1981; Goldman, 1990; Lopez-Colome et al., 1993; Henschel and Miller, 1992). Because many neuropeptides are expressed by retinal neurons, these molecules may serve as signaling candidates for inducing glycogen breakdown. Indeed, this suggestion is supported by the observation that vasoactive intestinal peptide (VIP) increases cAMP levels in Müller cell cultures (Koh and Roberge, 1989). Additionally, adenosine and K^+, two known glycogenolytic agents released by all activated neurons, could act as potential signals for glycogen breakdown in glial cells (Magistretti et al., 1986; Hof et al., 1988; Osborne, 1989; Pull and McIlwain, 1972; Sykova, 1983).

Because the Müller cell is the major repository of glycogen in the mammalian retina, it has been suggested that under conditions of energy stress, Müller cell glycogen is mobilized to provide neurons with a local source of glucose (Kuwabara and Cogan, 1959, 1961). Although a seemingly attractive hypothesis, there is little experimental evidence to support this notion. Indeed, the amount of glycogen stored is so small that it could not meet the energy demands of retinal neurons for more than a few minutes (Clarke and Sokoloff, 1994). Rather, the glucose produced by glycogen breakdown is probably used to meet the energy demands of the glial cell itself, and may not be transferred to neurons.

Experimental evidence for "metabolic trafficking" in retina is not as strong as in brain. Considering the anatomy of the retina, one might wonder whether such a process is essential. In the brain, astrocytic processes tightly surround capillaries and hence trafficking through glial cells may be necessary. In the retina, however, glucose can flow directly from choriod to photoreceptors, the major sites of energy metabolism, without having to pass across glial (Müller) cells. Also, photoreceptors express glucose transporters, and hence can take up glucose directly. Several 2-DG uptake studies support direct uptake of glucose by photoreceptors (Bensinger et al., 1979; Witkovsky and Yang, 1982; Sperling et al., 1982). Accordingly, metabolic trafficking may be superfluous for photoreceptors. Nevertheless, transfer via glial cells may be necessary to provide energy to inner retinal neurons.

Another view is that the basal activity of neurons, including photoreceptors, is supported by direct uptake of glucose. Under conditions of heightened activity, however, the energy needs of photoreceptors are met by lactate derived from metabolic trafficking between Müller cells and photoreceptors. This explanation is consistent with all available ^{3}H-2-DG uptake data (Basinger et al., 1979; Witkovsky and Yang, 1982; Sperling et al., 1982; Poitry-Yamate and Tsacopoulos, 1991, 1992) and recent studies on lactate trafficking (Poitry-Yamate et al., 1995).

3.2. GLUTAMATE METABOLISM

Glutamate is the major excitatory neurotransmitter in the retina and is involved in synaptic transmission from photoreceptors to bipolar cells, and from bipolar to ganglion cells. Glutamate is also a well-known excitotoxin associated with neuronal damage in conditions such as retinal ischemia. Therefore, it is not surprising that extracellular glutamate levels in retina are tightly regulated by the action of potent uptake systems and degradative pathways present in neurons and Müller cells. As discussed in Chapter 5, neurons and Müller cells express high affinity glutamate transporters that can rapidly clear glutamate from the extracellular environment. Within neurons, glutamate can be recycled, but in Müller cells glutamate appears to be inactivated by conversion to metabolites.

Glutamate metabolism in the CNS involves a large group of enzymes: aminotransferases, glutamate dehydrogenase, phosphate-activated glutaminase, and glutamine synthetase (Kugler, 1993). These enzymes are not uniformly distributed among neurons and glial cells. Aspartate aminotransferase and glutaminase are enriched in neurons, whereas glutamate dehydrogenase and glutamine synthetase are compartmentalized in glial cells. In the retina, glutamine synthetase is found exclusively in Müller cells, but not in neurons. Thus, the presence of high affinity glutamate uptake systems and glutamine synthetase provides the Müller cell with two different means to regulate glutamate levels in the retina. Acting in concert, these systems keep the extracellular glutamate levels in the retina quite low (Attwell et al., 1993; Müller et al., 1997).

3.2.1. Glutamine Synthetase

Although the localization and properties of glutamate dehydrogenase and glutamine synthetase have been well studied in the brain, very little is known about glutamate dehydrogenase in Müller cells. On the other hand, glutamine synthetase has been studied extensively in the retina. Glutamine synthetase is recognized as a key enzyme for maintenance of nitrogen homeostasis because it replenishes L-glutamine, an amino acid that is a component of many proteins, and also serves as a source of nitrogen for a number of important metabolites.

Glutamine synthetase catalyzes the reversible formation of glutamine from glutamate, ammonia, and ATP. As Fig. 3.8 shows, conversion of glutamate to glutamine involves fixation of ammonia in Müller cells (Meister, 1974). Since the retina (like the brain) does not possess a complete urea cycle, the ammonia generated is largely detoxified by formation of nontoxic glutamine. This process is likely to be important since high levels of ammonia appear to impair brain function and can lead to convulsions and seizures (Cooper and Plum, 1987). Similarly, accumulation of ammonia might be adverse for the normal functioning of the retina. Hence, glutamine synthetase serves two important roles in the retina: ammonia detoxification and amination of glutamate to glutamine.

Glutamine synthetase has been purified from brain as well as retina (Meister, 1974; Pahuja and Reid, 1985; Pahuja et al., 1985). It has been found

$$\text{L-glutamate} + \text{ATP} + NH_3 \xrightarrow{\text{Glutamine synthetase}} \text{L-glutamine} + \text{ADP} + P_i$$

Figure 3.8. Conversion of glutamate to glutamine is catalyzed by glutamine synthetase. This reaction is involved in ammonia detoxification and amination of glutamate to glutamine.

to exist as an octomer with a Mr ~360–390 kDa with subunits of Mr ~43–49 kDa and has four Mn^{2+} ions tightly associated with the enzyme. Moreover, maximum enzyme activity requires a divalent cation such as Mg^{2+}, Mn^{2+}, or Co^{2+}. The enzyme is activated by α-ketoglutarate and inhibited by glutamine. The cellular level of glutamine synthetase is regulated by a number of factors such as glucocorticoids, cAMP, insulin, triiodotyronine and norepinephrine (Kugler, 1993).

Immunocytochemical studies with glutamine synthetase-specific antibodies have shown that the enzyme is primarily associated with brain astrocytes (Norenberg and Martinez-Hernandez, 1979). Similar studies in retina show that it is expressed only in Müller cells (Fig. 3.9) (Riepe and Norenberg, 1977; Linser and Moscona, 1983). Moreover, glutamine synthetase expression in Müller cells is strongly dependent on contact interaction with neurons (see Chapter 2). The genetic elements involved in regulation of the glutamine synthetase gene have been extensively studied by Lily Vardimon, Anthony Young, and their associates (e.g., Li et al., 1995; Avisar et al., 1999).

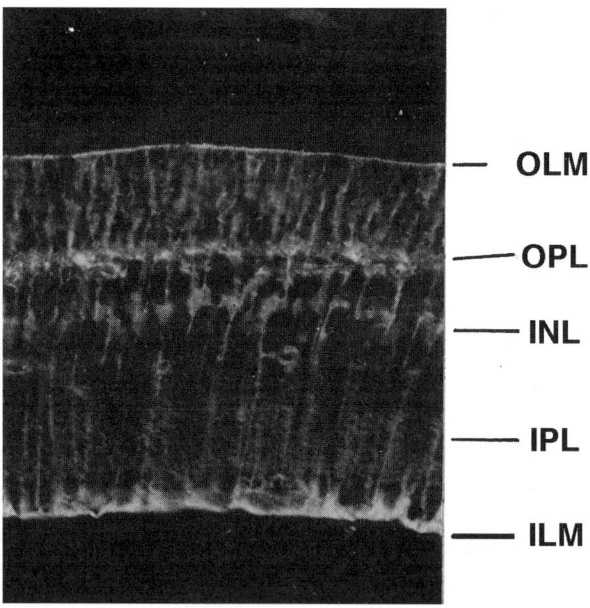

Figure. 3.9. Glutamine synthetase is localized in Müller cells. Immunocytochemical localization studies in chick retina show that labeling is present in the radial processes of Müller cells. Kindly provided by P. Linser.

3.2.2. Glutamate–Glutamine Cycle

The glutamate–glutamine cycle represents a good example of the complex metabolic interactions that exist between neurons and glial cells. The cycle was formulated to explain compartmentalization of glutamate in the brain (Benjamin and Quastel, 1974; van den Berg et al., 1969; cf. Shank and Aprison, 1981). Tracer studies with ^{14}C-acetate and ^{14}C-glucose indicated these substrates were handled differently. Label from ^{14}C-acetate was quickly incorporated into glutamine at a higher specific activity than into glutamate, whereas label from ^{14}C-glucose was incorporated into glutamate at a higher specific activity than into glutamine. Because the labeled carbon atoms from acetate and glucose must first enter the TCA cycle before conversion to glutamate and glutamine, the results pointed to the presence of at least two TCA cycles in the brain. Moreover, ^{14}C-acetate labeling was found associated with a small, rapidly-turning over pool of glutamate which was preferentially converted to glutamine, while ^{14}C-glucose appeared to label a larger pool of glutamate which was metabolized to a variety of products (including glutamine), again suggesting the existence of two pools of glutamate.

The cellular localization of the small, rapidly turning over glutamate pool was established when glutamine synthetase was shown to be present only in glial cells. Therefore, it appeared that glial cells were the main sites for rapid glutamine synthesis. About the same time, Hamberger et al. (1979) showed that glutamine could serve as a precursor for the synthesis of glutamate and GABA. These observations provided further support for the existence of a glutamate–glutamine cycle.

When glutamate and GABA are released from nerve terminals, some of the transmitter molecules are taken up by astrocytes via Na^+-dependent, high affinity transporter systems. Inside astrocytes, glutamate is converted to glutamine and released into the extracellular fluid where it is subsequently taken up by glutamatergic- and GABAergic neurons (Fig. 3.10). In the nerve terminal, glutamine is hydrolyzed by neuronal glutaminases into glutamate and ammonia. The glutamate generated is used either as a transmitter in glutamatergic terminals or is subsequently decarboxylated to GABA in GABAergic terminals.

Experimental evidence suggests a significant fraction of extracellular glutamine is not derived from glutamate taken up by glia. Therefore, some of the glial glutamine appears to come from *de novo* synthesis. Because glia contain pyruvate carboxylase, they are able to carry out net synthesis of glutamine by the anaplerotic pathway while maintaining a normal TCA cycle (Shank et al., 1985). *In vitro* studies with astrocytes cultures suggest that a large fraction of glial glutamine is derived from anaplerosis (Hertz,

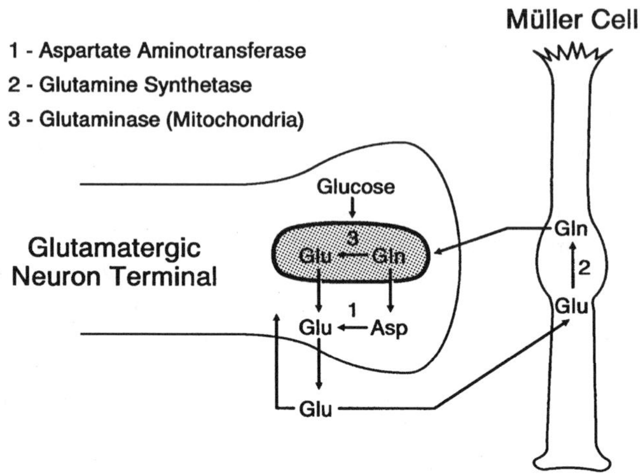

Figure 3.10. A scheme for recycling of glutamate between neurons and Müller cells (the glutamate–glutamine cycle). Glutamate released by neurons is taken up by the Müller cell, where it is converted to glutamine by glutamine synthetase. Glutamine released by the Müller cell is taken up by glutamatergic neurons and enters the mitochondria where it is converted to glutamine by glutaminase. The stippled structure is the mitochondrion where glutaminase converts glutamine to glutamate. Some extracellular glutamate is taken up by the neuronal terminal.

1980); however, estimates from *in vivo* NMR studies suggest that about 90% of glial glutamine is generated from the glutamate/glutamine cycle. Also, glutamine derived from anaplerotic pathway appears to be in agreement with the amount needed for ammonia detoxification (Sibson et al., 1998a; Shen et al., 1999). Nevertheless, these studies show that glia can readily provide neurons with adequate amounts of precursor for glutamate and GABA synthesis.

The existence of the glutamate–glutamine cycle is supported by a large body of evidence (cf. Westgaard et al., 1995). For example, treatment with gliotoxins such as fluroacetate or fluorocitrate lowers brain glutamine levels and eventually results in decreased glutamate and GABA content (Clark et al., 1970). An attractive feature of the glutamine cycle is that it conserves nitrogen atoms and also replenishes the glutamate pool of amino-acidergic neurons in the mammalian brain.

A similar scheme has been proposed for the retina. By following the metabolic fate of intravitreally injected glutamine, Voaden (1978) showed that glutamine is a precursor for glutamate and GABA, suggesting that the glutamate–glutamine cycle exists also in the retina. Many essential components of the cycle can be found in retina: Glutamate and GABA are major retinal neurotransmitters and glutamine is found extracellularly in retina (Massey, 1990; Yazulla, 1986; Müller et al., 1997). In addition, glutamate uptake and glutamine synthetase activity have been demonstrated in Müller cells (Riepe and Norenburg, 1977; Sarthy and Lam, 1979; Linser and Moscona, 1979), whereas glutaminase is present in synaptic layers of the retina (Ross et al., 1987).

Additional evidence for a retinal glutamate–glutamine cycle has come from immunocytochemical experiments. In normal retina, treatment with glutamate-specific antibodies does not lead to immunostaining of Müller cells (Pow and Robinson, 1994). Although there is avid glutamate uptake into Müller cells, glutamate levels are kept low presumably by the action of glutamine synthetase. However, in retinas treated with the glutamine synthetase inhibitor, methionine sulfoxide, there is strong glutamate-immunostaining in Müller cells, reflecting the increased level of intracellular glutamate (Fig. 3.11). Consistent with this observation, immunostaining retinas with glutamine-specific antibodies reveals high levels of glutamine in Müller cells (Kalloniatis et al., 1994; Pow and Robinson, 1994).

Quantitative NMR measurements show a 1:1 relationship between rate of glutamate–glutamine cycle and glucose oxidation under physiological conditions (Sibson et al., 1998a, b). Magistretti et al. (1999) proposed that ~85% of brain glucose utilization is attributable to energy requirements of glutamatergic neurotransmission. They argue that uptake of neuronally derived glutamate and its conversion to glutamine by glial cells uses up energy from glycolysis. This energy requirement is met by increased glucose uptake by glia, and the lactate generated by glycolysis is released for oxidation by neurons.

3.2.3. Other Metabolic Intermediates

Although the glutamate–glutamine cycle appears to be a major source of neuronal glutamate, experimental results indicate that glutamine alone cannot replace all the neurotransmitter pools of glutamate and GABA. Experiments with neuron/glia cultures show that a large amount of glutamate is transferred from neurons to glia, but a much smaller amount of glutamine is returned to neurons (Hertz et al., 1980; Schousboe et al., 1997). In addition, only a minor portion of the glutamate taken up into astrocytes is converted to glutamine while most of it is oxidized. Therefore, it has been suggested that other "shuttle" substrates such as citrate, lactate, malate, α-ketoglutarate and alanine may be transferred from glial cells to neurons to replenish the precursor supply (Schousboe et al., 1997). These data are,

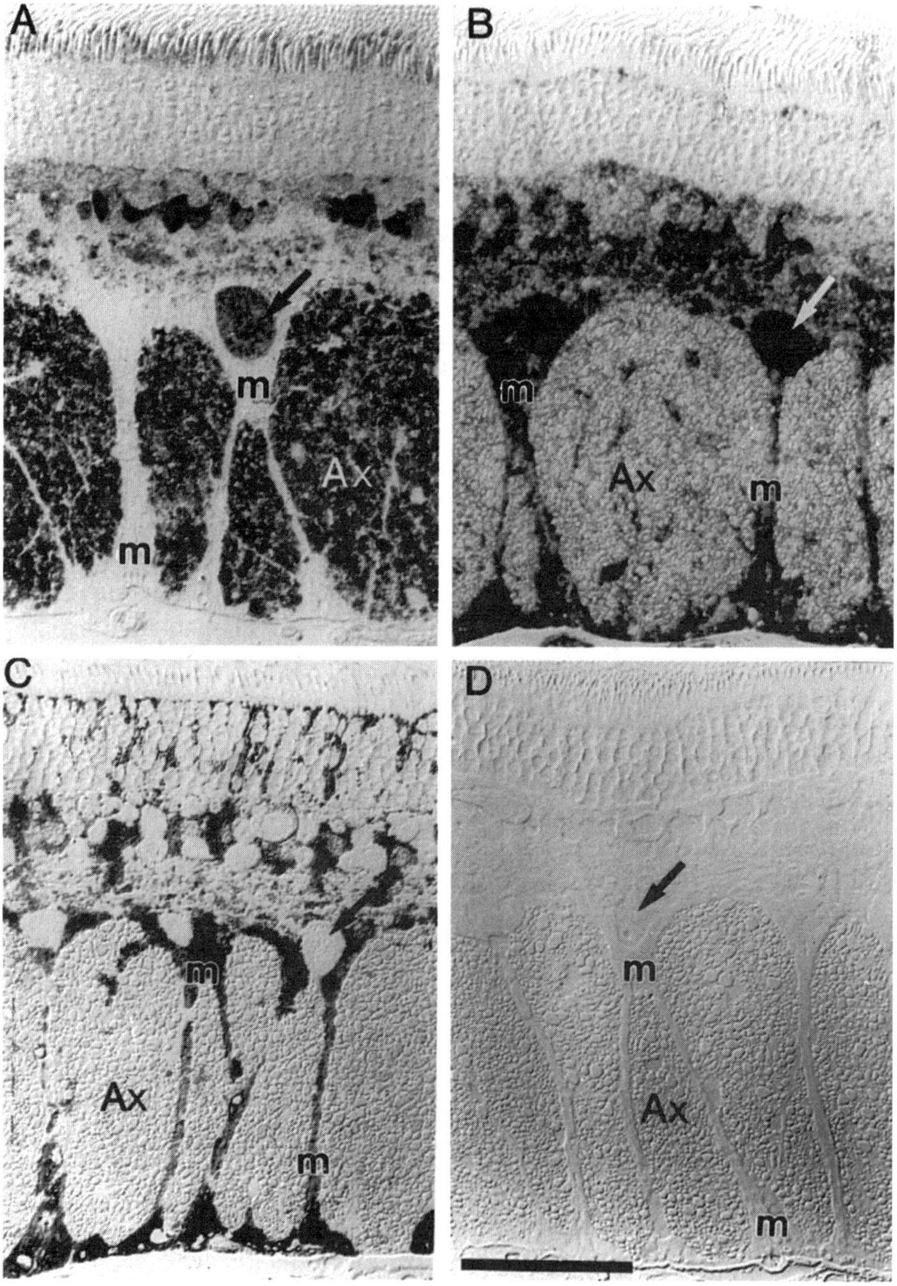
A
m
m
Ax
B
m
Ax
m
C
m
Ax
m
D
m
Ax
m

however, not in agreement with *in vivo* results from ^{15}N and ^{13}C NMR studies (Magistretti et al., 1999).

3.2.4. Gliotoxins

A popular approach to dissect the contribution of glial cells in retinal metabolism is to use gliotoxins to destroy glial cells in the experimental preparation. Many metabolic inhibitors have been used to selectively suppress glial cell activity in the nervous system. Among these, the most commonly used toxicants are: α-amino adipic acid (α-AAA), methionine sulfoximine (MSO), fluoroacetate, and flurocitrate (Martin and Winiewski, 1996).

α-AAA is a glutamate homologue that has been reported to selectively destroy brain astrocytes and retinal Müller cells when injected into animals (Olney et al., 1971; Lund Karlsen et al., 1982). In the retina, infusion of α-AAA strongly depresses the ERG b-wave and reduces the uptake of β-alanine, an indication of Müller cell dysfunction (Bonaventura et al., 1981; Szamier et al., 1981; Lund Karlsen, 1978; Kato et al., 1990). In most cases, L-α-AAA appears to be a more potent gliotoxin than D-α-AA. However, the racemic mixture of L- and D-forms appears to be more selective in its action. A somewhat surprising finding is that although the morphology of Müller cells changes quickly following injection of α-AAA, retinal structure is relatively well preserved (Pedersen and Lund Karlsen, 1979; Sugawara et al., 1980; Lund Karlsen, 1978). Similarly, in cultures of chick retina, treatment with α-AAA alters Müller cell morphology rapidly, but the cells recover completely when the drug is removed (Casper and Reif-Lehrer, 1983).

α-AAA does not appear to exert its toxic effect by activating excitatory amino acid receptors. Instead, α-AAA must be taken up by a Na^+-dependent mechanism to exert its action (Huck et al., 1984). Although the mechanism of α-AAA toxicity is not clearly established, there is some experimental evidence that α-AAA acts by depleting cells of glutathione, resulting in subsequent oxidative damage (Kato et al., 1993; Reichelt et al., 1997b; Szamier et al., 1981). Finally, a word of caution: α-AAA has been generally assumed to be a gliotoxin, yet there has been evidence of some neuronal damage in eyes treated with the toxin. It remains to be determined whether the impairment

Figure 3.11. Glutamate accumulates in Müller cells when glutamine synthetase is inhibited. Light micrographs of rabbit retina sections stained with an antibody to glutamate or glutamine. A. Retina stained with glutamate antibody; B. Retina stained with glutamine antibody; C. Glutamine synthetase inhibitor-treated retina immunostained with glutamate antibody; and D. glutamine synthetase inhibitor-treated retina labeled with glutamine antibody. Note glutamate-immunostaining of Müller cells in inhibitor-treated retina. Ax, axon bundle; and m, Müller cell (Pow and Robinson, 1994). (Copyright 1994 Elsevier Science, reprinted with permission.)

of neuronal function is due to the direct action of α-AAA on neurons or is a secondary consequence of glial damage.

Methionine sulfoximine (MSO) exerts its toxic effect by irreversibly inhibiting glutamine synthetase (Griffith and Meister, 1978). In the presence of MSO, glutamine levels fall, and the supply of precursors for glutamate and GABA synthesis becomes limiting (Martin and Winiewski, 1996). Indeed, convulsions produced by MSO are brought on by inhibition of glutamine synthetase. Last, fluoroacetate and fluorocitrate are effective toxins that appear to act by inhibiting the TCA cycle in glial cells (Fonnum et al., 1997). Fluoroacetate is metabolized to fluorocitrate, which inhibits aconitase and blocks the mitochondrial citrate carrier. Inhibition of the TCA cycle by fluorocitrate reduces glutamine production and leads to its gliotoxic effect. Since the TCA cycle operates in both glial cells and neurons, the glial selectivity of these agents remains enigmatic.

3.3. GABA METABOLISM

GABA, a major inhibitory neurotransmitter in the retina (Yazulla, 1986), is taken up by Müller cells in mammalian retinas (see Chapter 4). Inside the Müller cell, GABA is degraded through the action of GABA-transaminase (GABA-T) and succinic semialdehyde dehydrogenase, enzymes shown histochemically to be present in Müller cells (Hyde and Robinson, 1974; Moore and Gruberg, 1974). The demonstration of GABA uptake and localization of GABA-metabolizing enzymes suggest that Müller cells are likely to participate in regulating the extracellular concentration of GABA.

The idea that Müller cells may regulate GABA levels in retina is well supported by other studies. If GABA metabolism occurs mainly in Müller cells, inhibition of GABA-T activity can be expected to result in enhanced GABA accumulation in Müller cells but not in neurons (Fig. 3.12). Indeed, GABA-immunocytochemical studies show that although little GABA is found normally in Müller cells, there is an increase in GABA-immunoreactivity in Müller cells after the retina is treated with the GABA-T inhibitor, vigabatrin (Cubells et al., 1988; Neal et al., 1989). Since Müller cells cannot synthesize GABA, the GABA in Müller cells must come from its uptake and accumulation.

3.3.1. GABA-transaminase

GABA degradation is catalyzed by two mitochondrial enzymes: GABA-transaminase and succinic acid semialdehyde dehydrogenase (Kugler,

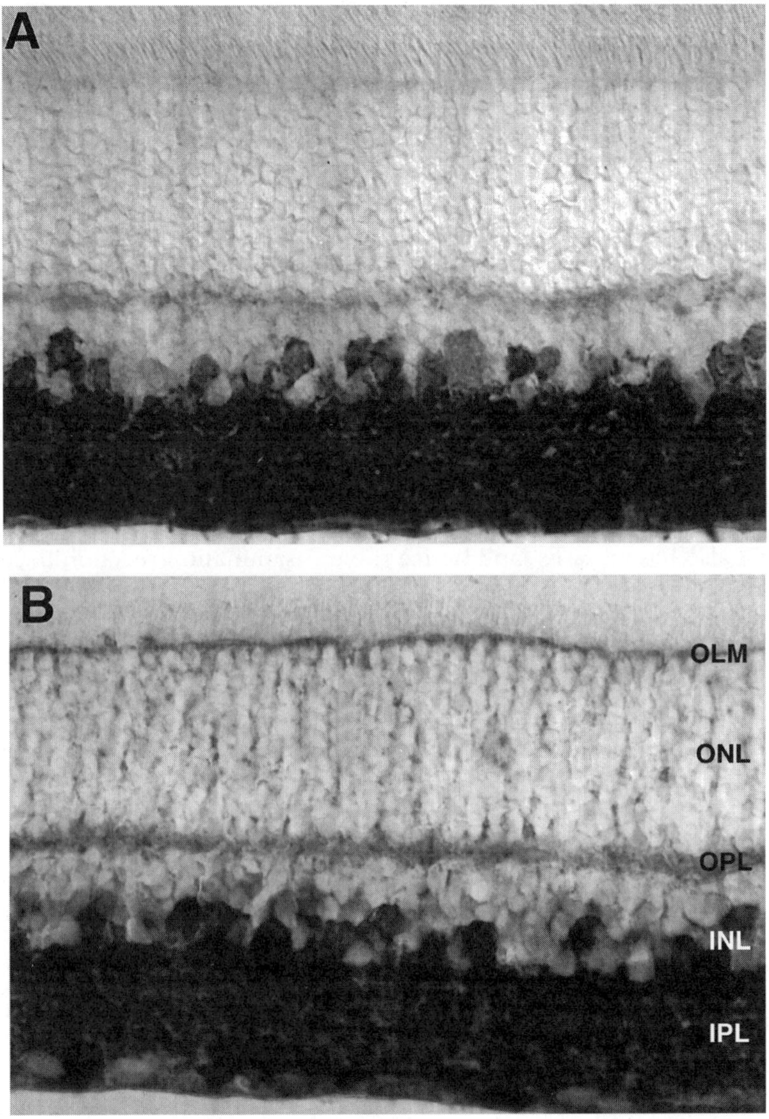

Figure 3.12. GABA-T inhibition elevates GABA level in Müller cells. The figure shows GABA-immunostaining pattern in normal rat retina (A) and rat retina-treated with GABA-transaminase inhibitor, γ-vinyl GABA (B). Although there are many GABA-positive cells in both retinal sections, GABA-immunoreactivity was seen in the radial process of Müller cells only in retinas treated with the GABA-T inhibitor (Neal et al., 1989). (Copyright 1989 Elsevier Science, reprinted with permission.)

GABA-T

GABA + α-Ketoglutarate ←------→ glutamate + succinic semialdehyde

SSADH

Succinic semialhehyde + NAD^+ -------→ succinate + NADH + H^+

Figure 3.13. GABA-transaminase catalyzes the breakdown of GABA. GABA-T, GABA transaminase; SSADH, succinic semialdehyde dehydrogenase.

1993). GABA-transaminase, a pyridoxal phosphate-dependent enzyme with high substrate specificity for GABA and α-ketoglutarate, carries out the initial step involving transamination of α-ketoglutarate to glutamate (Fig. 3.13). In the next step, succinic acid semialdehyde dehydrogenase oxidizes succinic semialdehyde to succinate. Succinate then enters the TCA cycle through succinate dehydrogenase. Succinic semialdehyde is not found in the nervous tissue because it is rapidly oxidized, and the activity of succinic acid semialdehyde dehydrogenase (SSADH) is normally higher than that of GABA-T making transamination the rate limiting step (Kugler, 1993). SSADH also has a low K_m and hence the transamination reaction becomes essentially irreversible *in vivo.*

3.3.2. GABA Shunt

The GABA shunt is a closed pathway whose purpose is to produce and conserve the supply of GABA (Fig. 3.14). The first step in the process is the

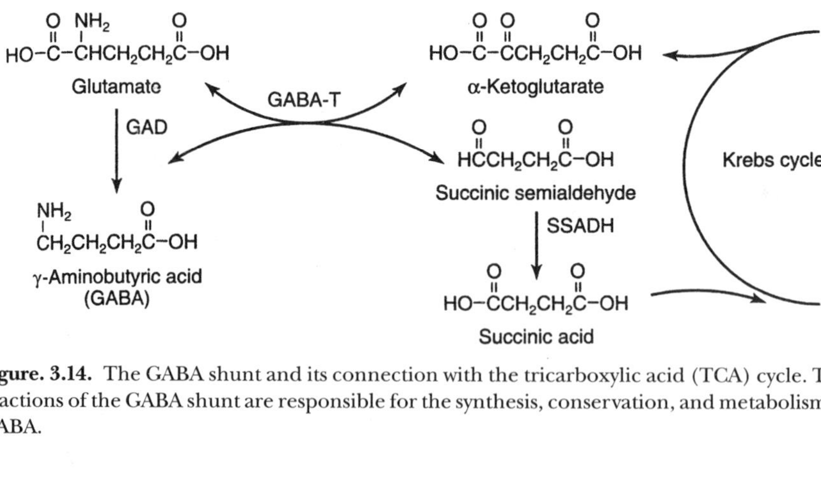

Figure. 3.14. The GABA shunt and its connection with the tricarboxylic acid (TCA) cycle. The reactions of the GABA shunt are responsible for the synthesis, conservation, and metabolism of GABA.

transamination of α-ketoglutarate by GABA-T to generate glutamate, which is subsequently converted to GABA by glutamic acid decarboxylase. Alternatively, GABA could be derived from GABA uptake. As previously described, GABA is metabolized by GABA-T to succinic acid semialdehyde. To conserve the GABA supply, transamination usually occurs when α-ketoglutarate is present to accept the NH_2 group from GABA, thus leading to formation of glutamate. The other product of the reaction, succinic acid semialdehyde, is converted to succinic acid, and can reenter the Krebs cycle, thus completing the loop.

Although GABA taken up by neurons can be recycled, the GABA in glial cells is metabolized and cannot be converted back to GABA because glia do not have glutamic acid decarboxylase. GABA, however, can be recovered through an indirect route involving the Krebs cycle. In glia, glutamate is converted to glutamine, which is transferred back to neurons where it is converted to glutamate by glutaminase, and thus can enter the GABA shunt. The main function of the shunt appears to be in the production of GABA and not in the TCA cycle since less than 10% of the metabolic flow occurs through the shunt (Kugler, 1993). More important, the shunt is not universal since glutamic acid decarboxylase (GAD) is almost exclusively found in GABAergic neurons and never in glial cells.

In summary, there is a tight metabolic link between retinal neurons and Müller cells in the turnover and metabolism of glutamate and GABA. Müller cells contain high levels of glutamate dehydrogenase and glutamine synthetase, and their products, α-ketoglutarate and glutamine, have been proposed to be involved in supplementing the neurotransmitter pool in glutamatergic and GABAergic neurons. Fig. 3.15 gives a summary diagram of the metabolic interactions of glutamatergic and GABAergic neurons with Müller cells.

3.4. ACID–BASE REGULATION

Because of its high rate of respiration, particularly in the dark, the retina produces significant amounts of CO_2 (Sickel, 1972). CO_2 has a high diffusion coefficient and is very soluble; therefore, most of the CO_2 generated can be expected to diffuse into adjacent retinal or choroidal vessels. Some of the CO_2 diffusing into the retina will pass into the neighboring Müller cells where it is converted to bicarbonate by the action of carbonic anhydrase. Müller cells are known to contain high levels of carbonic anhydrase which is likely to be involved in CO_2 fixation in the retina and the maintenance of a normal acid-base balance (Musser and Rosen, 1973; Sarthy and Lam, 1978; Linser and Moscona, 1981).

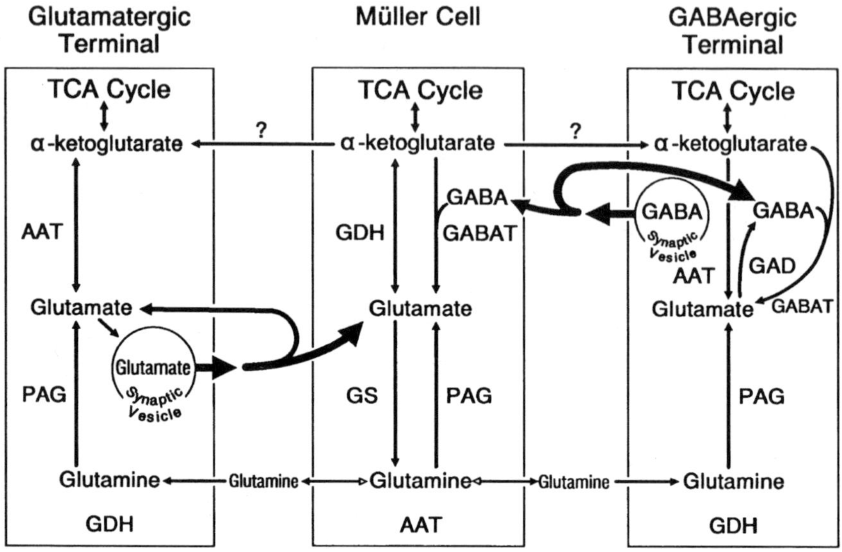

Figure 3.15. A schematic diagram of metabolic interactions among glutamatergic and GABA-ergic neurons and Müller cells. The enzymes aspartate aminotransferase (AAT), phosphate-activated glutaminase (PAG), and glutamate decarboxylase (GAD) are strongly enriched in neurons, whereas glutamate dehydrogenase (GDH), glutamine synthetase (GS), and GABA-transaminase (GABA-T) are strongly enriched in Müller cells.

3.4.1. Carbonic Anhydrase

The enzyme, carbonic anhydrase, catalyzes the conversion of CO_2 and water to carbonic acid, which dissociates spontaneously to yield HCO_3^- and H^+ (Fig. 3.16). Carbonic anhydrase is believed to provide H^+ and HCO_3^- ions for rapid intracellular buffering or for exchange of other ions resulting in movement of ions and fluids across the plasma membrane (Maren, 1967).

Mammalian carbonic anhydrase exists as seven isozymes: four cytoplasmic (CAI, CAII, CAIII, and CA IV); one membrane bound (CA IV); one secretory (CA VI); and one mitochondrial (CA V). The isozymes have broad structural similarity but are distinguished by their differential tissue distribution, membrane association, and catalytic activity (Tashian et al., 1991). CAII is the predominant isozyme in retina, and other isozymes make up a smaller fraction of the total CA content (Wistrand et al., 1986; Ridderstrale et al., 1994).

According to several lines of evidence, CA is localized almost exclusively in Müller cells in the vertebrate retina (Musser and Rosen, 1973; Korhonen and Korhonen, 1965; Hansson, 1967; Sarthy and Lam, 1979a; Wistrand et al.,

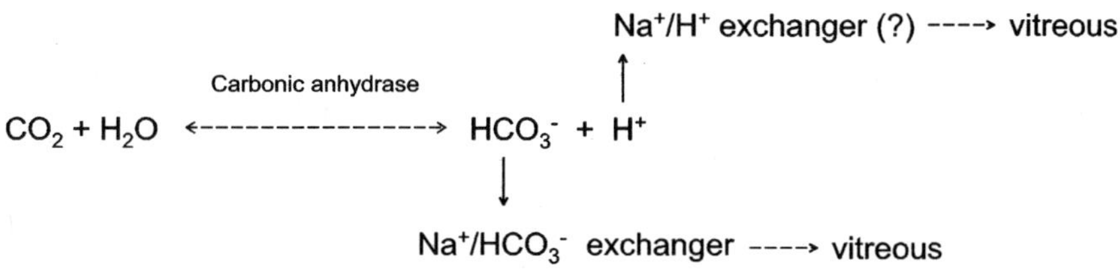

Figure 3.16. Carbonic anhydrase converts CO_2 to bicarbonate. The enzyme is almost exclusively localized in Müller cells in the vertebrate retina. Some of the bicarbonate is released into the vitreous through the sodium, bicarbonate exchanger. The protons generated in the reactions may also end up in the vitreous via a sodium–proton exchanger.

1986; Luten-Drecoll at al., 1983; Linser and Moscona, 1984; Ridderstrale et al., 1964). Histochemical methods as well as immunocytochemical studies using CA II-specific antibodies show that the radially oriented Müller cells are strongly labeled in many vertebrate retinas (Musser and Rosen, 1973; Linser and Moscona, 1984). Moreover, a comparison of CA levels in isolated Müller cells with values for the whole retina suggest that more than 90% of the retinal carbonic anhydrase activity can be attributed to Müller cells (Sarthy and Lam, 1978). The presence of high levels of CA in Müller cells may mean they play an active role in regulating acid-base balance in the retina.

3.4.2. Bicarbonate Exchange

There is good experimental evidence that neuronal activity leads to changes in intracellular and extracellular pH (Chesler, 1990; Brookes, 1997). The magnitude of the pH change can be expected to depend on intracellular and extracellular buffering systems as well as the activities of acid and base transporters (and exchangers) in retinal neurons and Müller cells. In CNS glial cells, a Na^+/HCO_3^- exchanger has been implicated in regulation of internal pH (Ritchie, 1987; Chesler, 1990).

A bicarbonate exchanger, present in Müller cells, could serve a similar function. Newman (1991) demonstrated electrogenic Na^+/HCO_3^- exchange in isolated Müller cells from the salamander retina (Fig. 3.17). The Na^+/HCO_3^- exchanger has a stoichiometry of ~3:1 (HCO_3^-:Na^+), and is predominantly localized to the endfoot region of the Müller cell. It is likely that the exchanger is involved in extruding HCO_3^- from Müller cells into the vitreous body. Therefore, when CO_2 is metabolized in Müller cells, a rise in $[HCO_3^-]_i$ occurs which would result in an efflux of HCO_3^- into the vitreous through the exchanger (Newman and Astion, 1991).

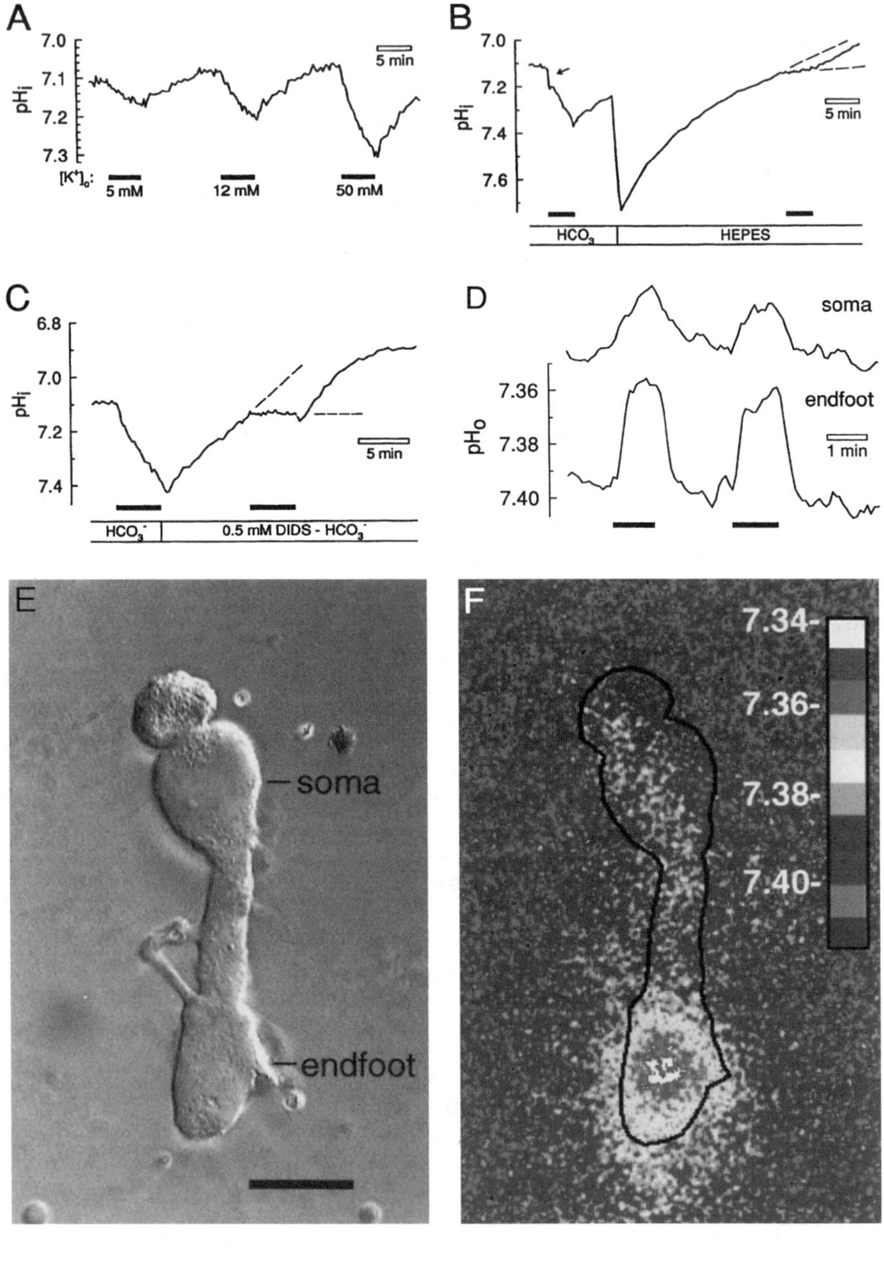
A
7.0
7.1
7.2
7.3
pHi
5 min
[K+]o:
5 mM
12 mM
50 mM
B
7.0
7.2
7.4
7.6
pHi
5 min
HCO3
HEPES
C
6.8
7.0
7.2
7.4
pHi
5 min
HCO3-
0.5 mM DIDS - HCO3-
D
soma
endfoot
7.36
7.38
7.40
pHo
1 min
E
soma
endfoot
F
7.34-
7.36-
7.38-
7.40-

Bicarbonate transport is known to be mediated by anion exchangers present in the plasma membrane (Kopito, 1990). Anion exchangers (AE) belong to a multigene family involved in the regulation of intracellular pH and chloride concentration in many tissues (Kopito, 1990). In the rat retina, two different isoforms of the AE3 gene product have been described (Kobayashi et al., 1994): A 125 kDa form is expressed in horizontal cells whereas a 165 kDa form is found in Müller cells (Fig. 3.18). AE3 in Müller cells is found predominantly in the endfoot region, an observation in agreement with the localization of the Na^+/HCO_3^- exchanger to the Müller cell endfoot region. In developing retina, AE3 is expressed at very low levels at embryonic stages but increases steadily during postnatal development as the retina becomes functionally mature (Kobayashi et al., 1994).

If the carbonic anhydrase present in Müller cells is involved in regulating pH levels in the retina, one would expect carbonic anhydrase inhibitors, such as acetazolamide and methazolamide, to have strong effects on extracellular pH. As shown in Fig. 3.19, there is experimental evidence that in the presence of the inhibitors, there is an increase in the light-evoked acidification in the inner plexiform layer and in the subretinal space (Borgula et al., 1989; Oakley and Wen, 1989).

As mentioned before, one consequence of CO_2 fixation is the lowering of intracellular pH in Müller cells. What happens to H^+ inside Müller cells? It has been proposed that the excess protons inside are released into the capillaries close to Müller cells, or into the interstitial fluid, or even into the vitreous through the endfoot (Newman, 1991). A Na^+/H^+ exchanger involved in H^+ extrusion has been localized in astrocytes (Kimelberg et al., 1979), and it seems likely that a Na^+/H^+ exchanger serves a similar function in the Müller cell plasma membrane.

The intimate association of Müller cells with the retinal vasculature and the influence of extracellular ionic changes on blood flow indicate that ionic activities of Müller cells may be linked to changes in retinal metabolism. In this regard, an attractive idea is that HCO_3^- efflux through the

←

Figure 3.17. Bicarbonate transport in Müller cells. The three figures show potassium-evoked intracellular alkalinization in Müller cells isolated from salamander retina. A. The rate and extent of alkalinization is greater for larger increases in $[K^+]_o$. B. substitution of HEPES for HCO_3 reduces the rate of alkalinization. C. Addition of DIDS, a Na^+–HCO_3^- blocker, reduces the rate of alkalinization. D. pHo was recorded simultaneously beneath the endfoot and soma of a single cell in HCO_3^- Ringer. E. Nomarski image of a Müller cell. F. Difference ratio image of pHo changes evoked by raising $[K^+]_o$ from .5 to 50 mM in HCO_3^- Ringer. Increasing $[K^+]_o$ evokes an acidification that is largest beneath the endfoot (outline) (Newman, 1996). (Copyright 1991 *The Journal of Neuroscience*, reprinted with permission.)

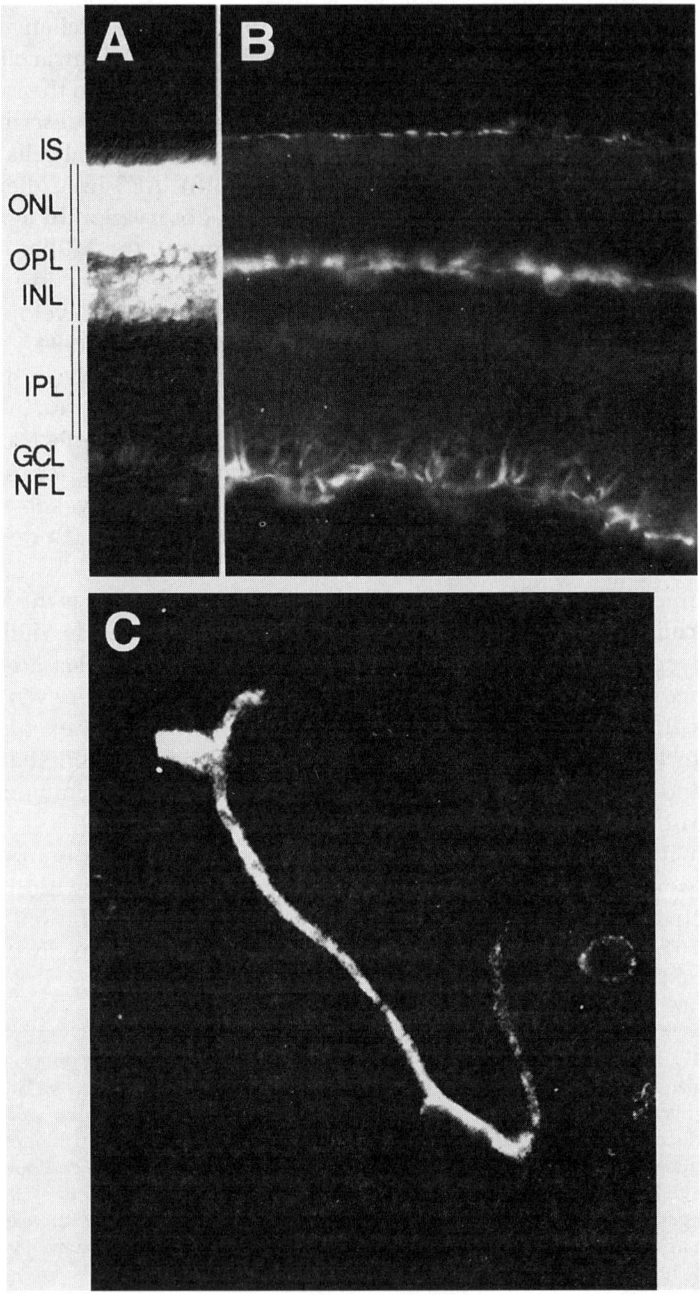
A
B
IS
ONL
OPL
INL
IPL
GCL
NFL
C

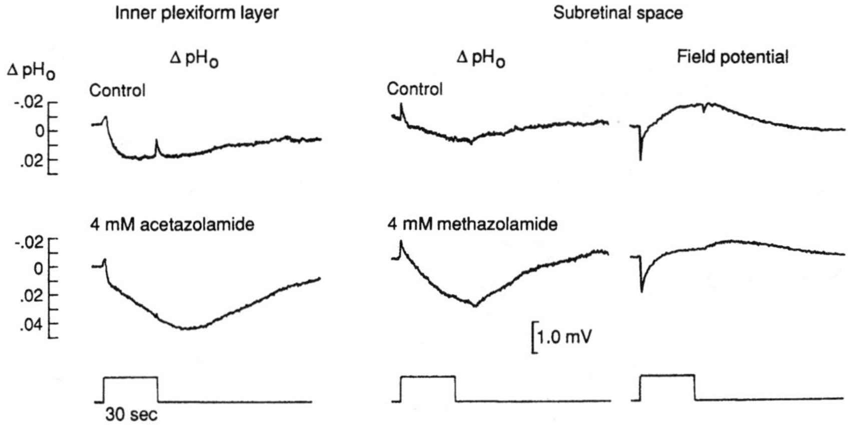

Figure 3.19. Effect of carbonic anhydrase inhibitors on retinal pH. The figure shows pHo from inner plexiform layer and subretinal space, with the simultaneously recorded field potential (right). Control responses (top) were obtained prior to superfusion with either acetazolamide or methazolamide (Borgula et al., 1989). (Copyright 1989 Elsevier Science, reprinted with permission.)

Na^+/HCO_3^- exchanger, H^+ efflux through a Na^+/H^+ exchanger, and K^+ efflux through potassium channels in Müller cell membrane can all serve to regulate retinal blood flow in response to variations in neuronal activity (Newman, 1991).

3.5. RETINOID METABOLISM

Vitamin A and its metabolites (retinoids) are the *sine qua non* for vision in all vertebrates. The essential role of vitamin A (retinol) in vision has been known since George Wald, Ruth Hubbard, and their coworkers (Wald, 1968) demonstrated that the 11-*cis* isomer of retinaldehyde (11-*cis* retinal) is the chromophore that binds to opsin within the disc membranes of rod outer segments to form the light-sensitive photopigment, rhodopsin. Light

Figure 3.18. Immunocytochemical localization of AE3 in Müller cells. Sections of rat retina (A,B) or dissociated Müller cells were stained with an antibody against AE3. A. Histological section of retina showing different layers; B. retinal section stained with anti-AE3. Immunoreactivity can be seen in the end feet, apical region and radial processes of Müller cells. C. Solitary Müller cell stained with anti-AE3 (Kobayashi et al., 1994). (Copyright 1994 *The Journal of Neuroscience*, reprinted with permission.)

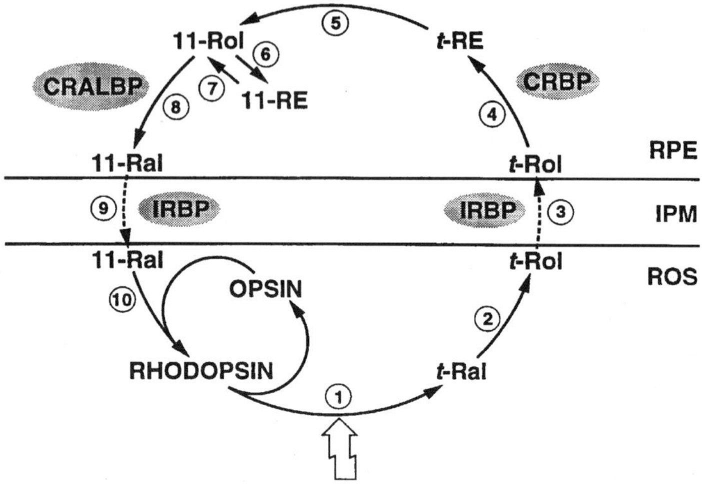

Figure 3.20. Schematic of the vitamin A cycle showing steps with potential involvement of retinoid binding proteins. Note the absence of binding proteins in the rod outer segment (ROS). CRALBP, cellular retinaldehyde-binding protein; CRBP, cellular retinol-binding protein; IRBP, interstial retinoid-binding protein; Ral, retinaldehyde; Rol, retinol; RE, retinyl ester; RPE, retinal pigment epithelium; and IPM, interphotoreceptor matrix. Kindly provided by John Saari.

exposure produces an instantaneous isomerization from 11-*cis* retinal to all-*trans* retinal, which subsequently degrades thermally through a series of intermediate stages, leading ultimately to the formation of all-*trans* retinol free of the opsin moiety. In this form, the retinoid is transferred to the RPE, where it is esterified, and eventually isomerized and oxidized to 11-*cis* retinal for return to the photoreceptors and the reformation of the visual pigment (Fig. 3.20). This series of reactions, from the bleaching of the photopigment to its reformation, is called the "visual cycle."

3.5.1. Retinoid-binding Proteins

Because of their potential toxicity and hydrophobic nature, retinoids are almost never found free outside or inside cells, but are generally stored as retinyl esters or are associated with binding proteins (Saari, 1994). Thus, extracellular retinoids are associated with IRBP or a retinol-binding protein (sRBP), and within retinal cells, retinoids are bound to a group of cellular retinoid-binding proteins: cellular retinaldehyde-binding protein (CRALBP), cellular retinol-binding protein (CRABP), and cellular retinoic acid binding protein (CRBP) (Saari, 1994). Although direct experimental evidence is lacking, it is generally assumed that the intracellular retinoid-binding proteins are involved in cytoplasmic transport and routing of retinoids, but they may serve also as substrate carriers for enzymatic reactions (Saari and Bredberg, 1982; Ong et al., 1994; Napoli, 1991).

More definitive clues to the functions of retinoid-binding proteins have come from studies that have examined the phenotypes of knockout mice. In the case of CRBP I, the protein appears to be indispensable for efficient retinyl ester synthesis and storage, and its absence results in 6-fold faster turnover of retinol (Ghyselinck et al., 1999). The mice appear normal, which could mean CRBP is not involved in regulating retinoic acid synthesis. The RBP-knockout mice have a markedly impaired ERG during the first few months of life but develop a normal response when fed a vitamin A-sufficient diet (Quandro et al., 1999). They are otherwise normal in appearance. Liver retinyl ester stores accumulate but cannot be mobilized. Thus, the animals are dependent on a regular vitamin A intake. The results suggest that an alternate pathway of delivery of retinol to tissues (including RPE) exists but it is inefficient *in utero.*

In CRALBP-knockout mice, visual pigment synthesis is delayed by about 15-fold (J. Saari, personal communication). HPLC analysis shows that retinyl esters accumulate during the delay, suggesting that the delay occurs at the isomerohydrolase step. Finally, the kinetics of visual pigment renewal

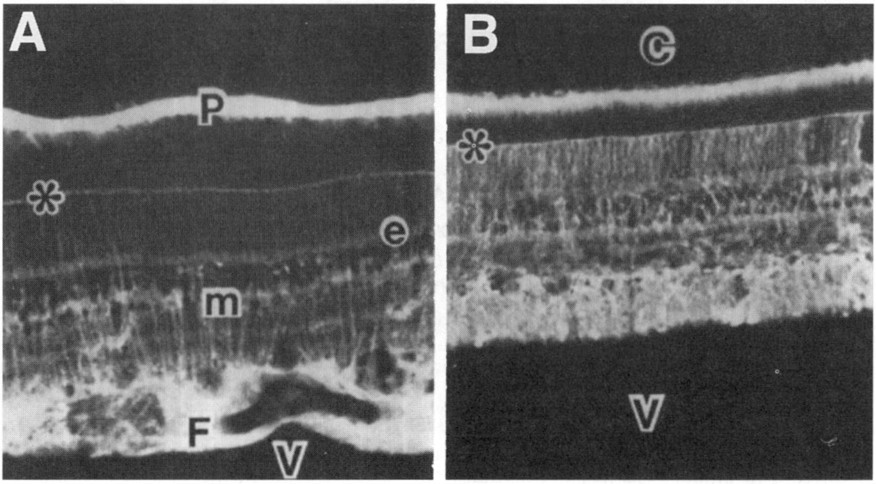

Figure 3.21. CRALBP is localized in Müller cells. Antibodies to CRALBP stain radial fibers of Müller cells in bovine (A) and monkey (B) retinas. P, Pigment epithelium; m, Müller cell; V, vitreous; C, choroid (Bunt-Milam and Saari, 1983). (Copyright 1983 The Rockefeller University Press, reprinted with permission.)

are normal in IRBP-knockout mice (Palczewski et al., 1999; Ripps et al., 2000). However, photoreceptors continue to die even when the animals are raised in the dark.

The availability of highly specific polyclonal and monoclonal antibodies has made it possible to investigate the cellular localization of retinoid binding proteins (Saari, 1994). Using this approach, Bunt-Milam and Saari (1983) made the surprising observation that CRALBP was localized in Müller cells as well as in the RPE of bovine retina (Fig. 3.21). When isolated from retina, this protein displays a high affinity for 11-*cis* retinaldehyde as

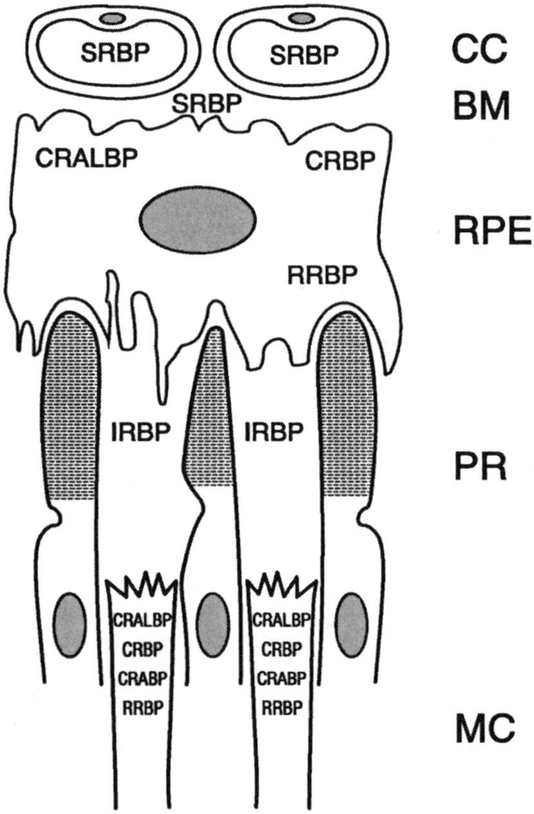

Figure 3.22. Localization of cellular retinoid binding proteins in the eye. CRALBP appears to be present in Müller cells in all vertebrate retinas. CRBP and CRABP have been found in Müller cells only in some retinas. RRBP has been reported in the Müller cells in some retinas. CC, choriocapillaris; BM, Bruch's membrane; RPE, retinal pigment epithelium; PR, photoreceptor; MC, Müller cell.

well as for 11-*cis* retinol (Saari et al., 1978, 1982). Since it had been generally assumed that Müller cells are not involved in any aspect of vitamin A metabolism associated with the visual cycle, the observation was totally unexpected.

Subsequent immunocytochemical studies showed that CRBP, which binds all-*trans* retinol to form a substrate for esterification, is also localized to Müller cells (Bok et al., 1984; Eisenfeld et al., 1985; De Leeuw et al., 1990). Thus, binding proteins for two of the key retinoids that participate in the visual cycle are found in Müller cells of all species examined (Fig. 3.22). Even more intriguing, two additional retinoid-binding proteins are found in Müller cells. In the primate retina, CRABP has been localized to Müller cells. In addition, a seven-transmembrane, microsomal protein with all-*trans* retinaldehyde-binding activity has been localized to Müller cells, where it probably acts as a substrate carrier in the oxidation of retinaldehyde to retinoic acid (Pandey et al., 1994).

3.5.2. Müller Cells and the Visual Cycle

Although retinoid-binding proteins have been found in Müller cells, no studies, to our knowledge, have directly addressed the question of

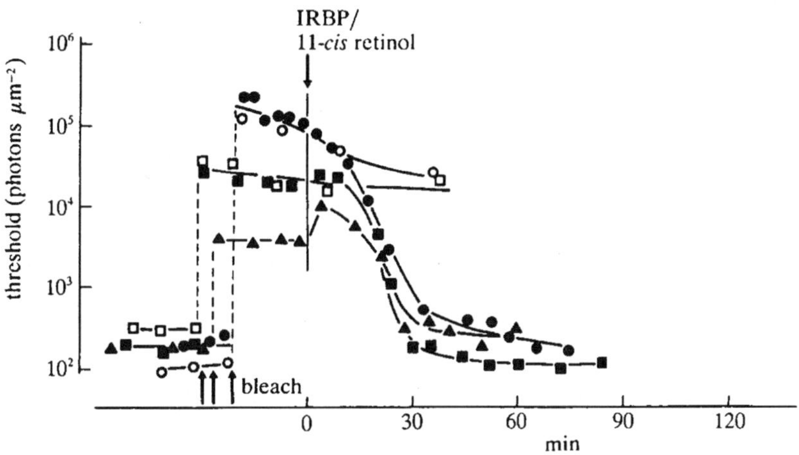

Figure 3.23. Recovery of sensitivity after bleaching in salamander cone photoreceptors. Solid symbols represent thresholds for three cones, before and after bleaching and during recovery with 11-*cis*-retinol with IRBP as carrier. Open symbols represent controls with bleached cones exposed to IRBP without retinoid (Jones et al., 1989). (Copyright 1989 G. J. Jones, reprinted with permission.)

whether Müller cells are involved in the visual cycle. There is, however, evidence that Müller cells are capable of synthesizing retinoids. Edwards et al. (1992) showed that Müller cell cultures, incubated with radioactive [^{3}H]-all-*trans* retinol, contain small amounts of retinyl esters, and are able to synthesize and release retinoic acid (See Fig. 2.15). Although 13-*cis* and all-*trans* retinal were found in the tissue, little were seen in the medium.

Another study that relates to this issue (Das et al., 1992) reported that Müller cells from the predominantly cone retina of the chicken are able to convert all *trans*-retinol into all-*trans* and 11-*cis*-retinyl palmitate, and to then release 11-*cis* retinol into the culture medium. Because cone photoreceptors, but not rods, are able to use 11-*cis* retinol to regenerate their photopigments (Jones et al., 1989) it is conceivable that the Müller cell plays a role in cone pigment regeneration (Fig. 3.23). This intriguing, but highly speculative, possibility brings to mind earlier studies indicating that cone pigments can be regenerated after bleaching in the isolated retina, i.e., in the absence of the RPE (Goldstein, 1967, 1970), but rod pigments do not regenerate. Based on these considerations, it would be hardly surprising if future studies proved Müller cells participate in certain aspects of the visual cycle.

Neuron–Glia Signaling Pathways 4

Interactions between neurons and glial cells occur throughout the nervous system during development, in synapse formation, and in many other functions under both normal and pathological conditions (Barres, 1991; Murphy et al., 1993; Nedergaard, 1994; Pfrieger and Barres, 1996; McCall et al., 1996; Pfrieger and Barres, 1997). Although there is no histological or physiological evidence of gap junctional pathways for direct communication between neurons and glial cells, there is a surprising array of intercellular interactions mediated by ions, metabolites, and various signaling molecules. This chapter describes some of the neuron–glia interactions mediated by neurotransmitters, calcium, and nitric oxide.

In addition to their conventional roles at nerve and muscle synapses, neurotransmitters provide a chemical pathway linking the activities of neurons and glial cells. In the vertebrate retina and throughout the CNS, neurotransmitter amino acids constitute one of the principal classes of molecule mediating intercellular communication. Because precise regulation of the magnitude and duration of transmitter activity is required, there is a need to inactivate or remove the transmitter in an efficient and timely manner from synaptic sites. Unlike the mechanism whereby acetylcholine is inactivated, degradative enzymes for amino acids and other neuroactive compounds are not present extracellularly, and termination of transmitter action depends largely on reuptake into neurons and/or transport into glial cells (cf. Kanner, 1994). After uptake, the transmitter may be degraded, transported into synaptic vesicles for reuse (Kanner and Sharon, 1978; Kish et al., 1989), or converted to a precursor for subsequent transmitter synthesis (cf. Paulsen and Fonnum, 1990). The importance of membrane transporters for the viability and functional integrity of neural tissue cannot be overemphasized (cf. Müller et al., 1998); even cholinergic neurons re quire an active transport process to recapture choline following acetylcholine hydrolysis (Kuhar and Murrin, 1978).

Glial cells possess high-affinity, carrier-mediated transport systems, and it is now firmly established that glia are major players in the active removal of neuroactive substances (Henn and Hamberger, 1971; Henn et al., 1974; Erecinska, 1987; Hansson and Rönnbäck, 1991). Furthermore, their biochemical properties suggest that glial cells may be involved in transmitter degradation and recycling, and the electrophysiological properties of their transporters indicate that they are able to discharge neurotransmitter into the extracellular space (cf. Nicholls and Attwell, 1990; Martin, 1992). Indeed, the glutamate released from glial cells has been shown to provide a signaling pathway by which neuronal NMDA receptors can be activated (Parpura et al. 1994, 1995). The transport systems of Müller cells exhibit many of the same features as other forms of glia, and it is likely these cells are engaged in similar activities in the retina (cf. Linser et al., 1984; Sarthy, 1983; Brew and Attwell, 1987; Szatkowski et al., 1990).

Glial cell membranes also contain voltage-gated (Chapter 5) and ligand-gated channels that behave very much like their neuronal counterparts (MacVicar et al., 1989; Wyllie et al., 1991; Clark and Mobbs, 1992). Here, too, the Müller cell has proven a good model for study. The findings have provided new insights into the functional and pharmacological properties of glial receptors and amino acid transporters (Brew and Attwell, 1987; Nicholls and Attwell, 1990; Malchow et al., 1989; Qian et al., 1993, 1996; Biedermann et al., 1995; Billups et al., 1996). Although the retina uses a multitude of neurotransmitters (cf. Adler, 1983; Pourcho et al., 1984; Biedermann et al., 1995), this chapter will focus primarily on neuron–glia interactions mediated by GABA and glutamate, the main inhibitory and excitatory amino acids, respectively, in retina and CNS. In addition, there is clear evidence that fluctuations in intracellular calcium, induced by activation of transmitter-gated ion channels, provide an important mechanism for neuron–glia signaling. Lastly, the observation that nitric oxide (NO) can serve as an intercellular messenger in neural tissues suggests yet another pathway for communication between neurons and glial cells (Garthwaite, 1991). Both cell types have the capacity to make NO, but whether glial cell NO affects neuronal function—and vice versa—is an issue that remains difficult to resolve. However, the emergence of NO as a major signaling molecule warrants consideration when discussing possible pathways for neuron–glia interactions.

4.1. MEMBRANE TRANSPORT

The rapid removal of amino acid transmitters is accomplished by transport proteins located in the plasma membranes of neurons and glial cells. The active transport process is typically Na^+-dependent, and driven by the transmembrane Na^+ gradient that is maintained by Na^+/K^+ ATPase activity. Thus, agents such as cyanide that block ATP synthesis, or substances such as ouabain that interfere with establishing an ionic gradient, inhibit transmitter uptake (cf. Kanner, 1983). In addition to Na^+, the coupling of other ions, e.g., K_i^+ or Cl_o^-, is often required for transport. GABA uptake, for example, is catalyzed by a Na^+ and Cl^--dependent transporter, whereas the glutamate carrier requires the countertransport of K^+ as well as Na^+ uptake. The stoichiometry suggests that for both neurotransmitters, the transport process is electrogenic—i.e., it is capable of generating a transmembrane current when activated (see Section 4.2).

4.1.1. Uptake of Radioactive Tracers

Tracer flux measurement is frequently used to determine the rate at which cells accumulate exogenously applied amino acids (cf. Keynan and Kanner, 1988). This technique enables visualization of the translocation of label to intracellular sites, and is particularly useful when the transport process does not involve a net transfer of electrical charge. Neurotransmitter uptake by Müller cells and retinal neurons has been frequently demonstrated using this approach. Although both cell types are able to accumulate radiolabeled amino acids, there are significant interspecies differences in the identity of the cells that take up the transmitter (cf. Marshall and Voaden, 1975; Ehinger, 1977). Autoradiographic studies of the cellular uptake of tritium-labeled GABA provide a good example of this variability (Table 4.1).

Table 4.1. Species Comparison of Retinal Cell Types Accumulating ^{3}H-GABA

Species	Cell type(s)	Reference
Frog	Horizontal, amacrine	Voaden et al., 1974
Goldfish	Horizontal, amacrine	Lam and Steinman, 1971; Marshall and Voaden, 1975
Pigeon, chicken	Horizontal, amacrine, ganglion	Marshall and Voaden, 1974b
Rabbit	Müller, amacrine	Ehinger, 1970, 1977
Cat, baboon	Müller	Marshall and Voaden, 1975
Guinea pig, goat	Müller	Marshall and Voaden, 1975
Mouse	Müller [amacrine, ganglion][a]	Blanks and Roffler-Tarlov, 1982
Rat	Müller	Neal and Iverson, 1972
Skate	Müller	Lam, 1975; Ripps and Witkovsky, 1985

[a]After prolonged (60 min) incubation.

In teleost, avian, and amphibian retinas, ^{3}H-labeled GABA appears to accumulate exclusively in neurons of the inner nuclear and ganglion cell layers (Lam and Steinman, 1971; Marshall and Voaden, 1974a; Voaden et al., 1974; Pourcho et al., 1984). On the other hand, ^{3}H-GABA is taken up primarily by the Müller cells in elasmobranch (Lam, 1975; Ripps and Witkovsky, 1985) and most mammalian retinas (Fig. 4.1). However, in rabbit and

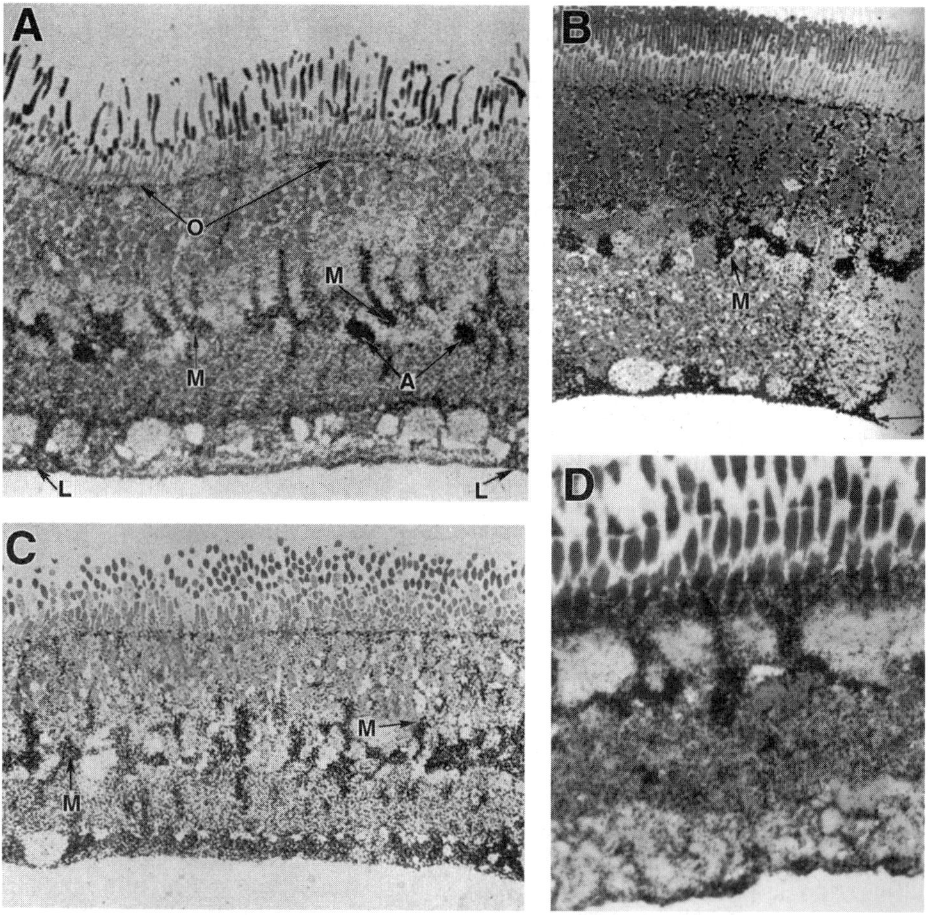

Figure 4.1. Light microscope autoradiographs of ^{3}H-GABA uptake in the retinas of several vertebrate species. Activity appears confined to the Müller cell fibers in rat (B), guinea pig (C), and skate (D) retinas. In rabbit (A), amacrine cells as well as Müller cells are labeled (Marshall and Voaden, 1974b, 1975; Ripps and Witkovsky 1985). ([B] copyright 1974 Academic Press, Inc., reprinted with permission. [A], [C] copyright 1975 Elsevier Science, [D] copyright 1985 Elsevier Science, reprinted with permission.)

mouse retinas, amacrine cells as well as Müller cells accumulate GABA (Marshall and Voaden, 1974b, 1975; Ehinger, 1977; Neal and Iverson, 1972; Blanks and Roffler-Tarlov, 1982); with prolonged incubation, mouse ganglion cells also take up radiolabeled GABA (Blanks and Roffler-Tarlov, 1982). Thus, GABA transport into Müller cells and into GABAergic nerve terminals can be expected to limit the spread of GABA from its release sites and thereby regulate GABA concentration in the synaptic clefts and extracellular space.

The identification of cells engaged in neurotransmitter uptake is sometimes complicated by inherent technical limitations. Because Müller cell endfeet occupy almost the entire surface area of the inner retina, caution needs to be exercised in interpreting results obtained when radiolabeled agents are applied to the vitreal surface of the intact retina. This is a particularly serious problem when a transmitter is delivered *in vivo* by intravitreal injection (cf. Ishida and Fain, 1981). The avid uptake of amino acids into Müller cells can mask uptake into neurons that would otherwise have accumulated these substances. This limitation will become more apparent when comparing some of the conclusions drawn from autoradiography with results obtained by electrophysiological methods (Section 4.2).

A second source of uncertainty is that the radiolabeled transmitter may be metabolized and lost to the incubation medium. This problem can often be minimized by pretreatment with compounds that inhibit degradation of the transmitter, and result in greater retention of radioactivity. Thus, the use of agents, e.g., hydroxylamine or amino-oxyacetic acid (AOAA), that block the activity of the degradative enzyme GABA-T, has revealed that the magnitude of transmitter uptake may be grossly underestimated because of degradation (Martin, 1976; Neal and Starr, 1973; Ehinger, 1977).

Although most of the GABA taken up by Müller cells is probably rapidly metabolized (see Chapter 3), it is important to recognize that carrier-mediated transport processes subserving the cellular uptake of amino acids can operate also in the reverse direction to discharge these substances (cf. Levi and Raiteri, 1993). In goldfish, for example, [^{3}H]GABA efflux can be evoked from horizontal cells, both *in vitro* and *in situ* (Ayoub and Lam, 1984; Yazulla, 1985), and a similar phenomenon has been reported in mammalian Müller cells. When rat retina is preloaded with radioactive GABA, the neurotransmitter appears to be taken up mainly by Müller cells (Neal and Iverson, 1972). It can then be released by electrical stimulation as well as by the application of solutions containing high K^+, veratridine, or the GABA mimetic ethylenediamine (cf. Voaden and Starr, 1972; Sarthy, 1983). However, the assumption that Müller cells alone take up exogenous GABA in rat retina is questionable, and other cell types may also be responsible for its discharge. Interestingly, both Ca^{2+}-dependent and Ca^{2+}-independent mechanisms have been implicated (Sarthy, 1983), the latter presumably mediated by reversal of the uptake mechanism. More definitive information on the voltage- and ionic-dependence of "reverse transport" has been obtained with electrophysiological methods, and results obtained with this approach leave little doubt that Müller cells are capable of transmitter release (cf. Szatkowski et al., 1990). It has yet to be determined whether GABA is released under normal conditions, and in sufficient concentration to affect neuronal excitability.

The uptake of glutamate, the major excitatory transmitter in the retina, has also been well studied. Fewer species have been examined for glutamate uptake using radioactive tracers, but several studies demonstrated that glutamate is accumulated by Müller cells and photoreceptors in the mammalian retina by a Na^+-dependent, high-affinity uptake mechanism (White and Neal, 1976; Ehinger, 1977; Sarthy et al., 1985). However, the use of electrophysiological methods in the analysis of membrane transport can often provide greater insight into the ionic dependence of the process. A more complete description of the glutamate uptake system is provided in Section 4.2.

4.1.2. GABA and Glutamate Transporters

Considerable progress has been made in elucidating the molecular structure and diversity of neurotransmitter transporters. Briefly, the GABA transporters belong to a large gene family that includes transporters for dopamine, serotonin, and several other neuroactive substances (Guastella et al., 1990, 1992; Clark et al., 1992; Kanner, 1994). Several glutamate transporters have also been cloned and characterized (Storck et al., 1992; Kanai and Hediger, 1992; Pines et al., 1992; Fairman et al., 1995; Arriza et al., 1997; Eliasof et al., 1998). Their molecular structure indicates they constitute a distinct family, unrelated to the GABA family of transport proteins. Schematic drawings representative of the two families of neurotransmitter transporters illustrate their predicted membrane topology (Fig. 4.2), and a number of excellent reviews provide detailed descriptions of the structural features, putative regulatory mechanisms, and ionic requirements of these transport proteins (cf. Uhl, 1992; Amara and Kuhar, 1993; Kanner, 1994; Worrall and Williams, 1994; Malandro and Kilberg, 1996).

Within each family of transporters, there are several subtypes, which are often distributed differentially among neurons and glial cells. Regions of divergence in the amino acid sequences of the various transport proteins probably determine their specificity for a particular transmitter, whereas conserved regions may account for a feature common to most high-affinity plasma membrane transporters, namely, their sodium dependence. Using the electrochemical gradient of sodium ions, these transporters can accumulate neurotransmitter against significant concentration gradients.

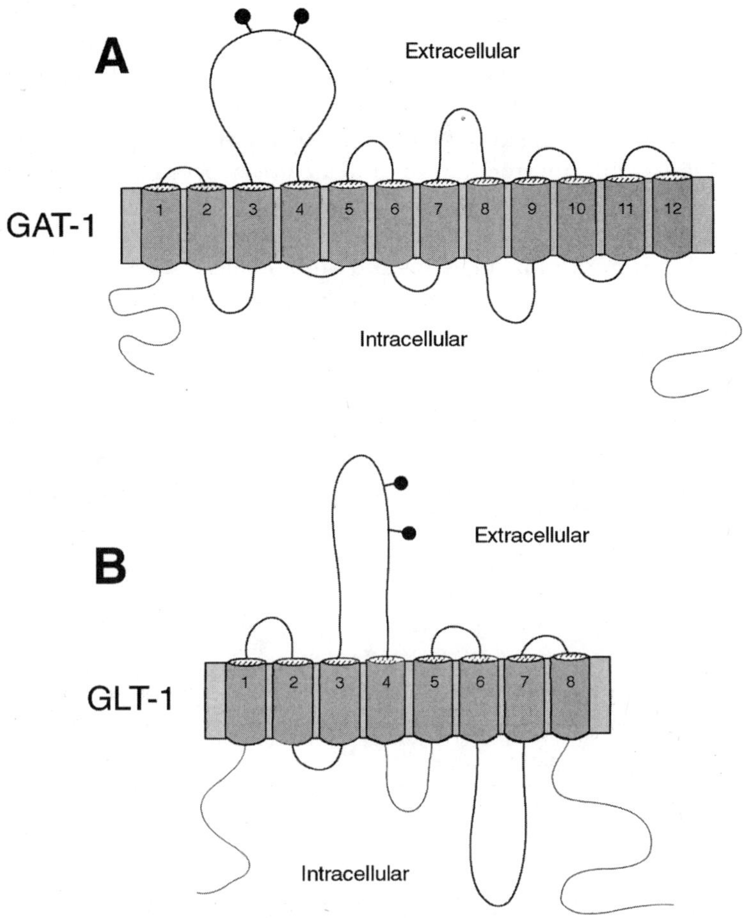

Figure 4.2. Schematic representation of GABA (GAT-1) and glutamate (GLT-1) transporters. Hydropathy analysis reveals 12 putative transmembrane domains for the GABA transporter (A), and 8 for the glutamate transporter (B). Filled circles indicate potential glycosylation sites (Worral and Williams, 1994). (Copyright 1994 The Biochemical Society, reprinted with permission.)

4.1.3. Localization of GABA Transporters

Studies of neuronal and glial preparations showed that GABA uptake systems for these different cell types exhibit different pharmacological properties (Amara and Kuhar, 1993). It was assumed, therefore, that there were at least two GABA transporters (GATs) expressed in the CNS: one specific to neurons and another specific to glial cells. In agreement with this expectation, several GABA transport proteins have been cloned and characterized (Guastella et al., 1990; Borden et al., 1992). In the mammalian retina, three GABA transporters (GAT-1, GAT-2, and GAT-3), which differ in peptide sequence, structure, and pharmacologic properties, have been reported (Ruiz et al., 1994; Brecha and Weigmann, 1994; Honda et al., 1995; Johnson et al., 1996). Immunocytochemical and *in situ* hybridization studies show that GAT-1 and GAT-3 are expressed in the neural retina, whereas GAT-2 is confined to RPE and the ciliary epithelia. However, the expression of GAT-1 and GAT-3 by inner retinal neurons and Müller cells illustrates the complexity of GABA uptake pharmacology in the retina. GAT-1 is highly expressed by amacrines, displaced amacrines, and some ganglion cells, but is found only at a low level in Müller cells (Fig. 4.3) (Ruiz et al., 1994; Honda et al., 1995;

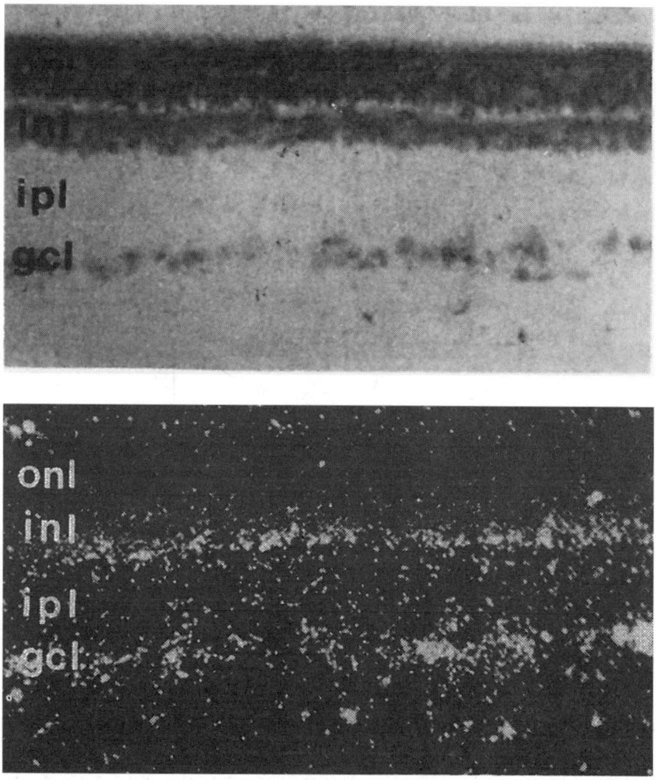

Figure 4.3. Localization of GABA transporter, GAT-1, in the mouse retina. (A) and (B) are bright- and dark-field micrographs, respectively, showing GAT-1 mRNA localization by *in situ* hybridization using 35S-labeled probes. Silver grains are localized to cell bodies in the inner nuclear and ganglion cell layers (Ruiz et al., 1994). (Copyright 1994 Association for Research in Vision and Ophthalmology, reprinted with permission.)

Brecha and Wegmann, 1994). In contrast, GAT-3 is found predominantly in Müller cells, as well as in some amacrines in the proximal INL. GAT-3 immunoreactivity is found throughout the Müller cell, including regions where there is no evidence for GABAergic synapses, e.g., the outer nuclear layer. This result is similar to GAT-3 localization in astrocytic processes that are not in the vicinity of GABAergic synapses (Minelli et al., 1996).

Perhaps the most intriguing outcome of the localization studies is the finding that nonmammalian retinal horizontal cells, which are known to avidly take up GABA, do not express either GAT-1 or GAT-3, suggesting that another plasma membrane transporter mediates GABA uptake by these cells (Yang et al., 1997). Moreover, a particular retinal cell could express more than one GAT, and many neurons that synthesize GABA may not express a GABA transporter. It appears, therefore, that GABA uptake may not be an ideal marker for identifying GABAergic neurons.

4.1.4. Localization of Glutamate Transporters

Based on DNA sequence, pharmacology, and channel properties, five types of glutamate transporter (also called excitatory amino acid transporter, EAAT)—EAAT1 (GLAST), EAAT2 (GLT-1), EAAT3 (EAAC1), EAAT4, and EAAT5—have been identified (Seal and Amara, 1999). EAAC1 is localized mostly to neurons whereas GLT-1 and GLAST appear to be the predominant glutamate transporters in astrocytes (Seal and Amara, 1999). EAAT4 is found only in the cerebellum whereas EAAT5 appears to be predominantly in the retina (Fairman et al., 1995; Arriza et al., 1997; Eliasof et al., 1998). Interestingly, EAAT4 and EAAT5 appear to function both as transporters and as ion channels permeable to Cl^- (see Section 4.2.2). Immunocytochemical studies suggest that EAAC1 is present in many neurons in the INL and the ganglion cell layer (GCL) (Rauen et al., 1996; Schultz and Stell, 1996) whereas EAAT5 is localized to both neurons and Müller cells (Arriza et al., 1997; Eliasof et al., 1998). Surprisingly, GLT-1 is localized mainly to cones and bipolar cells and is absent from both Müller cells and astrocytes (Rauen and Kanner, 1994). GLAST, however, has been localized exclusively to Müller cells (Fig. 4.4) (Rauen et al., 1996; Derouiche and Rauen, 1995; Lehre et al., 1997; Pow and Barnett, 1999; Wang et al., 1999). These results are in accord with earlier biochemical and electrophysiological data which suggested that Müller cells express a Na^+-dependent, high-affinity glutamate transporter (Neal and White, 1971; Sarthy et al., 1986; Brew and Attwell, 1987; Schwartz and Tachibana, 1990).

Recently, GLT-1 and GLAST knockout mice have been generated (Tanaka et al., 1997; Harada et al., 1998). Although the ERG in mice lacking GLT-1 appear to be normal, the b-wave and oscillatory potentials are depressed in GLAST-knockout mice. In addition, lack of GLAST leads to

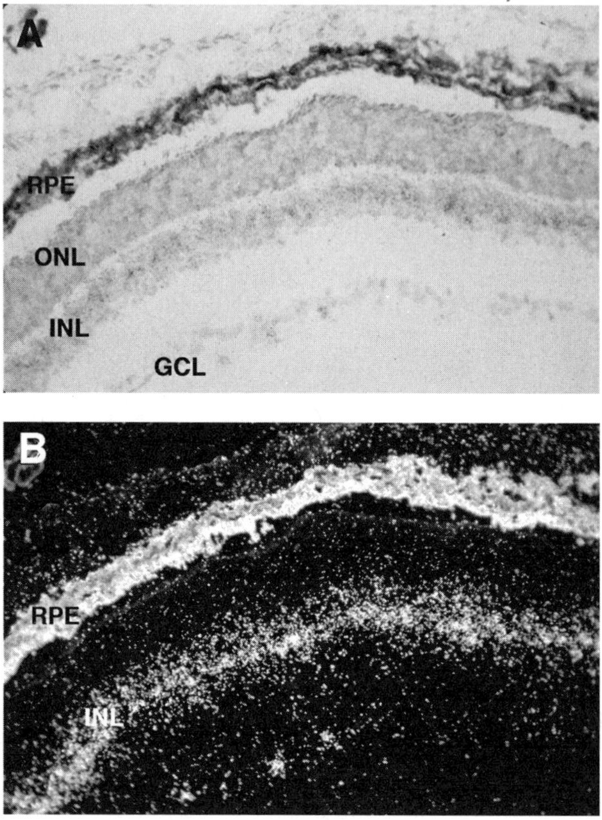

Figure 4.4. Localization of glutamate transporter, GLAST, in the mouse retina. (A) and (B) are bright- and dark-field micrographs, respectively, showing GLAST mRNA localization by *in situ* hybridization using ^{35}S-labeled riboprobes. Silver grains are localized to cell bodies in the inner nuclear and ganglion cell layers.

elevated extracellular glutamate, and neurodegenerative changes characteristic of excitotoxicity. These observations suggest that glutamate transport by Müller cells may play an important role in modulating synaptic signaling in the retina as well as serving to protect the retina from chronic glutamate neurotoxicity.

4.2. ELECTROPHYSIOLOGY OF MEMBRANE TRANSPORT

Because of the electrogenic nature of the transport process, electrophysiological methods have been useful for studying membrane transporters and ion exchangers. Simply stated, the cotransport of ions across the cell membrane typically results in a net charge displacement that gives rise to a transmembrane current. A good example of this approach can be found in the work of Attwell and associates who applied the whole-cell patch clamp method (Hamill et al., 1981) to study transport currents in isolated Müller cells (Brew and Attwell, 1987; Attwell et al., 1989). As pointed out by Attwell et al. (1989), electrophysiological techniques have several advantages over conventional autoradiographic procedures for investigating certain aspects of the transport machinery: (1) cells can be held at different potentials to determine the voltage dependence of the uptake mechanism; (2) the composition of the solutions inside and outside the cell can be controlled for pharmacological study; and (3) the kinetics of the carrier mechanism can be examined on a rapid time scale that minimizes the effects of intracellular metabolism. Studies dealing with the uptake of glutamate and GABA by Müller cells provide good examples of the power of this approach, but one should be aware that electroneutral transport, or the masking of small uptake currents by ligand-gated channel activity, may cause an uptake process to go undetected.

4.2.1. Glutamate Transport

A unique feature of the vertebrate photoreceptor is that, in darkness, there is a sustained cationic current carried mainly by Na^+ influx through cGMP-gated channels in the receptor outer segments (cf. Fesenko et al., 1985; Baylor, 1987). This current keeps the plasma membrane depolarized, resulting in continued release of the excitatory amino acid, glutamate. Thus, neurotransmitter discharge from the receptor terminal is maximal in darkness and is suppressed by the hyperpolarizing effect of photic stimulation (Trifonov, 1968; Dowling and Ripps, 1973; Schacher et al., 1974; Ripps et al., 1974, 1976; Schmitz and Witkovsky, 1996). Because glutamate released from photoreceptors serves, in turn, to induce a further intraretinal release of glutamate from OFF bipolar cells, it is evident that the removal of glutamate is essential not only to regulate synaptic transmission, but also to keep the glutamate concentration below neurotoxic levels (Olney, 1978; Rothman and Olney, 1986; Velasco et al., 1996). It is not surprising, therefore, that glutamate transporters are present on Müller cells as well as on retinal neurons (Brew and Attwell, 1987; Cammack and Schwartz, 1993; Eliasof and Werblin, 1993; Grant and Dowling, 1995).

The electrogenic transport mechanism mediating glutamate uptake has been studied in detail on Müller cells that were enzymatically isolated from amphibian retinas (Brew and Attwell, 1987), and several key findings have been replicated on isolated Müller cells from a mammalian (rabbit) retina (Attwell et al., 1989). As Fig. 4.5A shows, application of L-glutamate

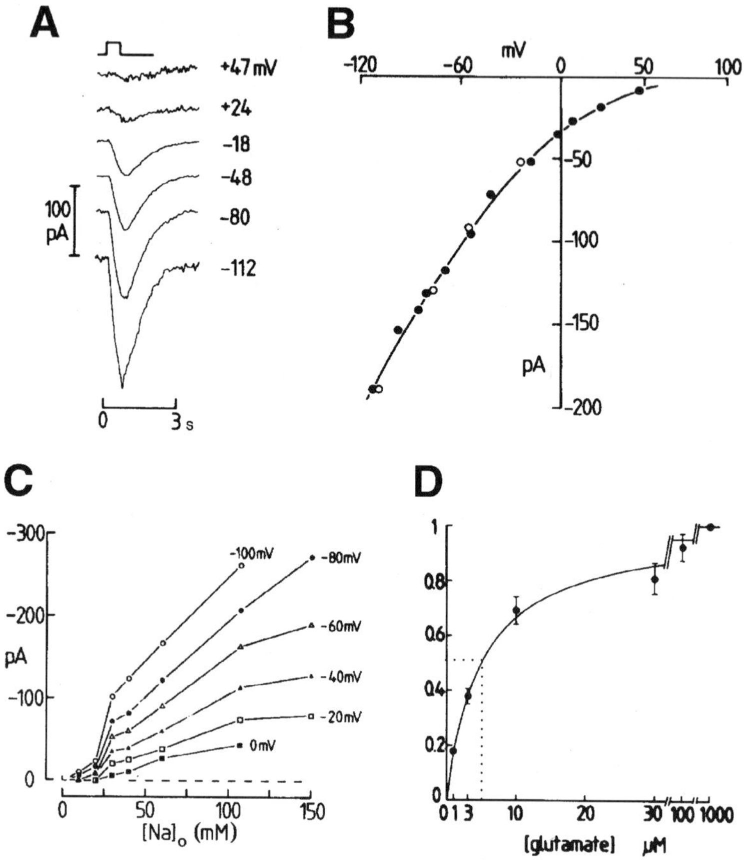

Figure 4.5. Glutamate uptake currents recorded from isolated Müller cells of the tiger salamander retina. A. The amplitude of the inward current increases with membrane hyperpolarization. B. The I-V curve shows no reversal as the cell is voltage-clamped to depolarizing potentials. C. Sodium-dependence of the transport current measured at holding potentials from 0–100 mV (indicated to the right of the data points). Note that the greater the hyperpolarization and the lower the $[Na]_o$, the smaller is the current response to 100 μM glutamate; in the absence of extracellular Na^+ (0 $[Na]_o$ on the scale of abscissae) the responses to glutamate are blocked at all holding potentials. D. Glutamate-induced uptake responses of rabbit Müller cells are mediated by a high affinity transporter with a K_m of 5 μM (Brew and Attwell, 1987; Sarantis and Attwell, 1990). (A, B, C copyright 1987 Macmillan Magazines Ltd., reprinted with permission. D copyright 1990 Elsevier Science, reprinted with permission.)

to axolotl Müller cells elicits an inward current that decreases in amplitude as the holding potential is moved from −120 mV to +47 mV (Brew and Attwell, 1987). However, even with the cell depolarized to +47 mV, there is no sign of reversal to an outward current (Fig. 4.5B), which would typically occur near 0 mV for glutamate-gated ion channels (cf. Wyllie et al., 1991). These findings, and the observation that the glutamate-induced responses are not accompanied by an increase in membrane current noise, argue for the presence of an electrogenic glutamate uptake carrier rather than the opening of ion channels through activation of glutamate receptors. Consistent with this interpretation are results indicating that removal of extracellular Na^+ completely abolishes the glutamate-evoked current (Fig. 4.5C), and that glutamate analogs are far less effective than L-glutamate in inducing a membrane current (cf. Brew and Attwell, 1987; Schwartz and Tachibana, 1990; Barbour et al., 1991).

Na^+-dependent glutamate uptake is mediated by a high affinity transporter with a K_m of about 20 M for axolotl (Brew and Attwell, 1987) and tiger salamander Müller cells (Schwartz and Tachibana, 1990; Barbour et al., 1991), and by one of even higher affinity (K_m = 5 μM) in rabbit Müller cells (Fig. 4.5D) (Sarantis and Attwell, 1990). Because glutamate is a negatively charged amino acid, the fact that transporter activity is associated with an inward current indicates that Na^+, or other cationic species, contributes a net positive charge that accompanies glutamate transport into the cell.

It appears now that in addition to its dependence upon sodium, the carrier requires the outward movement of intracellular K^+. As Fig. 4.6A shows, removing intracellular K^+ (or raising $[K^+]_o$) greatly reduces the glutamate-induced current (Sarantis and Attwell, 1990; Barbour et al., 1988, 1991; Szatkowski et al., 1991). Thus, there is an ordered process of ion exchange by the glutamate transporter: After sodium and glutamate are released into the cell, potassium binds to the carrier and is transported outward before a new cycle begins (Kanner and Bendahan, 1982).

However, the stoichiometry of the glutamate transporter in Müller cells appears to be even more complex. The previously cited data, together with the results obtained from measurements of intra- and extracellular pH (Bouvier et al., 1992), and a consideration of the carrier-associated anion conductance (Billups et al., 1996; see Section 4.2.2), strongly suggest that glutamate uptake induces an intracellular acidification by transporting H^+ into the cell. Accordingly, for each cycle of the carrier, the inward current derives from cotransport into the cell of one glutamate anion together with two sodium ions and one proton, and the countertransport of one potassium ion (Fig. 4.6B). As a result, there is a net transport of one positive charge per glutamate anion; i.e., the electrogenic current is proportional to the number of glutamate ions entering the cell.

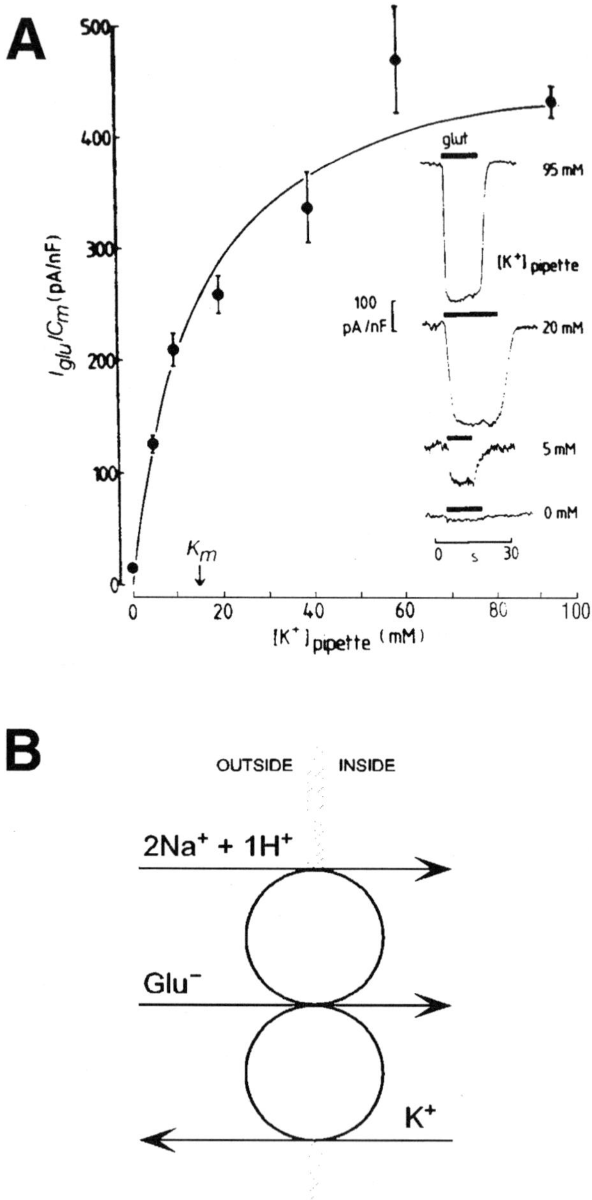
A
I_{glu}/C_m (pA/nF)
500
400
300
200
100
0
K_m
0
20
40
60
80
100
$[K^+]_{pipette}$ (mM)
glut
95 mM
$[K^+]_{pipette}$
100
pA/nF
20 mM
5 mM
0 mM
0
s
30
B
OUTSIDE
INSIDE
2Na⁺ + 1H⁺
Glu⁻
K⁺

4.2.2. Anion Conductance and Glutamate Transport

An unusual property of the glutamate transporter is its association with an anionic conductance carried by Cl^- (Fairman et al., 1995). This glutamate-activated chloride current has been recorded from Müller cells (Billups et al., 1996; Eliasof and Jahr, 1996), from oocytes expressing several subtypes of excitatory amino acid transporters (Arriza et al., 1997; Eliasof et al., 1998), as well as from retinal neurons (e.g. cone photoreceptors, ON bipolar cells) that exhibit glutamate transport activity (Eliasof and Werblin, 1993; Grant and Dowling, 1995). In each case, the anion conductance, activated when the transporter binds glutamate and Na^+, accounts for a significant fraction of the glutamate-elicited current.

Patch-clamp recordings from isolated salamander Müller cells (Fig. 4.7) show that in the presence of normal extracellular chloride, exposure to 100 μM L-glutamate evokes a large inwardly rectifying current. The current is inward at hyperpolarizing potentials and reverses to an outward current at potentials more positive than the chloride equilibrium potential ($E_{Cl} \approx 0$ mV). Removing chloride from the bath (replacement with gluconate) eliminates the outward current, but has no effect on the transporter-activated inward current. Thus, transport activity is not dependent upon activation of the anionic conductance. Moreover, the outward current occurs at depolarizing potentials for which the transporter is inactive, indicating the anionic conductance can occur independently of net glutamate transport. Nevertheless, both response components require extracellular sodium.

The cloning of excitatory amino acid transporters and their functional characterization in the oocyte expression system has enabled extension of these observations to the rapidly expanding subfamily of glutamate transporters. cDNAs encoding a variety of excitatory amino acid transporter (EAAT) subtypes were identified and shown by immunocytochemistry to be present in amphibian and mammalian retina (cf. Rauen et al., 1996; Arriza et al., 1997; Eliasof et al., 1998). Five distinct subtypes were isolated from a

Figure 4.6. The potassium dependence of glutamate transport. A. Removing intracellular K^+ by perfusion through the whole-cell patch pipette reduces glutamate-induced currents in a concentration-dependent fashion (bracketed values are the number of experiments for each data point). Inset shows actual recordings of the responses to 100 μM glutamate obtained at different values of $[K^+]_i$. Glutamate does not produce an inward current in the absence of intracellular K^+ (Barbour et al., 1988). B. Stoichiometry of electrogenic glutamate transport in Müller cells. A net transfer of positive charge results from the inward movement of one glutamate anion, two sodium ions, and one proton, coupled to the outward movement of one potassium ion (Barbour et al., 1988). (Copyright 1988 Macmillan Magazines, Ltd., reprinted with permission.)

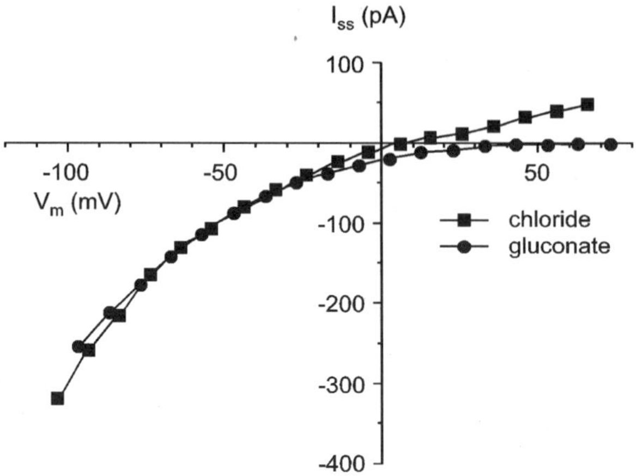

Figure 4.7. The association of the glutamate transporter with an anionic conductance is evident in current-voltage data from tiger salamander Müller cells. Steady-state responses (I_{ss}) were elicited by 100 μM L-glutamate in normal (high) chloride (■) and in chloride-free (●) solutions; the internal (pipette) solution also contained high chloride. The removal of $[Cl^-]_o$ eliminates the outward current carried predominantly by the influx of external chloride ions, but has no effect on the inward (transport) current (Eliasof and Jahr, 1996.) (Copyright 1996 National Academy of Sciences, U.S.A, reprinted with permission.)

cDNA library prepared from the tiger salamander retina, and at least four of these (sEAAT1, sEAAT2A, sEAAT5A, and sEAAT5B) were shown to be present in Müller cells (Eliasof et al., 1998). It is difficult to explain why so many different glutamate transporters exist in the Müller cell of a single species. Although the various subtypes exhibit somewhat different properties when expressed in *Xenopus* oocytes, they share one significant feature: The transport of glutamate is combined with a mechanism for increasing chloride permeability. It has been suggested that this anionic conductance may be important for preventing a reduction in the rate of transport due to the depolarization that would occur as a result of electrogenic glutamate uptake (Eliasof and Jahr, 1996).

4.2.3. GABA Transport

The electrophysiology of the Müller cell GABA transporter provides a number of parallels, as well as differences, when compared with results obtained for glutamate transport. In both skate and mammalian retina, radiolabeled GABA accumulates almost exclusively in Müller cells (Section 4.1.1). GABA uptake is known to be mediated by a sodium-dependent, electrogenic mechanism (Keynan and Kanner, 1988; Malchow and Ripps, 1990; Qian et al., 1993; Faude et al., 1995), and studies in *Xenopus* oocytes expressing the GABA transporter GAT-1 indicate that each GABA molecule is cotransported with two sodium ions and one chloride ion (Mager et al., 1996).

The current generated by GABA transport in Müller cells appears often to be much smaller than that resulting from glutamate uptake. For example, application of 100 μM GABA to Müller cells from guinea pig retina elicits an inward current of about 12 pA with the cells voltage-clamped at −80 mV (Biedermann et al., 1994); this is about 6-fold less than the current induced by only 20 μM glutamate at a comparable holding potential in rabbit Müller cells (Sarantis and Attwell, 1990). It is not known whether this is due to fewer GABA transport molecules on Müller cell membranes, to the differential loss of protein as a result of the dissociation procedure, or simply to species differences.

In the skate, a more unusual situation is encountered—a good example of the care that must be exercised in identifying the cell types involved in transmitter uptake. We noted earlier that based on autoradiographic findings, ^{3}H-GABA uptake in skate retina occurs almost exclusively in Müller cells, with little or no uptake into horizontal cells or other neurons (see Fig. 4.1). On the other hand, voltage-clamp recordings from skate horizontal and Müller cells give diametrically opposite results. The GABA-induced membrane currents of Müller cells appear to result solely from activation of $GABA_A$ receptors (Fig. 4.8C,D; Malchow et al., 1989), whereas the GABA currents of horizontal cells are mediated by a high affinity electrogenic uptake mechanism (Fig. 4.8A,B; Malchow and Ripps, 1990).

It is not easy to reconcile these findings. The apparent lack of GABA uptake into horizontal cells in the intact retina could possibly be explained by their limited access to exogenous GABA, i.e., most of the applied ^{3}H-GABA may have been removed by Müller cells at the site of application. Indeed, if the autoradiographic findings are correct, GABA uptake must surely occur in skate Müller cells. Why then do we fail to record GABA-induced currents that are consistent with the presence of a transport mechanism? As it turned out, the large currents generated by activation of the $GABA_A$ receptors on Müller cells completely masked the small current elicited by the GABA transport process. Using a clever experimental approach that exploits the sodium dependence of the GABA transporter (cf. Akaike et al., 1987), it is possible to demonstrate electrophysiologically the GABA transport activity of Müller cells (Qian et al., 1993).

The strategy is quite straightforward. Applying very low concentrations of GABA, and comparing responses obtained in the presence and in the absence of sodium, serves to expose the competition between the GABA receptor and the GABA transporter for the available GABA. As illustrated in

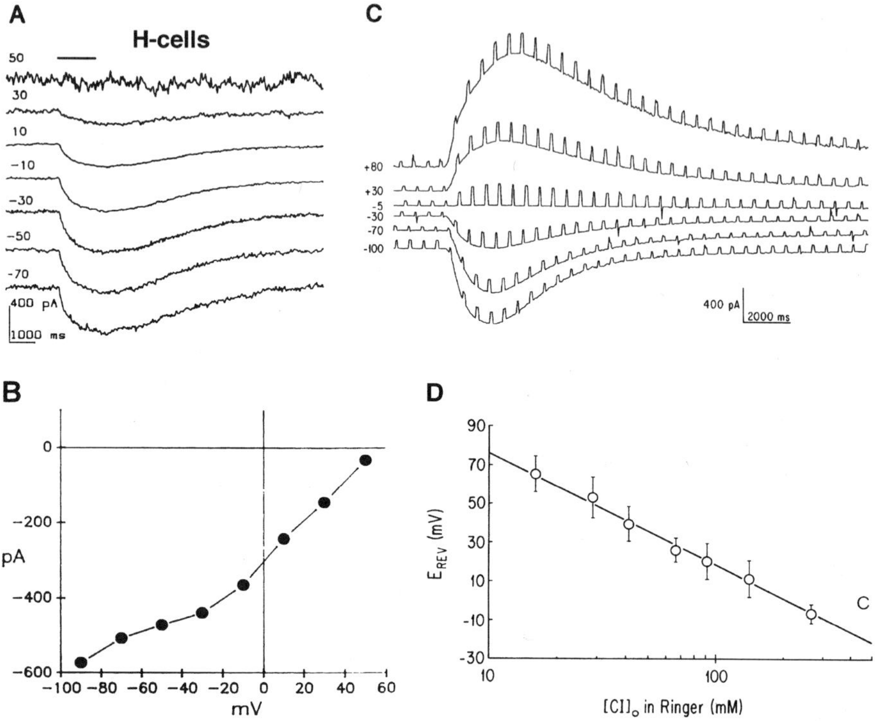

Figure 4.8. The current responses resulting from activation of the GABA transporter on horizontal cells, and $GABA_A$ receptors on Müller cells. A. GABA-induced inward currents of skate horizontal cells are mediated by an electrogenic carrier mechanism, and as shown in B, the resultant I–V curve shows no sign of a reversal potential. C. The GABA-induced currents of skate Müller cells result from activation of $GABA_A$ receptors. The voltage-dependence of the response to GABA exhibits a reversal potential (at ~0 mV) as the holding potential is stepped from −100 mV to +80 mV. The increase in the current responses to brief 10 mV depolarizations superimposed on each trace reflects the GABA-induced increase in membrane conductance. D. Dependence of the reversal potential on [Cl^-] follows the Nernst equation (Malchow et al., 1989; Malchow and Ripps, 1990). (Copyright 1989, 1990 National Academy of Sciences, U.S.A., reprinted with permission.)

Fig. 4.9A, low concentrations of GABA induce current responses that are significantly enhanced when sodium ions are eliminated from the bathing solution. Thus, inhibiting GABA transport by removing extracellular Na^+ allows GABA to act solely on the $GABA_A$ receptor, resulting in currents of larger amplitude. But as the GABA concentration is increased, the current contributed by the GABA transporter is proportionately reduced until, with GABA concentrations $>2\ \mu M$, removing Na^+ no longer enhances the

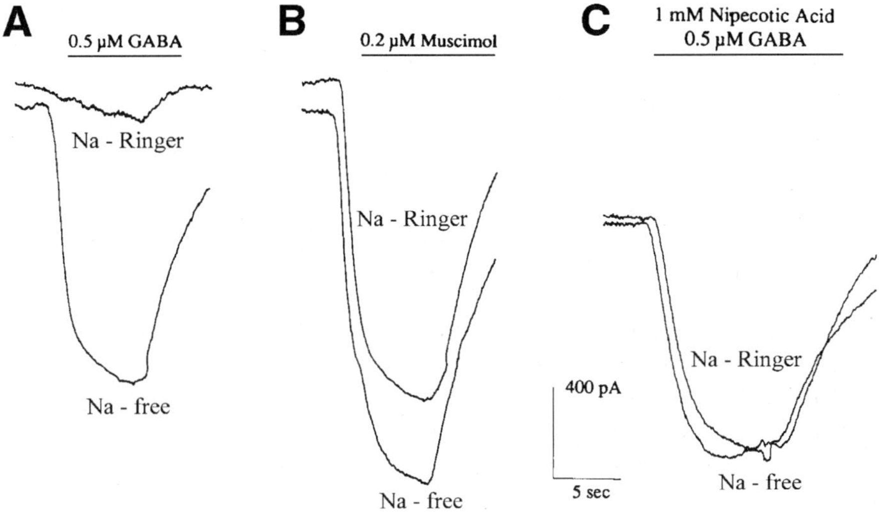

Figure 4.9. $GABA_A$ receptor-mediated currents can be used to unmask the small GABA uptake current of skate Müller cells. A. Removal of extracellular sodium suppresses the uptake current, and enhances the $GABA_A$-mediated currents in response to a low concentration (0.5 µM) of GABA. B. Removing extracellular Na^+ has no effect on the response to the $GABA_A$ agonist muscimol, which does not activate the GABA transporter, nor is the response enhanced when the uptake current is blocked by nipecotic acid (C) (Qian et al., 1993). (Copyright 1993 Plenum Publishing Corp., reprinted with permission.)

GABA-induced response (not shown). Results obtained with the $GABA_A$ receptor agonist muscimol (Johnston et al., 1968) and the GABA uptake blocker nipecotic acid (Schousboe et al., 1978) are consistent with this interpretation. Because muscimol activates the $GABA_A$ receptor, but is a poor ligand for the GABA transporter (DeFeudis, 1980), removing Na^+ does not affect the muscimol-induced response (Fig. 4.9B). When GABA responses are elicited from Müller cells bathed in nipecotic acid, the GABA-induced currents are similar with and without sodium in the Ringer solution (Fig. 4.9C). Thus, the electrophysiological findings confirm the autoradiographic evidence of a GABA transporter on skate Müller cells.

4.3. TRANSMITTER-OPERATED ION CHANNELS

In the previous section, we saw the power of electrophysiological techniques in the analysis of amino acid membrane transporters. As already shown, these methods have led to the remarkable discovery that functional, "neuronal-like" neurotransmitter receptors are expressed on glial cell membranes. The striking similarities in channel properties to those found on their neuronal partners initially raised concerns that the glial receptors might represent an artifact of the cell cultures, and not normally present in the intact tissue. Although culture conditions can undoubtedly influence the expression or functional properties of ion channels (cf. Barres et al., 1990), results obtained from acutely isolated Müller cells (Malchow et al. 1989), and *in situ* recordings from astrocytes in hippocampal slices (MacVicar et al., 1989) and retinal astrocytes in flat mount preparations (Clark and Mobbs, 1992), have largely dispelled these concerns. Indeed, there is abundant evidence that glial cells throughout the nervous system express receptors for a broad range of neurotransmitters and neuromodulators (Murphy and Pearce, 1987; Barres et al., 1990; Gallo and Russell, 1995). Here we describe key features of the membrane currents elicited by activation of GABA and glutamate receptors on Müller cells.

4.3.1. GABA Receptors

Three pharmacologically distinct classes of GABA receptor (GABA*R*) have been identified: the $GABA_A$, $GABA_B$, and $GABA_C$ receptors (cf. Sieghart, 1992; Bowery, 1989; Bormann and Feigenspan, 1995). $GABA_A$*R*s and $GABA_C$*R*s are ligand-gated chloride channels. They differ, however, in that the $GABA_A$*R*s are competitively inhibited by bicuculline and allosterically modulated by barbiturates, whereas $GABA_C$*R*s are bicuculline-insensitive and unresponsive to barbiturates. $GABA_B$*R*s, on the other hand, are coupled to potassium and calcium channels through G-protein and intracellular second-messenger pathways, are selectively activated by baclofen, and are antagonized by phaclofen and related compounds.

Many neurons of the vertebrate retina express one or more of these receptor types (cf. Slaughter and Bai, 1989; Qian and Dowling, 1993, 1995), but thus far only $GABA_A$*R*s have been demonstrated conclusively on Müller cells—and surprisingly, only in skate (Malchow et al., 1989) and primate (baboon and human) retinas (Reichelt et al., 1996). Like their neuronal counterparts, the $GABA_A$*R*-mediated currents of skate Müller cells are carried by Cl^- (Fig. 4.8D), and as Fig. 4.10 shows, the GABA-induced currents are highly sensitive to GABA, blocked by bicuculline in a dose-dependent fashion, and enhanced by pentobarbital and diazepam.

Despite the similarities to neuronal $GABA_A$-*R*, there are some unusual features associated with the $GABA_A$*R*s of skate and baboon Müller cells. Most notable is the response to the divalent cation zinc, which is found in the glutamatergic synaptic terminals of photoreceptors (Wu et al., 1993; Qian et al., 1997) as well as neurons in other parts of the CNS, and is

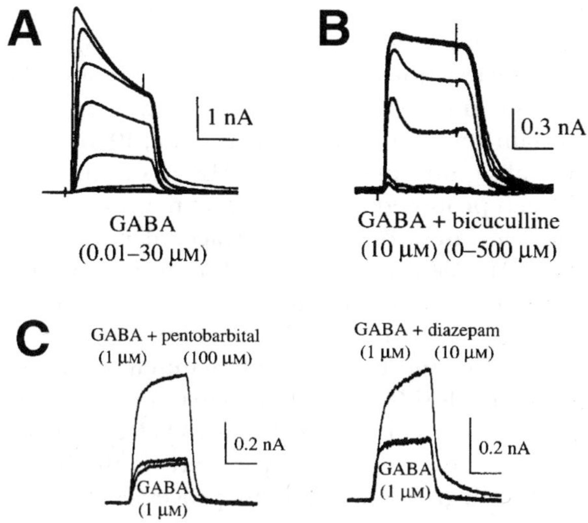

Figure 4.10. Isolated skate Müller cells possess functional $GABA_AR$s that activate membrane chloride channels A. Concentration dependence of the GABA-induced currents. B. The $GABA_A$ blocker bicuculline suppresses the GABA responses in a dose-dependent fashion. C. Diazepam (which increases the frequency of Cl^- channel openings) and pentobarbital (which increases the average time of channel openings) enhance the actions of GABA at the $GABA_AR$ (Qian et al., 1996). (Copyright 1996 The Royal Society, reprinted with permission.)

presumably coreleased with glutamate (Howell et al., 1984; Pérez-Clausell and Danscher, 1985; Beaulieu et al., 1992. Typically, Zn^{2+} either suppresses or exerts no effect on GABA-induced currents. By contrast, Zn^{2+} enhances significantly the $GABA_AR$-mediated currents of skate Müller cells (Fig. 4.11), and a similar effect is seen with the $GABA_AR$ of baboon Müller cells (Reich-

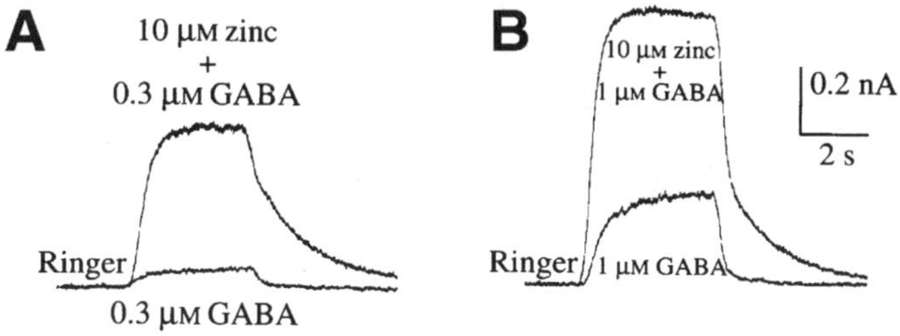

Figure 4.11. Coapplication of 10 μM zinc with 0.3 μM GABA (A) or 1 μM GABA (B) upregulates the $GABA_AR$-mediated currents of skate Müller cells (Qian et al., 1996). (Copyright 1996 The Royal Society, reprinted with permission.)

elt et al., 1997a; Qian and Ripps, 1999). As with other membrane receptors, drug interactions are dependent upon their subunit composition (Draghun et al., 1990); thus, the reactions to zinc suggest that the subunit composition of $GABA_AR$ on skate and baboon Müller cells differs from that identified previously in certain retinal neurons (Qian et al., 1996).

The functional significance of glial $GABA_ARs$ is not fully understood, although it has been proposed that they affect neuronal activity by regulating the chloride concentration of the extracellular space (Bormann and Kettenmann, 1988; MacVicar et al., 1989). Moreover, activation of $GABA_ARs$ has been shown to induce an efflux of HCO_3^- from astrocytes (Kaila et al., 1991); the resultant extracellular alkalinization can affect pH-sensitive mechanisms on voltage- and ligand-gated channels, e.g., enhancement of the calcium-activated chloride current in cone photoreceptors (Barnes and Bui, 1991). Thus, upregulation by zinc of $GABA_AR$ activity on Müller cells could enhance further the influence they exert on the behavior of neighboring neurons. The degree to which these processes affect signal transmission in the vertebrate retina, and the efficacy of regulation by zinc, may be resolved when it becomes possible to examine selectively the GABA-mediated activity of Müller cells within their normal environment, e.g., in a retinal slice preparation.

4.3.2. Glutamate Receptors

Patch-clamp studies of isolated glial cells and of glial cells *in situ* have revealed the presence of glutamate receptors on CNS astrocytes (cf. Steinhäuser and Gallo, 1996). Of the three classes of ionotropic glutamate receptor classes present on neurons, only non-NMDA activated channels, exhibiting many of the pharmacological properties associated with the AMPA receptor subtype, have been consistently demonstrated on glial cells thus far (Usowicz et al., 1989; Backus et al., 1989). Although NMDA-induced currents were recorded from Bergmann glial cells in mouse cerebellar slices (Müller et al., 1993), it was suggested that the responses may have resulted indirectly from neuronal stimulation, a release of glutamate, and subsequent activation of AMPA receptors (Gallo and Russell, 1995).

The retinal glia also express the AMPA subtype of glutamate receptor. Electrophysiological responses recorded *in situ* from rabbit retinal astrocytes provide strong evidence for the presence of non-NMDA glutamate receptors (Clark and Mobbs, 1992), and primary cultures of Müller cells from chick retina have been shown to express AMPA receptors (López-Colomé et al., 1991; López, 1994). The latter finding is consistent with fura-2 results obtained from cultured rabbit Müller cells by Wakakura and Yamamoto (1994), who found that transient increases in $[Ca^{2+}]_i$ are elicited by gluta-

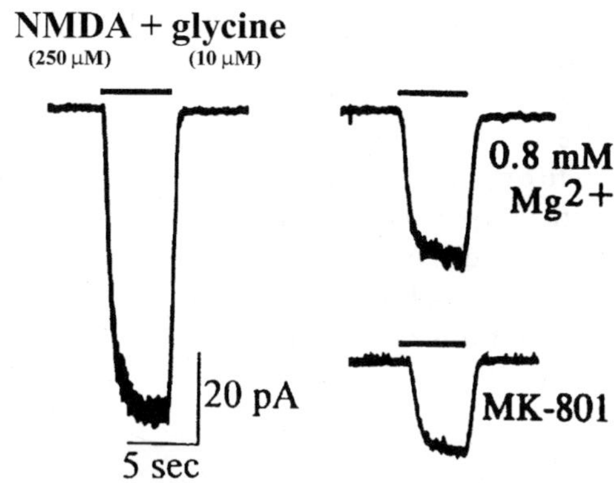

Figure 4.12. NMDA-activated current recordings from a Müller cell in culture; activation of NMDA*R*s requires coapplication of glycine. The inward current elicited by NMDA (250 μM)/glycine (10 μM) is partially blocked by extracellular magnesium and the noncompetitive NMDA antagonist MK-801 (Puro et al., 1996). (Copyright 1996 Cambridge University Press, reprinted with permission.)

mate, AMPA, and kainate, but not by application of NMDA (cf. Hume et al., 1991). Removing extracellular Ca^{2+} or applying the non-NMDA glutamate receptor antagonist 6-cyano-7-nitroquinoxaline-2,3-dione (CNQX) blocked the kainate–AMPA-induced response, indicating that the rise in $[Ca^{2+}]_i$ resulted from an influx of Ca^{2+} through calcium-permeable channels regulated by non-NMDA glutamate receptors (but cf. Section 4.4.1). In contrast, there is evidence that in some species, notably human, NMDA receptors may be the principal type of glutamate receptor. Results of both immunocytochemistry and whole-cell current recordings (Fig. 4.12) show that human Müller cells express functional NMDA receptors (Puro et al., 1996).

As with other transmitter-activated glial cell receptors, the role of glutamate-operated channels on Müller cells is largely speculative. The cation selectivity of these channels (MacVicar et al., 1989; Bormann and Kettenmann, 1988) suggests that the Müller cell response to glutamate may affect the extracellular levels of Na^+, K^+, and Ca^{2+}, and thereby exert a significant effect on neuronal excitability. Moreover, the unusually high levels of glutamate that can occur in pathological conditions (cf. Chapter 6) could trigger a wide range of reactions that are likely to seriously affect glial and neuronal functions. For example, prolonged exposure to glutamate has been shown to induce a proliferative glial-cell response via activation of NMDA receptors (Uchihori and Puro, 1993). In addition, Schwartz (1993) discovered a metabotropic, quisqualate-sensitive receptor that inhibits the K^+ conductance of tiger salamander Müller cells through activation of adenylyl cyclase and its effect on intracellular cAMP. Thus, a pathological rise in extracellular glutamate may suppress the large K^+ conductance of Müller cells (Puro et al., 1996) and impair the buffering mechanisms by which Müller cells regulate the extracellular concentrations of K^+ and glutamate.

In summary, it is becoming increasingly apparent that Müller cells, like the neuroglia of the CNS, are targets for chemical signals released from neurons, and that these elements, previously considered to be passive, actively participate in the processing of visual signals. The presence on Müller cells of membrane transporters, transmitter-operated channels, and voltage-regulated ion channels (Chapter 5), suggests that neuron–glia interactions occur via a variety of pathways that link the actions of Müller cells with every class of neuron in the vertebrate retina.

4.4. A CALCIUM-BASED INTERCELLULAR SIGNAL PATHWAY

Another transmitter-mediated route for neuron–glia interaction is via changes in intracellular calcium. Fluorometric techniques, and the development of Ca^{2+}- indicator dyes such as fluo-3 and fura-2, have made possible simultaneous measurements of spatial and temporal changes in $[Ca^{2+}]_i$. As a result, calcium dynamics has been studied extensively, and it is now well established that many neurotransmitters can mobilize intracellular Ca^{2+}, and trigger Ca^{2+} oscillations in astrocytes both in culture and *in situ* (cf. Reiser et al., 1989; Cornell-Bell et al., 1990; Jensen and Chiu, 1990, 1991; Cornell-Bell and Finkbeiner, 1991; Dani et al., 1992; Nilsson et al., 1992, 1993; Finkbeiner, 1993; di Scala-Guenot et al., 1994). These Ca^{2+} fluxes may afford a means by which bidirectional signals pass between neurons and glial cells and thus provide an effective route for long-range intercellular communication (cf. Kriegler and Chiu, 1993; Nedergaard, 1994; Pasti et al., 1997).

A rise in $[Ca^{2+}]_i$ can occur in glial cells from different sources and by various mechanisms, e.g., activation of a Ca^{2+}-uptake process, or calcium influx through voltage-regulated Ca^{2+} channels (cf. Corvalen et al., 1990; Chapter 5). However, a great deal of interest has been aroused by evidence that activation of glutamate receptors produces a rise in $[Ca^{2+}]_i$ independent of changes in membrane voltage, the operation of glutamate uptake systems, or the presence of extracellular Ca^{2+} (Cornell-Bell et al., 1990). Although the increase in $[Ca^{2+}]_i$ is less robust and not well sustained in the absence of $[Ca^{2+}]_o$, it is now well established that glutamate is capable of inducing the hydrolysis of inositol lipids, resulting in the production of inositol 1,4,5-triphosphate (IP_3), and the liberation of calcium from intracellular stores (cf. Pearse et al., 1986; Jensen and Chiu, 1990).

Pharmacological analysis of the receptors mediating these effects revealed that the glutamate response is mediated primarily through two mechanisms (Fig. 4.13): (1) a voltage-independent, quisqualate-sensitive

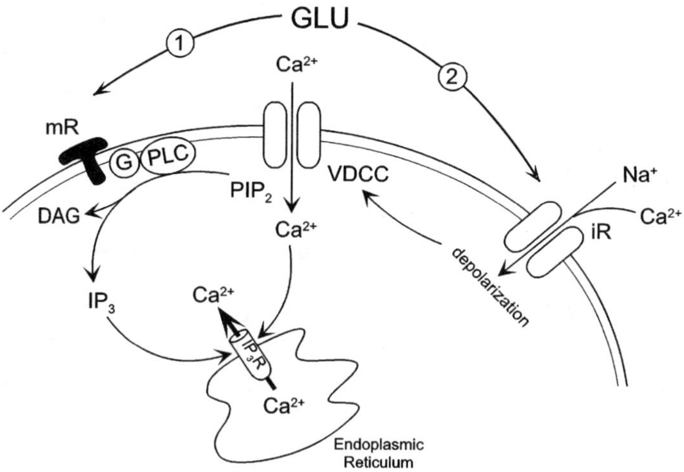

Figure 4.13. Pathways by which glutamate can effect a rise in $[Ca^{2+}]_i$ in glial cells. In pathway 1, activation of a metabotropic receptor (m*R*) induces G-protein stimulation of phospholipase C (PLC), which degrades phosphatidylinositol 4,5-bisphosphate (PIP_2) to form diacylglycerol and inositol 1,4,5-triphosphate (IP_3). The IP_3 activates a receptor (IP_3R) to trigger Ca^{2+} release from Ca^{2+}-storing organelles (e.g., endoplasmic reticulum). Pathway 2 depicts the rise in $[Ca^{2+}]_i$ that occurs when glutamate activates an ionotropic receptor (i*R*), producing an influx of cations. The resultant membrane depolarization activates voltage-dependent Ca^{2+} channels (VDCC) that allow the entry of Ca^{2+}. The rise in $[Ca^{2+}]_i$ through this route may also activate the IP_3R complex to release Ca^{2+} from intracellular storage sites.

second-messenger system that stimulates IP_3 production and the activation of Ca^{2+} channels on intracellular storage sites (Cornell-Bell et al., 1990; Jensen and Chiu, 1990); and (2) activation of an AMPA–kainate receptor subtype that induces membrane depolarization and raises $[Ca^{2+}]_i$ via entry through voltage-gated ion channels (see also Enkvist et al., 1989; Müller et al., 1992; Burnashev et al., 1992). Recent *in situ* studies support this view, showing that glutamate release from synaptic terminals induces a rise in $[Ca^{2+}]_i$ through activation of both metabotropic and ionotropic glutamate receptors (Porter and McCarthy, 1996). Interestingly, it is now thought that the depolarization-induced rise in $[Ca^{2+}]_i$ may be further enhanced by direct calcium activation of the IP_3R (Mignery et al., 1989). Thus far, no studies have conclusively demonstrated that a rise in $[Ca^{2+}]_i$ results from activation of the NMDA-sensitive subtype of glutamate receptor, but there are many other neuroactive substances which modulate Ca^{2+} fluxes, and other mechanisms by which $[Ca^{2+}]_i$ is regulated (cf. Hoth and Penner, 1992; Penner et al., 1993; Finkbeiner, 1993).

4.4.1. Ca^{2+} Waves in Retinal Glia

As previously noted (Section 4.3.2), glutamate-induced increases in $[Ca^{2+}]_i$ in rabbit Müller cells resulted from the entry of extracellular Ca^{2+} via activation of non-NMDA (AMPA–kainate) ionotropic receptors (Wakakura and Yamamoto, 1994). However, using fura-2 fluorescence to view calcium dynamics in individual Müller cells dissociated from tiger salamander retina, Keirstead and Miller (1997) showed that glutamate and various metabotropic receptor agonists elicit large increases in $[Ca^{2+}]_i$ in the absence of extracellular Ca^{2+}, presumably through activation of an IP_3-coupled release of Ca^{2+} from intracellular stores (cf. Fig. 4.13). Although drug application was via the bath solution, and therefore reached all parts of the cell more or less simultaneously, the rise in $[Ca^{2+}]_i$ frequently occurred first in the distal part of the cell and was propagated to the proximal (endfoot) region (Fig. 4.14). The authors speculated that this may be a mechanism by which signals are transmitted from the outer to the inner retinal layers.

Despite the evidence of glutamate-induced calcium waves in astrocytes and Müller cells, Newman and Zahs (1997) report that, when tested *in situ* in the acutely isolated rat retina, neither cell type shows a $[Ca^{2+}]_i$ increase in response to glutamate. Although glutamate alone was ineffective in evoking

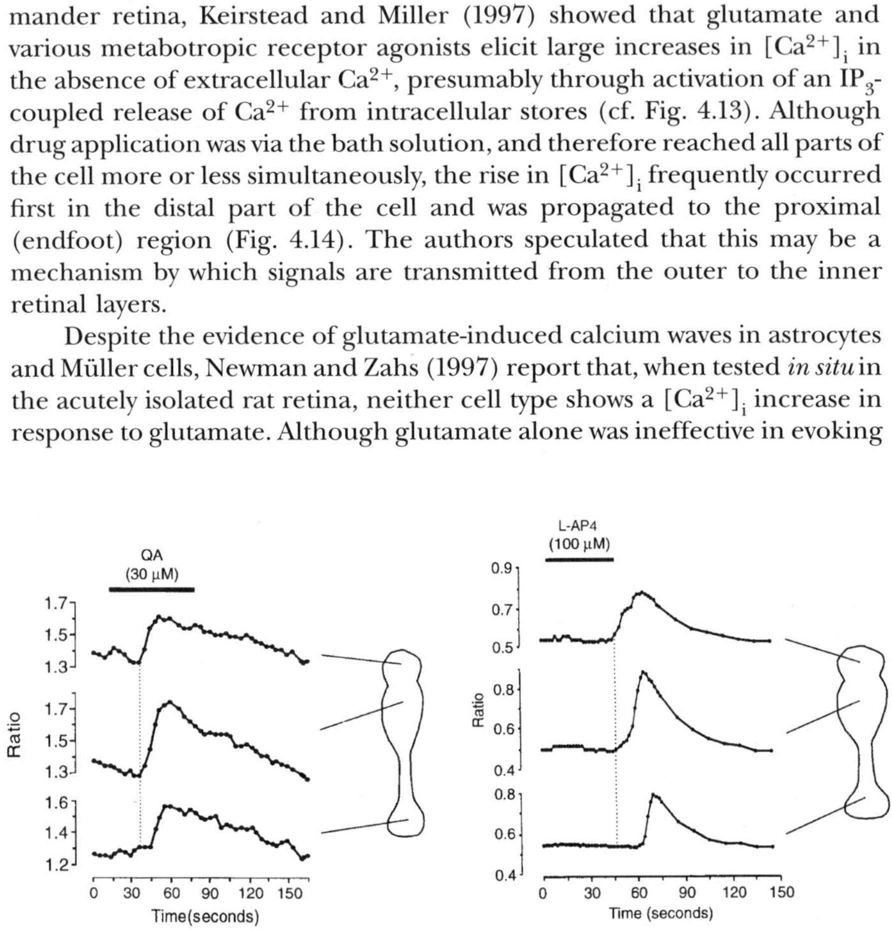

Figure 4.14. Activation of metabotropic glutamate receptors (mGlu*R*) on isolated salamander Müller cells release Ca^{2+} from intracellular stores. Measurements of Fura-2 fluorescence, obtained in the absence of extracellular Ca^{2+}, show the temporal changes in $[Ca^{2+}]_i$ induced by the application of the mGlu*R* agonists quisqualate and L-2-amino-4-phosphonobutyrate (L-AP4). Note that the calcium increase originates in the distal portion of the Müller cell, and spreads proximally through the soma to the endfoot (Keirstead and Miller, 1997). (Copyright 1997 Wiley-Liss, Inc., a subsidiary of John Wiley & Sons, Inc., reprinted with permission.)

a rise in $[Ca^{2+}]_i$, it did potentiate the $[Ca^{2+}]_i$ responses elicited by electrical and mechanical stimulation as well as by application of adenosine triphosphate. These stimuli consistently initiated Ca^{2+} waves in astrocytes that propagated via gap junctions to neighboring astrocytes and Müller cells. Particularly noteworthy was the observation that, irrespective of the type of stimulus, the Ca^{2+} waves were observed in the absence of extracellular Ca^{2+}. Thus, although unresponsive to glutamate, a rise in $[Ca^{2+}]_i$ can be derived from intracellular stores, mediated probably by activation of inositol 1,4,5-triphosphate (IP_3) receptors. Interestingly, the calcium release was not accompanied by significant changes in membrane potential and is therefore unlikely to be involved in the K^+ buffering activity of Müller cells (cf. Chapter 5). It is not clear whether the very different results obtained from rat glial cells *in situ* as compared to those from rabbit and salamander Müller cells in culture are a consequence of differences in tissue preparation or are the result of species variation. Nevertheless, the foregoing observations again demonstrate a potentially important signal pathway linking astrocytes and Müller cells. If it can be shown that this pathway influences neuronal activity, it will be important to learn how the information is transmitted intercellularly, the targets of the signals, and in what way neuronal processes are affected by the propagation of Ca^{2+} waves.

4.5. NITRIC OXIDE

The discovery that a toxic gas such as nitric oxide (NO) can serve as a neuronal messenger (cf. Garthwaite et al., 1988; Moncada et al., 1989) has generated widespread interest in the physiological actions of this molecule. There is a large body of literature on the cellular localization of NO, the biochemical reactions that mediate its cytological effects, its role in synaptic processes, and its involvement in pathological conditions. In the brain, NO has been implicated as a mediator of neurotoxicity, a retrograde signaling molecule, a regulator of blood flow in response to neuronal activity, a potent neuromodulatory agent, and in a diversity of higher-order neuronal processes ranging from long-term potentiation to long-term depression (cf. Hibbs et al., 1988; Gally et al., 1990; Schuman and Madison, 1991; Garthwaite, 1991; Moncada et al., 1991; Dawson et al., 1991; Snyder and Bredt, 1992; Bredt and Snyder, 1994; Schuman and Madison, 1994; Szabó, 1996). Our knowledge of how NO participates in this broad spectrum of activities is still only fragmentary, but the availability of NO donors such as sodium nitroprusside (SNP) and S-nitroso-N-acetylpenacillamine (SNAP), as well as drugs that inhibit nitric oxide synthase (e.g., analogs of L-arginine such as

L-NAME) or scavenge NO (e.g., hemoglobin) has helped provide insight into some of its actions.

4.5.1. General Features

NO, a reactive free radical with an estimated half-life of ~5–10 seconds, is a lipophilic molecule that can rapidly diffuse through cell membranes to exert its influence within a radius of ~200 μm from the site of generation (Wood and Garthwaite, 1994). NO is generated during the conversion of L-arginine to L-citrulline (Fig. 4.15). The reaction is catalyzed by nitric oxide synthase (NOS), a cytosolic enzyme that requires as cofactor nicotinamide adenine dinucleotide phosphate diaphorase (NADPHd). The transfer of electrons from NADPHd to tetrazolium salts, and the enzymatic reduction of soluble nitroblue tetrazolium salt to a visible formazan precipitate, provides a widely used histochemical marker for cells containing NOS (Dawson et al., 1991; Hope et al., 1991).

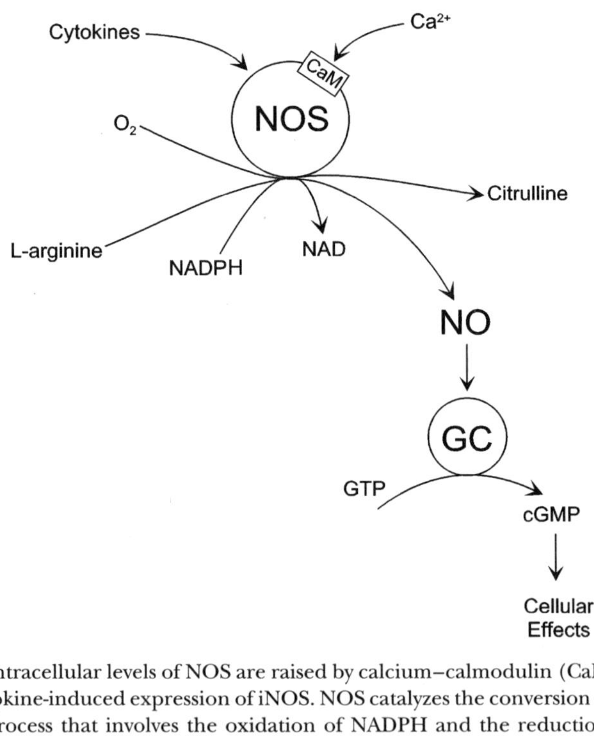

Figure 4.15. Intracellular levels of NOS are raised by calcium–calmodulin (CaM) activation of cNOS or a cytokine-induced expression of iNOS. NOS catalyzes the conversion of L-arginine to citrulline, a process that involves the oxidation of NADPH and the reduction of molecular oxygen. NO can diffuse intra- and intercellularly to activate a soluble guanylate cyclase (GC) which serves to catalyze the formation of cGMP.

There are two distinct classes of NO synthase: a constitutive form (cNOS) and an inducible form (iNOS), both of which have been cloned and sequenced. The two forms display 50% identity and 65% similarity at the amino acid level (Bredt et al., 1991; Xie et al., 1992; Lowenstein et al., 1992), but appear to be distinct genetically and antigenically (Lowenstein and Snyder, 1992). cNOS is Ca^{2+}–calmodulin dependent, relatively short-lived, and produces small quantities (picomoles) of NO; its two isoforms are found predominantly in the vascular endothelium and in neurons. Neuronal cNOS tends to be inactive at resting cytosolic calcium levels (Garthwaite, 1991), but a rise in $[Ca^{2+}]_i$, whether via entry through calcium channels or by release from intracellular stores, will enhance enzyme activity and lead to the production of NO.

iNOS, first characterized in macrophages, is typically calcium independent, and is expressed in many cell types in the presence of cytokines (Moncada et al., 1991; Nathan, 1992; Simmons and Murphy, 1992). Following its induction, iNOS can sustain NO production (in nanomolar quantities) for several hours. Agents that have proven effective in activating the iNOS gene include lipopolysaccharide, interferon γ, tumor necrosis factor α, interleukin 1β, and other cytokines (cf. Goureau et al., 1994b).

Glial cells are an important source of NO. Both the calcium-dependent and inducible forms of NOS are found in glial cells throughout the CNS (Agulló et al., 1995; Weikert et al., 1997; Simmons and Murphy, 1992; Kugler and Drenckhahn, 1996; Merrill et al., 1997). However, it is not known whether both forms co-exist in the same cell, and to what extent each contributes to NO production under physiological or pathological conditions. In the presence of L-arginine, cytokine-mediated induction of iNOS activity can continue in astrocytes for many hours, producing high levels of NO (Nomura and Kitamura, 1993). In contrast, formation of NO in astrocytes and cerebellar glia is mediated by cNOS, activated by the rise in intracellular calcium resulting from noradrenaline stimulation of α_1-adrenoreceptors (Agulló et al., 1995) or Ca^{2+}-permeable glutamate-operated ion channels (Müller et al., 1992). Regardless of its source, NO may subsequently diffuse to cerebral vessels to regulate vascular tone; pass to nerve cells, where it can disrupt essential metabolic pathways and induce cell death; or affect neuronal function through its ability to activate soluble guanylate cyclase (Fig. 4.15), the enzyme that catalyzes the production of cGMP (cf. Schmidt et al., 1993).

4.5.2. Nitric Oxide in Retina

NADPH/diaphorase histochemistry has shown that NOS is present throughout the visual system (Sandell, 1985, 1986; Cudeiro and Rivadulla, 1999). In the retina, the cellular localization of NOS has been examined

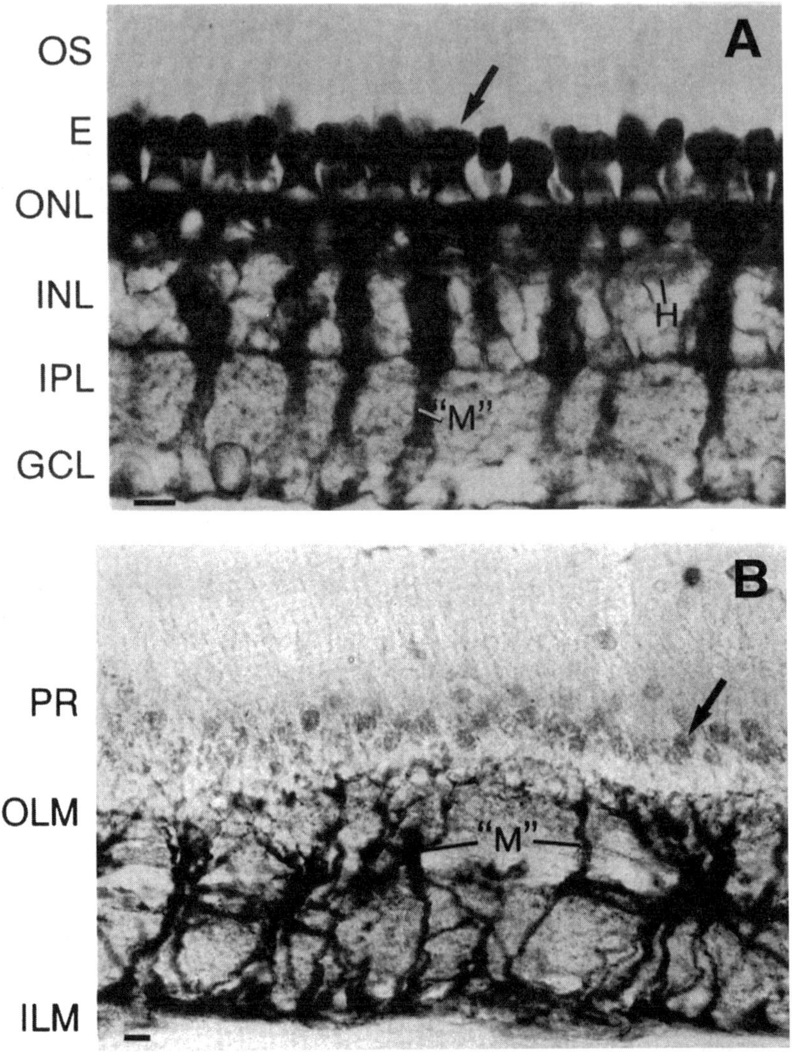

Figure 4.16. Localization of NOS in neurons and Müller cells of vertebrate retinas. NADPHd staining in salamander (A) is seen mainly in the ellipsoids (arrow) and nuclei (ONL) of photoreceptors, and in the thick radial processes of Müller cells (M). In catfish (B), the Müller cells are heavily stained for NADPHd, with less reaction product seen in the ellipsoid region of photoreceptors and in cell bodies of the ganglion cell layer (GCL) (Liepe et al., 1994). (Copyright 1994 *The Journal of Neuroscience*, reprinted with permission.)

both with NADPHd staining and NOS-selective antibodies (cf. Vaney and Young, 1988; Straznicky and Gábriel, 1991; Osborne et al., 1993; Weiler and Kewitz, 1993; Kurenni et al., 1995). Although there are significant differences among species (Fig. 4.16), NOS is expressed to a varying degree in the ellipsoids of photoreceptor inner segments, in every neuronal cell type of the inner nuclear layer, in ganglion cells, and notably, within Müller cells (Goureau et al., 1994b; Liepe et al., 1994; Darius et al., 1995; Huxlin, 1995; Kim et al., 1999). Interestingly, long before neuronal NADPHd activity was identified with NOS or associated with NO production, Toichiro Kuwabara, David Cogan, and their coworkers (Kuwabara et al., 1957; Cogan and Kuwabara, 1959) studying the retinal distribution of dehydrogenases reported that tetrazolium precipitate was present in human Müller cells under substrate conditions requiring activity of the diaphorase. In addition to cells of the neural retina, both human and bovine RPE cells produce NO in response to cytokines (Goureau et al., 1992, 1994a), a reaction that may result in inhibition of the important phagocytic activity performed by RPE cells (Becquet et al., 1994).

With so many potential sources of NO in the retina, questions arise as to which cellular functions are affected by NO, and which cells provide the NO that modulate these functions. Not surprisingly, the molecule has been implicated in a broad range of phenomena, including visual adaptation, modulation of gap junctions, and ischemic injury (reviewed by Goldstein et al., 1996). In isolated rod photoreceptors, for example, NO donors induce membrane depolarization in dark-adapted cells, and accelerate recovery of the dark current after photic stimulation (Schmidt et al., 1992; Tsuyama et al., 1993). Both reactions reflect changes in guanylate cyclase activity, cGMP production, and the opening of cGMP-gated cation channels. Activation of a cGMP-gated conductance by NO donors has been demonstrated also in ganglion (Ahmad et al., 1994) and bipolar cells (Shiells and Falk, 1992), and as illustrated in Fig. 4.17, agents that generate NO or activate endogenous guanylate cyclase close gap junction channels and significantly block electrical and dye coupling between horizontal cells (DeVries and Schwartz, 1989; Miyachi et al., 1991, 1994; Pottek et al., 1997). The effect on gap-junctional coupling may vary with cell type, considering results obtained from recordings within rat supraoptic nucleus, where NO-induced cGMP production *increased* neuronal coupling (Yang and Hatton, 1999).

Clearly retinal cells which express NOS and contain soluble guanylyl cyclase may autoregulate their activity via NO. However, the NO directly responsible for mediating a physiological response *in vivo* also could reach its targets by diffusion from neighboring cells (cf. Vincent and Hope, 1992). As we have seen, Müller cells contribute to NO production in the retina of many vertebrates (Goureau et al., 1994b; Liepe et al., 1994; Darius et al.,

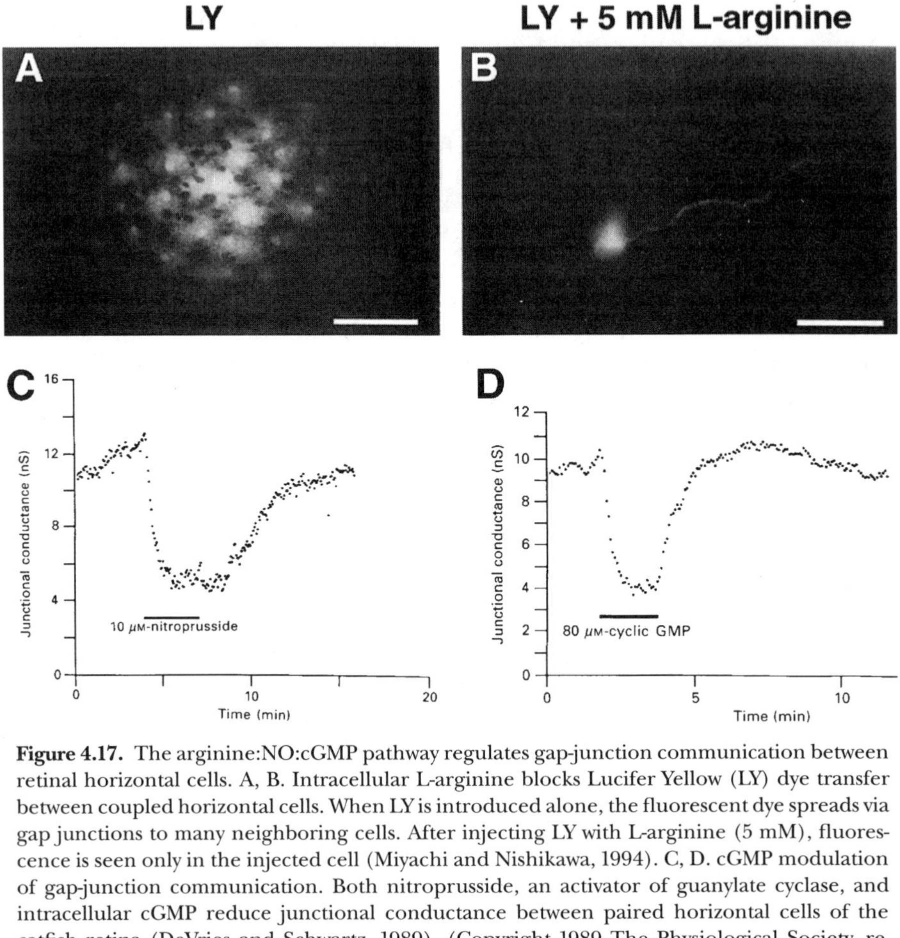

Figure 4.17. The arginine:NO:cGMP pathway regulates gap-junction communication between retinal horizontal cells. A, B. Intracellular L-arginine blocks Lucifer Yellow (LY) dye transfer between coupled horizontal cells. When LY is introduced alone, the fluorescent dye spreads via gap junctions to many neighboring cells. After injecting LY with L-arginine (5 mM), fluorescence is seen only in the injected cell (Miyachi and Nishikawa, 1994). C, D. cGMP modulation of gap-junction communication. Both nitroprusside, an activator of guanylate cyclase, and intracellular cGMP reduce junctional conductance between paired horizontal cells of the catfish retina (DeVries and Schwartz, 1989). (Copyright 1989 The Physiological Society, reprinted with permission.)

1995; Huxlin, 1995; Kim et al., 1999), raising the possibility that NO release from Müller cells may be involved in the regulation of neuronal responses. Although it had been thought that Müller cells express only the inducible form of NOS (Goureau et al., 1994b), recent studies showed that both iNOS and neuronal cNOS are present in mammalian Müller cells. By comparing electron micrographs of NADPHd activity in normal mouse retina with those from mice in which the neuronal NOS gene had been disrupted, Darius et al. (1995) demonstrated the loss of the constitutively-active neuronal cNOS in the "knockout" nNOS−/− mice. A more direct approach, combining NADPHd histochemistry with nNOS immunoreactivity, showed that the neuronal isoform is expressed also by Müller cells of the tree shrew retina (Cao et al., 1999).

Thus, activation of cNOS via calcium entry through the voltage- and ligand-regulated Ca^{2+} channels of Müller cells or cytokine-induced activation of iNOS, can lead to the formation of NO. The sustained release of high levels of NO from Müller cells could potentially have profound effects on neuronal excitability, retinal blood flow, and cellular reactions to injury. Recent *in vitro* experiments by Goureau et al. (1999) provide evidence that such an intercellular transfer can occur. Using cocultures of mouse retinal neurons and Müller cells, they demonstrated that cytokine-induced stimulation of glial iNOS can lead to neuronal cell death. The neurotoxic reaction was not seen in cocultures in which the Müller cells were derived from NOS-2 knockout mice, nor did it occur after the production of NOS or the release of NO was suppressed.

It is evident that we will gain a better understanding of the importance of NO to the physiology and pathology of the retina from detailed *in vivo* studies of knockout mice. Preliminary data on the ocular effects of gene deletion indicate that retinal morphology appears to be well-preserved in NOS-null mice (Huang et al., 1993; Darius et al., 1995). However, a careful analysis of retinal function has not yet been performed on these animals. In addition, the availability of selective inhibitors of constitutive (e.g., 7-nitroindazole) or inducible (e.g., L-N^6-l-iminoethyl lysine) forms of NOS should also prove useful in further elucidating the effects of Müller cell-derived NO on retinal physiology.

K^+ Dynamics, Ion Channels, and Transretinal Potentials

5

A stable internal environment (Claude Bernard's concept of *le maintenance de la milieu intérieur*) has long been recognized as an essential requirement for cells to carry out their normal activities. Nerve cells, for example, are particularly sensitive to fluctuations in extracellular potassium ions $[K^+]_o$ because even a small rise represents a large fractional change from the baseline level of about 3 mM. The increase in $[K^+]_o$ can be far greater under pathological conditions, resulting in disruption of neuronal function and eventual tissue damage (Szatkowski and Attwell, 1994).

A series of landmark studies by Stephen Kuffler and his colleagues on the electrophysiological properties of glial cells led to an appreciation of their importance in potassium homeostasis in the nervous system (Kuffler and Potter, 1964; Kuffler et al., 1966; Orkand et al., 1966). Intracellular recordings from glial cells in the amphibian optic nerve showed that the glial cell membrane potential was almost completely dependent upon the extracellular concentration of potassium ions $[K^+]_o$. As a result, resting potentials were close to −90 mV (about 10–20 mV more negative than in neurons), and over a large range of $[K^+]_o$, the concentration dependence was in accordance with the Nernst equation, i.e., a ten-fold change in K^+ (at T = 24°C) resulted in a 59 mV change in membrane potential.

Subsequently, this log–linear relationship was found also with frog Müller cells (Fig 5.1, continuous line). However, in both astrocytes and Müller cells, the membrane potential shows a departure from the Nernst equation at low potassium concentrations, due to a small but significant membrane permeability to Na^+ (Newman, 1985a; Conner et al., 1985; Reichenbach and Eberhardt, 1986). Thus, substituting a nonpermeant ion (e.g., choline) for sodium in the extracellular bathing solution extends the log–linear relation (Fig. 5.1, dotted line).

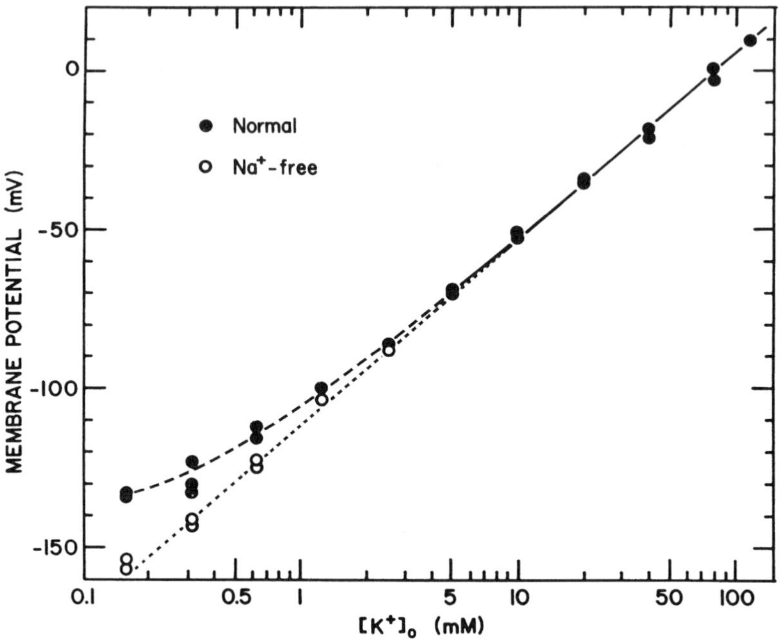

Figure 5.1. Müller cell membrane potential as a function of $[K^+]_o$; results from a single cell in the frog retinal slice preparation. Data obtained in normal Ringer solution (filled symbols) follow the Nernst equation (continuous line) for $[K^+]_o > 2$ mM. When $[K^+]_o$ falls below 2.0 mM, the membrane potential recordings depart from linearity. When choline is substituted for $[Na^+]_o$, the linear relation is restored (open symbols and dotted lines). However, the results obtained in normal Na^+-Ringer can be fit by the constant field equation (dashed line) in which the K^+:Na^+ permeability ratio is 435:1 (Newman, 1985a). (Copyright 1985 *The Journal of Neuroscience*, reprinted with permission.)

Alternatively, it is possible to fit the experimental data with the Goldman- Hodgkin-Katz constant field equation, which takes into account the effect of other permeant ions on membrane potential (cf., Goldman, 1943):

$$E_m = \frac{RT}{\psi F} \ln \frac{P_K[K]_o + P_{Na}[Na]_o + P_{Cl}[Cl]_i}{P_K[K]_i + P_{Na}[Na]_i + P_{Cl}[Cl]_o} \tag{1}$$

where E_m is the membrane potential, F is the Faraday constant (96,500 coulombs per mole), R is the universal gas constant (8.314 joules degree^{-1} mole^{-1}), T is the absolute temperature in degrees Kelvin, ψ is the valence of the ionic species, and P is the permeability of the individual ion and the bracketed value is its concentration.

Ignoring chloride because at physiological K^+ it carries no net current, and converting to common logarithms, yields the expression:

$$E_m \text{ (mV)} = 59 \log \frac{[K]_o + P_{Na}/P_K[Na]_o}{[K]_i + P_{Na}/P_K[Na]_i} \tag{2}$$

Assuming a P_{Na}:P_K of about 1:500 (Newman, 1985a), Equation 2 fits the whole range of experimental data including values of $[K^+]_o < 2$ mM (Fig. 5.1, dashed line).

5.1. REGULATION OF EXTRACELLULAR POTASSIUM

Owing to its almost exclusive dependence upon potassium, the glial membrane potential behaves as a K^+-sensitive electrode and can be used to monitor changes in extracellular K^+ under physiological conditions. This feature enabled Orkand et al. (1966) to demonstrate that, if not rapidly dissipated, K^+ accumulation within the intercellular clefts around active neurons could reduce their sensitivity and severely impair synaptic transmission. However, both neuronal and glial mechanisms participate to prevent the rise in $[K^+]_o$ from reaching abnormally high levels. Although it is essential that nerve cells ultimately recapture the K^+ they extrude, the clearance of excess K^+_o by re-uptake into neurons is a relatively slow process and accounts for only a small part of the removal system (Syskova, 1983). On the other hand, glial cells play a prominent role in controlling the potassium concentration of the neuronal microenvironment (cf. Nicholson 1980; Gardner-Medwin, 1981, 1983a, b). In the retina and CNS, the involvement of glial cells in the regulation of $[K^+]_o$ has been of particular interest because neither diffusion through the narrow and often tortuous pathways of the extracellular space (ECS), nor transport into the blood stream are likely to be effective routes for the rapid removal of K^+_o (Mutsuga et al., 1976; Hansen et al., 1977; Gardner-Medwin, 1983a, b).

One mechanism to dissipate the potassium generated by active neurons was postulated originally by Orkand et al. (1966), who reasoned that the currents generated by the K^+-induced glial cell depolarization will allow K^+ to enter the glial cells in a region of high K^+ concentration and be redistributed to regions where $[K^+]$ is low. This passive process, referred to as "spatial buffering," is most effective when K^+ is accumulated in a restricted (focal) area of neuronal activity. The establishment of a K^+ concentration gradient enables K^+ to enter the glial cells in a highly active region, be transferred via gap junctions to neighboring glia, and be released in an area that is relatively inactive. In this way, glial cells can serve to buffer extracellular K^+ without a net change in their own internal $[K^+]$.

5.1.1. K^+ Regulation in Retina

Müller cells do not seem well suited to participate in spatial buffering via K^+ transfer through gap junctions. Unlike the gap-junctional coupling seen between astrocytes of the inner retina (Quigley, 1977; Zahs and Newman, 1997), Müller cells are poorly coupled to each other (cf. Chapter 1). Müller cells may form gap junctions with astrocytes, but their channels are highly directional (from astrocytes to Müller cells) (Robinson et al., 1993; Zahs and Newman, 1997) and may not provide a useful pathway for redistributing K^+ across the retina. In addition to the anatomical restrictions to transcellular communication between glial cells, it is important to recognize that retinal illumination is often global. In these circumstances, K^+ efflux is more uniform across the retina, and it would be difficult to establish an intraretinal $[K^+]_o$ gradient that could provide a suitable pathway for translocating ions from regions of high to regions of low concentration.

However, a novel and effective system of K^+ buffering has evolved in the vertebrate retina. Newman, through an innovative series of experiments, showed the Müller cell utilizes a mechanism of spatial buffering that enables K^+ to be "siphoned" into the large body of vitreous at the inner surface of the retina (Newman, 1984, 1985a, 1987; Newman et al., 1984). The localization of K^+ channels along the Müller cell membrane is a major factor in the siphoning mechanism. Intracellular recordings from enzymatically isolated Müller cells of tiger salamander retina revealed a striking nonuniformity in the distribution of K^+ conductances (Fig. 5.2A) (Newman, 1984). When the endfoot was severed from the cell body, the membrane resistance (R) recorded from the resealed cell (Fig. 5.2C) increased by 16-fold. Thus, about 94% of the total membrane conductance (1/R) was localized in the Müller cell endfoot. The remaining 6% of the potassium conductance was present in more distal parts and is probably involved in K^+ entry into the cell. This highly asymmetric channel distribution was confirmed by recording the depolarizing responses of Müller cells when K^+ was ejected at various sites along the cell membrane (Fig. 5.2B). The Müller cell responded to the local increase in extracellular K^+ throughout its length, but the largest response was seen when the ejection pipette faced the endfoot.

Although highly suggestive, these findings do not demonstrate unequivocally that K^+ is actually translocated through the Müller cell to its endfoot, nor do they indicate that a light-evoked rise in $[K^+]_o$ results in an efflux of K^+ at the vitreal surface of the cell. However, both issues have been addressed, and the findings provide additional support for a K^+-siphoning mechanism in vertebrate retina. Figure 5.3 A, B shows the results of an experimental design that reveals the siphoning pathway in an isolated Müller cell (Newman et al., 1984). An ejection pipette filled with 85 mM KCl

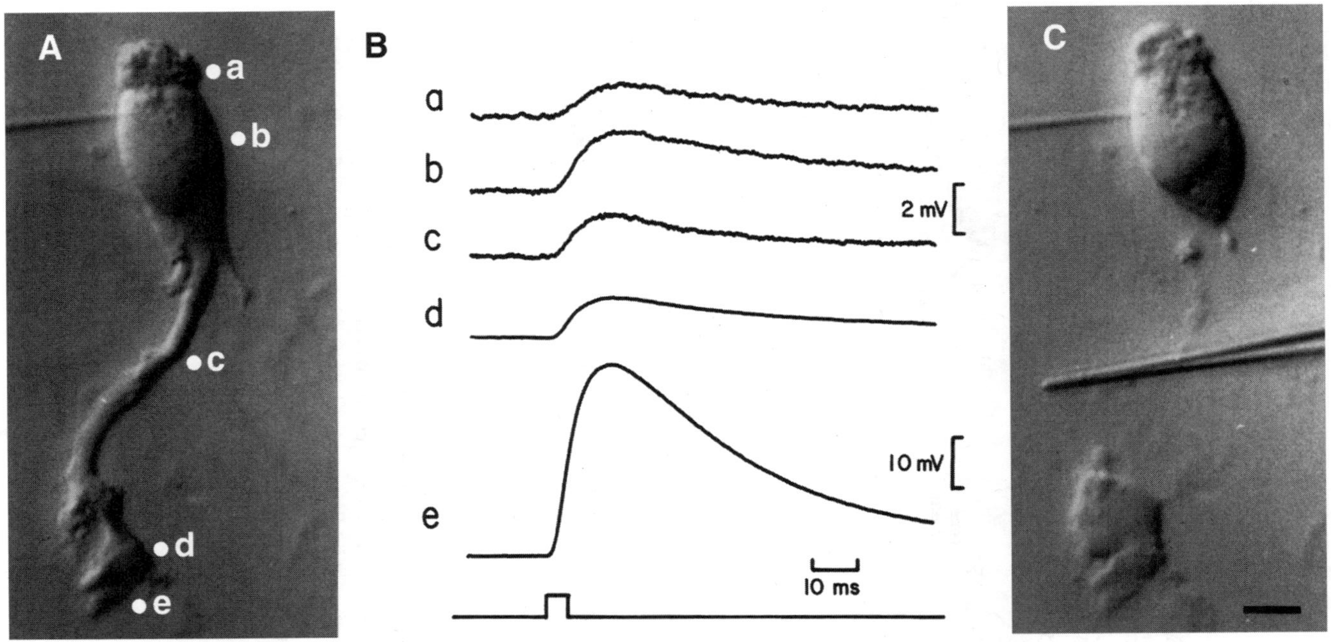

Figure 5.2. A. Photomicrograph of an isolated Müller cell from the salamander. A microelectrode (upper left) is seen penetrating the nuclear region of the cell. B. Intracellular recordings in response to K^+ ejections at various sites (a–e) along the cell membrane. C. When the endfoot was severed from the cell, the cell resistance increased 16-fold, i.e., 94% of the K^+ conductance was in the endfoot region (Newman, 1984). (Copyright 1984 Macmillan Magazines Ltd., reprinted with permission.)

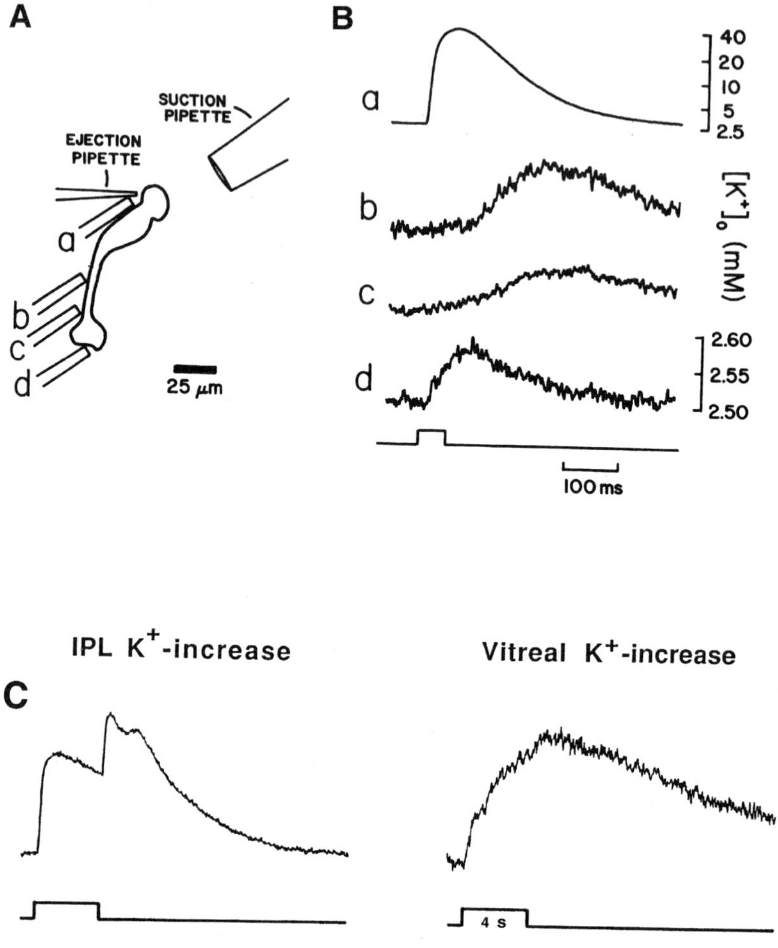

Figure 5.3. A. An ion-selective electrode placed at several sites (*a–d*) along the Müller cell measures the efflux of K^+ that occurs when K^+ is pressure ejected from a pipette located near the distal end of the cell body. A large suction pipette serves to minimize diffusion through the extracellular medium. B. Potassium fluxes derived from voltage recordings at the four positions of the ion-selective pipette (note that traces *b*, *c*, and *d* are expanded vertically relative to *a*). The large change at *a* results from the K^+ pulse (indicated by the deflection on the time trace below the recordings). The slowly developing K^+ increases at *b* and *c* are due to diffusion from the ejection site, whereas the short latency response at *d* is the efflux of K^+ that was siphoned intracellularly to the endfoot region of the Müller cell (Newman et al., 1984). (Copyright 1984 American Association for the Advancement of Science, reprinted with permission.) C. Ion-selective electrodes in the inner retina, and near the retinal border of the vitreous, record the $[K^+]_o$ changes evoked at these sites by a 4 sec exposure to diffuse illumination. Both the onset and offset of the photic stimulus elicit transient increases in $[K^+]_o$ in the IPL; the rise in $[K]^+$ is significantly smaller and slower as K^+ is transferred to the vitreous humor (Karwoski et al., 1989). (Copyright 1989 American Association for the Advancement of Science, reprinted with permission.)

was placed close to the distal end of the Müller cell, and a large suction pipette was positioned on the opposite side to draw away the solution and minimize diffusion away from the ejection site. With an ion-selective recording electrode placed sequentially at several loci (a–d) along the length of the cell, it was possible to record the K^+ flux resulting from the distal pulse of KCl. Aside from the expected K^+ change recorded at the ejection site (a), and the very slow K^+ increases attributable to diffusion at (b) and (c), the largest and most rapid efflux was seen at (d), the proximal face of the endfoot. Thus, the experimentally-induced rise in K^+ was quickly transferred from the distal to the proximal end of the Müller cell. Experiments by Karwoski et al. (1989) in eyecup preparations provide a nice demonstration that K^+ siphoning also occurs *in vivo* following photic stimulation. Using a K^+-selective electrode to measure changes in $[K^+]_o$, they were able to show that a light-evoked increase in intraretinal K^+ leads to a $[K^+]_o$ increase in the vitreous near the inner surface of the retina (Fig. 5.3C).

5.1.2. Directionality of Potassium Siphoning

The membrane distribution of K^+ channels, the properties of the channels mediating K^+ buffering, and the movement of K^+ through the Müller cell, appear now to be far more complex than originally envisioned. Although the high potassium conductance in the endfoot membrane has been confirmed both with patch-clamp recordings of individual ion channels in amphibian Müller cells (Brew et al., 1986) and with intracellular recordings from rabbit Müller cells (Reichenbach and Eberhardt, 1988), the K^+-channel distribution is highly species-dependent (Newman, 1987). In mouse and monkey, for example, the maximal conductance is located in midretina, whereas in cat, the conductance at the extreme distal end of the Müller cell is almost twice that of the endfoot (Fig. 5.4).

This variability provides a basis for the different ways in which potassium buffering is managed by different species. For example, results obtained in toad (Oakley et al., 1992) and cat (Frishman and Steinberg, 1989), not only confirmed spatial buffering currents transport K^+ to the vitreous body but also provided considerable evidence that a similar process leads to the movement of K^+ from the inner retina and vitreous to the distal retina, where it is extruded into the OPL and subretinal space. These observations led Oakley et al. (1992) to propose that K^+ remains largely within the retina, moving between the vitreous and more distal sites to buffer the transient photically induced increases and decreases in $[K^+]_o$ (see Section 5.6).

Moreover, it has become increasingly evident that K^+-buffering currents are not mediated primarily by passive ion channels. The magnitude of the increase in $[K^+]_o$ will determine the degree to which voltage-dependent channels, ligand-gated channels, and a Na^+-K^+-pump contribute to K^+

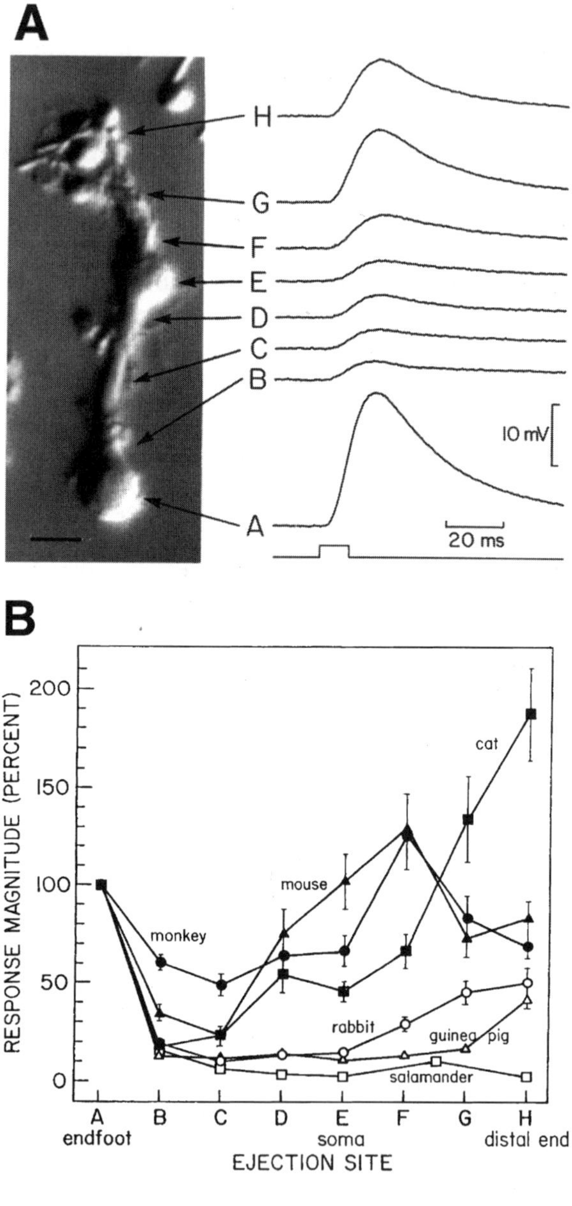
A
H
G
F
E
D
C
B
A
10 mV
20 ms
B
RESPONSE MAGNITUDE (PERCENT)
200
150
100
50
0
cat
mouse
monkey
rabbit
guinea pig
salamander
A
endfoot
B
C
D
E
soma
F
G
H
distal end
EJECTION SITE

buffering; Müller cells are well endowed with membrane proteins that subserve each of these activities (Newman, 1985b; Reichenbach et al., 1985, 1986, 1992; Brew et al., 1986; Nilius and Reichenbach, 1988; Reichelt et al., 1993; see also Chapter 4).

5.2. VOLTAGE-ACTIVATED ION CHANNELS

The discovery that glial cells possess voltage-gated ion channels (MacVicar, 1984; Bevan and Raff, 1985) opened a new era in the study of glial cell physiology throughout the nervous system. This observation was particularly surprising in view of the results of earlier studies on invertebrate and vertebrate glia in situ, which led to the conclusion that glial cell membranes are entirely passive, and unresponsive to changes in transmembrane voltage (Kuffler, 1967). Indeed, it was reported that altering the transmembrane potential by as much as 200 mV produced no apparent change in membrane resistance (Kuffler et al., 1966).

We know now that the seeming lack of voltage-dependent ion channels was almost certainly due to shunting of current from the site of injection to neighboring glial cells via intercellular gap junctions. With the advent of patch-clamp technology (Hamill et al., 1981), advances in cell isolation and culture procedures, and development of the tissue slice preparation, it has become possible to study in detail the membrane properties of individual glial cells. The results have been remarkable in demonstrating the presence of almost every major class of voltage-dependent Na^+, K^+, Ca^{2+}, and anion channel previously thought to be exclusively on excitable cells (cf. MacVicar, 1984; Ritchie, 1987; Sontheimer, 1994).

Channels with similar properties have been shown to be present also on Müller cells (Newman, 1985b; Brew et al., 1986; Nilius and Reichenbach, 1988; Reichelt et al., 1993; Chao et al., 1993, 1994a, 1997; Ishii et al., 1997). Several examples are illustrated in the following sections. The striking similarities in channel properties between glial and neuronal voltage-activated

Figure 5.4. A. Intracellular voltage responses of an isolated rabbit Müller cell to focal ejections of K^+ at the sites indicated. Note that the distribution of conductances is markedly different from that of the salamander Müller cell shown in Fig. 5.2. B. A comparison of the distribution of K^+ conductances along the Müller cells of several vertebrate species. Values represent the mean ± SEM of individual voltage responses expressed as a percentage of the response recorded at the endfoot of each cell. Note that in mouse and monkey retina, responses recorded in the region of the cell soma exceed that at the endfoot, whereas in cat, the K^+ conductance at the distal end of the cell is almost 2-fold greater than at the endfoot region (Newman, 1987). (Copyright 1987 *The Journal of Neuroscience*, reprinted with permission.)

channels have led to some provocative suggestions regarding their functional significance. However, a great deal of experimental work is necessary to provide a clear picture of the ways in which voltage-gated channels contribute to the activities of glial cells or to their interactions with neurons.

One problem in assigning a role to the voltage-sensitive channels is the high resting K^+ permeability of glia, which tends to hold the membrane potential near the K^+ equilibrium potential of −80 to −90 mV. For many of these ionic channels, there is little current flow until the cell is depolarized by 40 mV or more above its resting potential, a condition that is not thought to occur under normal physiological conditions. However, not all voltage-regulated channels require large membrane depolarizations, and those that do may be activated under pathological conditions.

5.2.1. Voltage-dependent Channels of Müller Cells

Newman (1984, 1985a) reported that the channels present in the endfoot membrane at the vitreous surface are primarily of one type, namely, a passive, time- and voltage-independent K^+ channel. However, when shorn of their endfeet, it was possible to identify four types of voltage-dependent ion channel (Fig. 5.5) in the Müller cell membrane distal to the endfoot (Newman, 1985b): a Ca^{2+} channel, a Ca^{2+}-activated K^+ channel (K_{Ca}), a rapidly inactivating A-type K^+ channel (K_A), and an inward rectifying K^+ channel (K_{IR}), each of which had been identified previously both in excitable and inexcitable cells. The different voltage sensitivities and ionic properties of the various K^+ channels are representative of but a few of the diverse potassium channel subtypes that have been characterized biophysically and pharmacologically (Rudy, 1988; Jan and Jan, 1990, 1994).

Surprisingly, single-channel patch recordings from axolotl Müller cells revealed only one type of K^+ channel: an inwardly-rectifying K^+ channel, both at the endfoot where channel density is high, as well as at the cell body where few channels are located (Brew et al., 1986). One would not have expected the K^+-channel population of Müller cells from salamander and axolotl to be so different, and indeed, more recent data from tiger salamander, in which the recording conditions were optimized to provide an adequate space clamp, confirmed the overwhelming preponderance of inward-rectifying K^+ channels throughout the Müller cell membrane (Newman, 1993).

Although there is mounting evidence that the inward-rectifying K^+ channels of Müller cells provide the principal avenue for K^+ fluxes in mammalian as well as amphibian retinae, it would appear that the types of K^+ channel expressed by Müller cells may vary considerably among species. An example of this diversity can be seen in the results obtained from

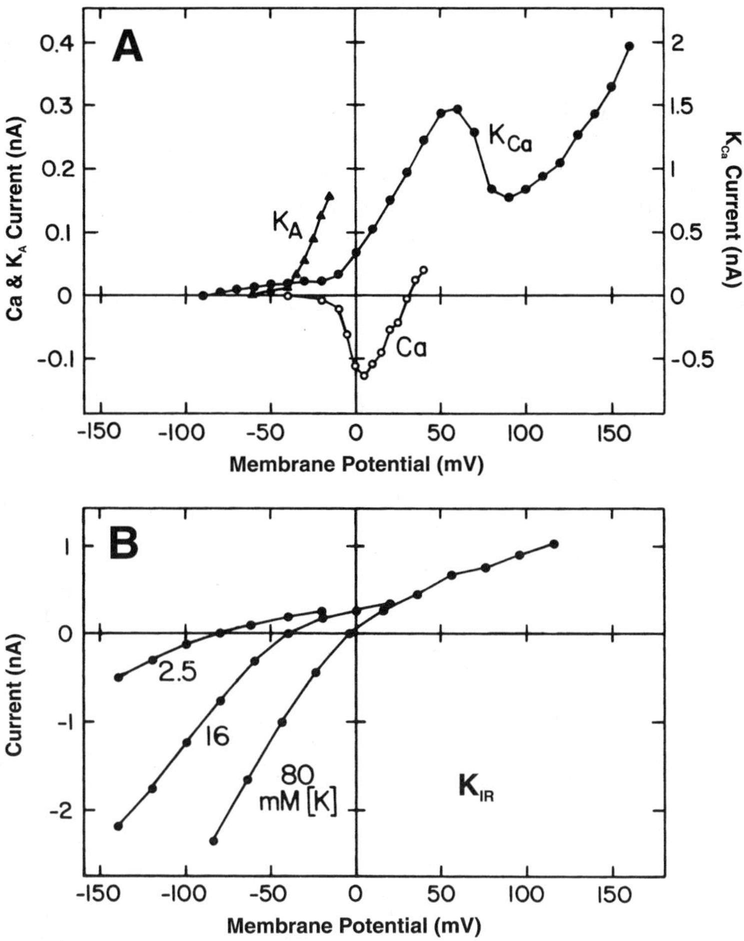

Figure 5.5. Current–voltage relations obtained from whole-cell patch clamp recordings reveal four types of voltage-regulated channels on salamander Müller cells. A. Current responses for the inward calcium currents (Ca) and A-type potassium currents (K_A) are shown on the left hand scale of ordinates; the larger calcium-activated potassium currents (K_{Ca}) are on the right. Note that activation of these currents occurs when cells are depolarized to between −40 and −60 mV. B. Potassium currents through the inward rectifier (K_{IR}) recorded from a cell shorn of its endfoot, and bathed sequentially in solutions containing 2.5, 16, and 80 mM K^+. This voltage-dependent channel is open under physiological conditions, and the channel conductance increases as the cell is hyperpolarized from the resting membrane potential. In addition, the I–V relation shifts to the right at increasing levels of $[K^+]_o$, producing a further rise in conductance in regions of the cell exposed to the high $[K^+]_o$ (Newman, 1985b). (Copyright 1985 Macmillan Magazines Ltd., reprinted with permission.)

mammalian Müller cells. In addition to the currents mediated by rapidly inactivating (K_A) and inwardly rectifying (K_{IR}) channels, voltage-regulated K^+ channels with markedly different conductances, as well as currents flowing through a delayed rectifier (K_{DR}) channel have been reported (cf. Nilius and Reichenbach, 1988; Reichenbach et al., 1992; Reichelt et al., 1993). Each of these channels, as well as the Ca^{2+}-activated K^+ channel, could participate to some degree in the regulation of K^+_o, although it has proven difficult to distinguish their individual contributions to the overall process. In addition, there is reason to believe that other mechanisms are also engaged directly or indirectly in K^+ buffering, all of which confirm the importance of K^+ homeostasis to the functional integrity of neural tissues (cf. Amédée et al., 1997).

5.2.2. Inward-Rectifying Potassium Channels

As already noted, establishing the functional significance of voltage-dependent channels can present a difficult challenge. However, K_{IR} channels are clearly a significant component of the K^+ buffering system of Müller cells. Strictly speaking, K_{IR} channels are functionally distinct from voltage-gated K^+ channels—i.e., they do not activate over a fixed range of membrane potential, and inward rectification is due mainly to the blockade of outward current by internal Mg^{2+} rather than a voltage-activated gating mechanism (Matsuda et al., 1987). In addition, unlike classical voltage-gated channels, K_{IR} is typically open at or near the cell's resting potential, its current regulation is highly dependent upon $[K^+]_o$, and for the most part it is unidirectional, allowing a much larger K^+ influx than efflux (cf. Fig. 5.5 B).

However, it is important to note that the outward K^+ current carried by $I_{K(IR)}$ is not insignificant, and as will be described, the channel can serve to extrude K^+ in regions of the cell membrane not exposed to high $[K^+]_o$. Fig. 5.5B shows that the current through K_{IR} channels is determined by the electrochemical driving force on K^+. Typically, the channel conductance is maximal at potentials below the potassium equilibrium potential (E_K), and when $[K^+]_o$ is raised, the resultant changes in (E_K), and the voltage-dependence of the channel, serve to effectively preserve inward rectification and enhance K^+ influx. Thus, a rise in $[K^+]_o$ may reduce E_K, but the membrane potential (E_M) remains "clamped" by those regions of the membrane not exposed to high K^+_o, and K^+ enters the cell.

There is now general consensus that K_{IR} channels play a prominent role in the K^+ buffering activity of Müller cells, but their properties and membrane distribution remain somewhat controversial. In mammalian (rabbit) retinae, two types of inward rectifier were identified: high conductance (105 pS) K_{IR} channels localized in the proximal processes of the

Müller cells, and low conductance (60 pS) K_{IR} channels in the membrane of the soma and distal processes (Nilius and Reichenbach, 1988). The endfoot membrane, on the other hand, was reportedly dominated by nonrectifying, large conductance channels (3.5 to 6-fold greater than the conductances of the inward rectifiers) which the authors contend are responsible for the time-and voltage-independent currents through K^+ channels at the vitreal surface of the retina (see also Reichelt et al., 1993). This channel distribution would evidently provide an effective means for K^+ siphoning to the vitreous, i.e., K^+ entry through inward rectifying channels within the plexiform (synaptic) layers of the retina, and extrusion through passive ion channels at the Müller cell endfoot.

However, a comprehensive investigation of K_{IR} channels of mammalian Müller cells with molecular biological, immunochemical, and electrophysiological techniques, failed to confirm these observations and brought into question the existence of multiple types of K_{IR} channels in the Müller cell membrane (Ishii et al., 1997). Whole-cell and cell-attached patch-clamp recordings revealed only a single population of K_{IR} channel currents in dissociated Müller cells from rabbit retina. This channel was identified as the $K_{IR}4.1$ member of the superfamily of functionally diverse inward-rectifier channels (Takumi et al., 1995; Fakler and Puppersberg, 1996). Moreover, it exhibited properties that matched almost precisely those of rat $K_{IR}4.1$ channels expressed in human embryonic kidney cells, and a glial-cell derived low-conductance $K_{IR}4.1$ channel expressed in *Xenopus* oocytes (Takumi et al., 1995). The unitary conductance of the channel (~ 25 pS), and its dependence upon intracellular ATP, are properties found also in $I_{K(IR)}$ channels of monkey Müller cells (Kusaka and Puro, 1997). Surprisingly, recordings from more than 200 cell-attached membrane patches of rabbit Müller cells failed to detect conductances consistent with any of the variety of K_{IR} channels reported previously (Nilius and Reichenbach, 1988). It will be interesting to see how this discrepancy is resolved, but it should be obvious that we have yet to identify the many factors that influence channel expression. Nevertheless, it has become increasingly apparent that the inward rectifying K^+ channel is the principal mediator of K^+ buffering currents in the cell membrane of amphibian and mammalian Müller cells.

In addition to their functional studies of rabbit Müller cells, Ishii et al. (1997) used a polyclonal antibody raised in rabbit against a C-terminal region of rat $K_{IR}4.1$ to study the distribution of these channels in rat retina. Confocal light-microscopic immunocytochemistry of retinal wholemounts and immunogold EM of retinal sections revealed that the channels were present in the cell body, processes, and endfoot of the Müller cell. Moreover, $K_{IR}4.1$ immunoreactivity in isolated Müller cells demonstrated that the channels tended to be grouped in clusters. This clustering, which probably enhances channel activity in regions where K^+ fluxes are maximal, was shown to be induced by insulin and laminin, two endogenous proteins of the extracellular matrix. It is likely that channel clustering also accounts for the wide variability in the numbers of channels recorded in cell-attached patches from particular regions of the Müller cell membrane.

5.2.3. K_{IR} Channels and K^+ Buffering Currents

As already indicated, the potassium conductance in various regions of the Müller cell membrane is highly species-dependent (Fig. 5.4), suggesting that there is more than one avenue for K^+-induced current flow. In amphibia, for example, there is an overwhelming preponderance of $I_{K(IR)}$ channels in the endfoot (Brew et al., 1986), whereas in rat, channel density is high both at the endfoot and at the distal ends (Ishii et al., 1997).

As shown in Fig. 5.5, a rise in extracellular K^+ increases channel conductance and shifts the reversal potential in a depolarizing direction (in accordance with the Nernst equation). Because current flow through K_{IR} channels is inward at resting potentials more negative than the potassium equilibrium potential (E_k), K^+ will enter the Müller cell in regions where neuronal activity produces a local increase in $[K^+]_o$, e.g., in the plexiform layers. In those regions where K^+ is unchanged (the vitreous) or reduced (the subretinal space) by neural activity, the resting potential is positive to E_k, and K^+ can exit the Müller cell via K_{IR} channels. Thus, potassium buffering currents will depend upon the membrane distribution of K_{IR} (Brew and Attwell, 1985) and can lead to the deposition of K^+ in the vitreous as well as in the distal retina (Fig. 5.6), where it may be actively transported across the RPE to the choroidal vasculature (Steinberg and Miller, 1979).

5.3. OTHER VOLTAGE-DEPENDENT ION CHANNELS

There has been a great deal of speculation as to the significance of voltage-dependent ion channels on glial cells. Much of this has been engendered by results obtained in studies on the current-voltage (I–V) relation of the individual channels, which suggest that many of them are inactive under normal physiological conditions. Because the glial cell membrane potential is highly dependent upon $[K^+]_o$, there is the assumption that the potassium efflux resulting from neuronal activity is too small to depolarize the glial cell membrane sufficiently to cause channel activation.

There is, however, an important caveat. Estimates of $[K^+]_o$ are often obtained from measurements in the extracellular space with relatively large K^+-sensitive electrodes. These and other ion-specific microelectrodes (ISM) typically taper to tip diameters of ~1.5 um, but create a large "dead" space

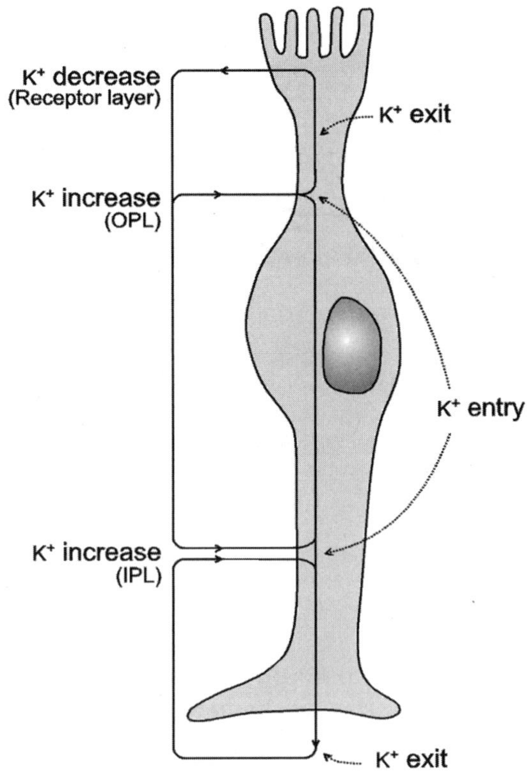

Figure 5.6. Schematic drawing of potassium buffering currents resulting from the light-evoked extrusion of K^+ at synaptic sites within the IPL and OPL. The current path will depend upon the membrane distribution of $I_{K(IR)}$, and can lead to the deposition of K^+ in the vitreous, as well as in the distal retina. In this diagram, K^+ enters the Müller cell in the plexiform layers, and exits in regions where K^+ is unchanged (the vitreous) or reduced by neural activity (the subretinal space).

around the electrode and cause damage as they penetrate the tissue to reach the recording site (often poorly defined). Moreover, the ISM records "net" changes in ion activity, and the measured values may be reduced by active uptake mechanisms. Collectively, these factors probably lead to gross underestimates of the changes in $[K^+]_o$ occurring within the confines of the synaptic clefts (cf. Orkand, 1980; Ripps and Wikovsky, 1985). Consequently, there may be large potential changes across local regions of the glial cell membrane that go undetected by the recording electrode.

5.3.1. The Delayed Rectifier Channel (K_{DR})

The outward current carried by the delayed rectifier potassium channel provides a good example of why there is confusion regarding the role of most voltage-activated channels in glial cells. First described in the squid giant axon, where it serves to repolarize the cell following discharge of the action potential, its presence on astrocytes both in culture and *in situ* has

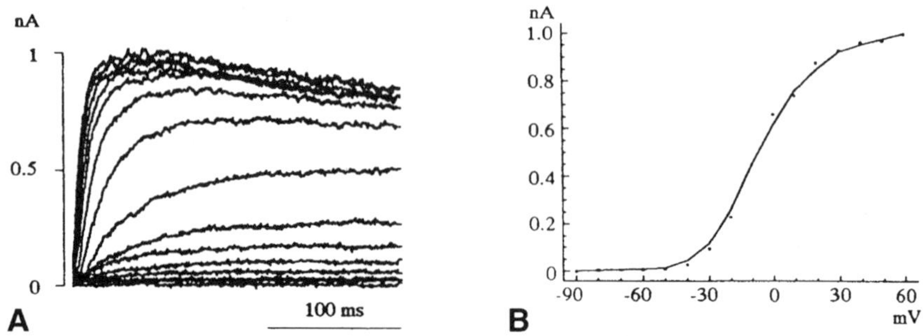

Figure 5.7. Whole cell recordings of a sustained (noninactivating), outwardly rectifying K^+ current in rabbit Müller cells. A. Slowly activating outward currents elicited by depolarizing voltage steps from −80 to +60 mV in 10 mV increments. B. The I–V curve for this delayed rectifier ($I_{K(DR)}$) shows that the threshold of activation is about −30 mV, well beyond the normal operating range of Müller cells (Chao et al., 1994a). (Copyright 1994 Springer-Verlag, reprinted with permission.)

been amply documented (Bevan and Raff, 1985; Bevan et al., 1985; Nowak et al., 1987; Tse et al., 1992; Steinhauser, 1993; Steinhauser et al., 1994). Delayed rectifier channels mediating the outward current in different cell types appear to be pharmacologically and structurally distinct (cf. Hille, 1992). However, the channels are typically inactive at the resting membrane potential of glial cells and activate when the membrane is depolarized to about −30 mV; with prolonged depolarization, the outwardly-rectifying K^+ current is well maintained.

Recently, the presence of a similar type of voltage-dependent K^+ channel was revealed with whole-cell voltage-clamp recordings from isolated mammalian Müller cells (Chao et al., 1994a, 1997; Reichelt et al., 1993). In rabbit, the K_{DR} channels required depolarizing voltages (~ −30 mV) for activation (Fig. 5.7), whereas in guinea pig and mouse Müller cells, the channels were activated when membrane depolarization was positive to −20 mV and 0 mV, respectively.

It is hard to speculate about a possible role for these channels in Müller cells. Although it has been suggested that they may be involved in promoting cell proliferation following neural injury (Chiu and Wilson, 1989; Chao et al., 1997), no plausible explanation of how this is accomplished has been put forth.

5.3.2. The A-type K^+ Channel (K_A)

A-type K^+ currents activate and inactivate quickly, properties that enable them to regulate the frequency of repetitive firing in spontaneously active neurons (Connor and Stevens, 1971). Once again, there appears to be no rational explanation for their presence on glial cells. Nevertheless, transient outward currents with characteristics similar to those of neuronal K_A have been described in salamander (Fig. 5.5), guinea pig, and rabbit Müller cells (Newman, 1985b; Chao et al., 1994a; Reichelt et al., 1993) as well as in other types of glia in the CNS (cf. Bevan et al., 1987; Barres et al., 1990; Ritchie, 1992).

It is not too difficult to distinguish between currents carried by K_A and those due to K_{DR} channels. Like the delayed rectifier, the "A" current is activated only at depolarized membrane potentials (~ −30 mV), but the K_A-mediated current is far more transient and rapidly inactivates. In Müller cells, the A current appears to be significantly less sensitive to the convulsant 4-aminopyridine (4-AP) and its derivatives, as well as to external tetraethylammonium (Chao et al., 1994a). Moreover, activation of K_A channels typically requires a brief hyperpolarization preceding the depolarization. Retinal glia are rarely exposed to patterns of activity that could induce activation of K_A, e.g., large, rapid fluctuations in K^+_o resulting from recurrent action potentials, and thus the functional significance of K_A channels on Müller cells is far from clear.

5.3.3. Ca^{2+} Channels and Ca^{2+}-activated K^+ Channels

Voltage-dependent Ca^{2+} channels resembling the L-type neuronal channel were first described in cultures of rat brain astrocytes (MacVicar, 1984), and subsequently shown to be present in Müller cells of salamander retina (Fig. 5.5; Newman, 1985b). More recently, two classes of Ca^{2+} channel, the L-type and T-type (Fig. 5.8), were identified in cultures of Müller cells from donor eyes (Puro and Mano, 1991a). These channels are readily distinguished by their kinetics and pharmacology. L-type currents are prolonged, and blocked by 1,4-dihydropyridines; T-type currents are transient, inactivate within tens of msec, and are insensitive to dihydropyridines. Interestingly, the expression of calcium channels in glia appears to be dependent upon extrinsic factors derived from serum (Barres et al., 1989b), cocultured neurons (Corvalan et al., 1990; Puro and Mano, 1991a; Puro, 1994a), or exposure to agents that increase intracellular cAMP (MacVicar, 1984).

Activation of both types of voltage-dependent Ca^{2+} channel requires membrane depolarization to about −50 mV, again seemingly beyond the normal operating range of Müller cells. However, there is evidence that the depolarization-induced rise in $[Ca^{2+}]_i$ seen in astrocytes *in situ* is an indirect effect, resulting from the activation of metabotropic receptors by the neuronal release of neurotransmitters, e.g., glutamate (Carmignoto et al., 1998). Regardless of how the rise in $[Ca^{2+}]_i$ is brought about, the

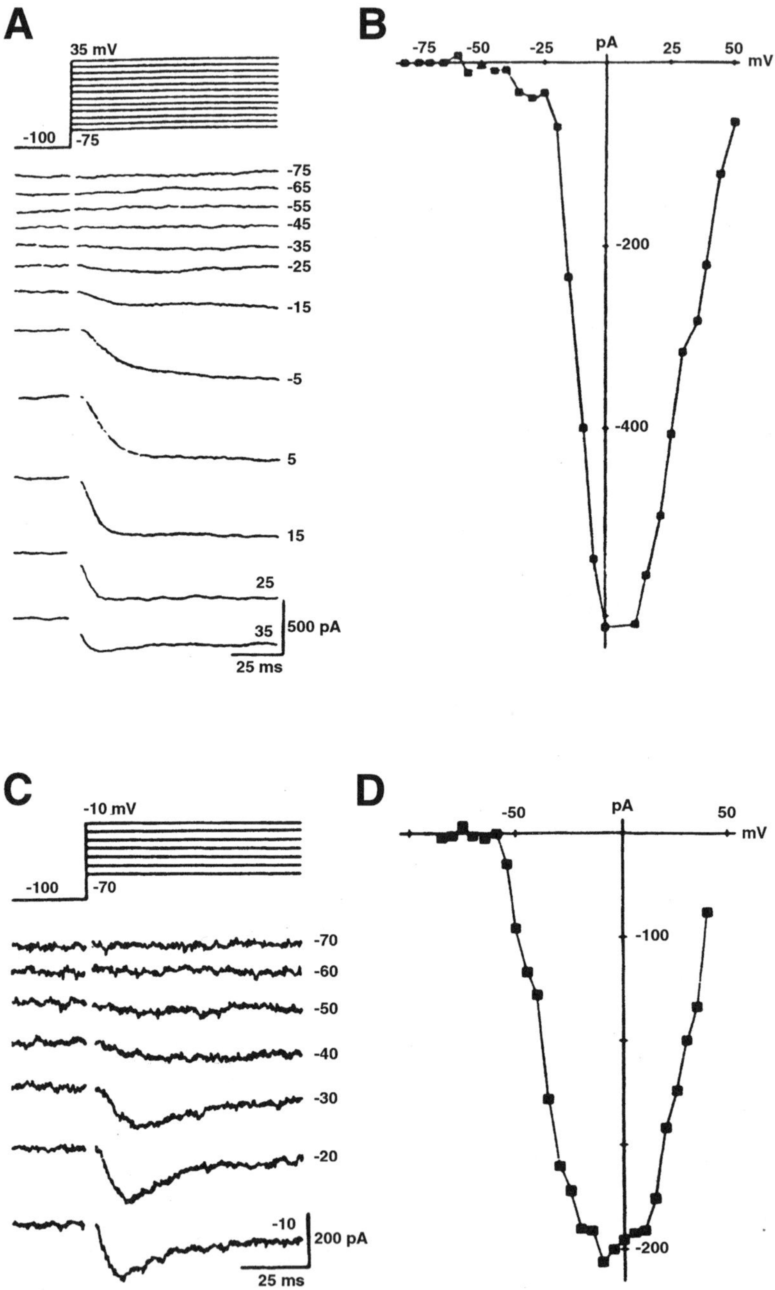
A
35 mV
-100
-75
-75
-65
-55
-45
-35
-25
-15
-5
5
15
25
35
500 pA
25 ms
B
-75
-50
-25
pA
25
50
mV
-200
-400
C
-10 mV
-100
-70
-70
-60
-50
-40
-30
-20
-10
200 pA
25 ms
D
-50
pA
50
mV
-100
-200

depolarization-induced influx of calcium may lead to activation of Ca^{2+}-sensitive potassium channels (Fig. 5.5) and thus contribute to the buffering capacity of glial cells (Quandt and MacVicar, 1986). Indeed, the process may be further enhanced in human Müller cells by a voltage-insensitive, calcium-activated, nonspecific cation channel that responds to the rise in $[Ca^{2+}]_i$ (Puro, 1991a).

5.3.4. Sodium Channels

Perhaps even less well understood is the finding that glial cells express voltage-activated, tetrodotoxin-sensitive Na^+ channels indistinguishable from those used by nerve cells to generate action potentials (cf. Bevan et al., 1985; Barres et al., 1989a; Sontheimer and Waxman, 1992). Their functional significance in glial cells remains a mystery, although various suggestions have been put forth. One popular idea is that the influx of Na^+ through these channels may fuel the (Na^+, K^+)-ATPase that subserves K^+ uptake. Another is that a rise in $[Na^+]_i$ could produce graded potentials that serve to activate various intracellular signaling cascades (Black and Waxman, 1996). There is little reason for thinking these events occur under normal conditions; the I–V relationship for astrocytes shows that the threshold for activation of the sodium current is about −40 mV, and the maximal current is elicited at approximately −20 mV (Bevan et al., 1985).

Similar voltage-dependent sodium channels have been found in Müller cells of human and other (but not all) mammalian species (Chao et al., 1993, 1994b). As with astrocytes, threshold activation occurs at about −40 mV, and the peak current is reached at −12 mV (Fig. 5.9). Because these levels of depolarization would not be expected under normal physiological conditions, the sodium channels on Müller cells are almost certainly not involved in impulse propagation, nor is there evidence that other types of glia are capable of establishing regenerative Na^+ spikes (cf. Ritchie, 1987,

Figure 5.8. Calcium currents obtained from whole-cell patch clamp recordings on isolated human Müller cells are predominantly a slowly-inactivating L-type current, or a transient T-type current. A. Currents evoked by voltage steps from a holding potential of −100 mV to the potentials indicated to the right of the current traces reveal the presence of a L-type inward current. B. Peak currents plotted against voltage for the cell depicted in A show that the threshold for activation is about −35 mV, and the maximum current is reached when the membrane potential is depolarized to approximately 0 mV. C. Another Müller cell responded to depolarizing pulses (from a holding potential of −100 mV) with transient T-type calcium currents. D. The peak current–voltage relation shows that the activation threshold is at about −50 mV, and I_{max} again occurs in the region of 0 mV (Puro and Mano, 1991a). (Copyright 1991 *The Journal of Neuroscience*, reprinted with permission.)

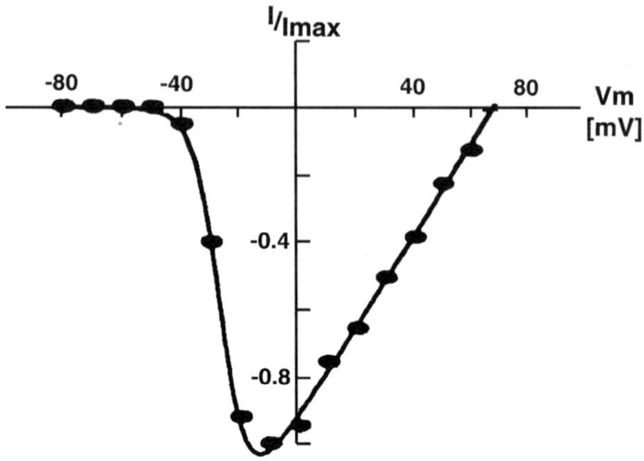

Figure 5.9. The current–voltage relation of the inward sodium current obtained with whole cell recording from an isolated mouse Müller cell. The threshold for activation was about −45 mV, and the maximum current was elicited at about −12 mV, after which there was a decline with further depolarization (Chao et al., 1994b). (Copyright 1994 Wiley-Liss, a subsidiary of John Wiley & Sons, Inc., reprinted with permission.)

1992). Nevertheless, the high level of K^+_o often associated with pathological conditions has led to speculation that the resultant depolarization might be sufficient to activate Na^+ channels. The influx of Na^+ could serve as a neuron–glial signal for G-protein activation and other metabolic processes (Pentreath and Kai-Kai, 1982; Cohen-Armon and Sokolovsky, 1993), to modulate sodium-dependent transport mechanisms (Martin, 1992), or to enhance K^+ uptake by stimulation of the cell membrane Na^+–K^+-ATPase (see below).

5.3.5. Chloride Channels

Voltage-activated, inwardly rectifying chloride channels reported initially in rat astrocytes (Bevan et al., 1985) have also been identified in retinal Müller cells. Assuming a low intracellular chloride concentration, and a Cl^- equilibrium potential more negative than the membrane potential, these channels could provide an adjunct to other K^+ buffering mechanisms—i.e., Cl^- entry would be accompanied by K^+ (driven by the extracellular K^+ gradient) and water. Although Cl^- channels that are open at the resting potential of glial cells have not been found, channel activity could occur under extreme conditions where high K^+ concentrations substantially depolarize the cells. The accompanying entry of water would result in the swelling of the Müller cells and a reduction in the ECS, events that are not without physiological importance (Puro, 1991b). Nevertheless, it would allow potassium to be stored as KCl in the glial cell, and thus be available for release and eventual reuptake by neurons.

In addition, the GABA-gated chloride channels of glia (Bormann and Kettenmann, 1988; cf. Chapter 4) may participate in K^+ homeostasis. Elevated $[K^+]_o$ can lead to a depolarization-induced release of GABA from retinal neurons, e.g. horizontal cells (Schwartz, 1987; Attwell et al., 1993), which would, in turn, activate Müller cell $GABA_A$ receptors (Malchow et al., 1989). The resultant GABA-mediated increase in chloride conductance will prevent the Müller cell membrane from reaching a new K^+ equilibrium potential, and by increasing the driving force on potassium, enhance the uptake of K^+ (along with Cl^-) (Boyle and Conway, 1941; Hodgkin and Horowicz, 1959).

5.4. THE Na^+–K^+ PUMP

In addition to their many voltage-regulated channels, glial cells possess a Na^+–K^+-ATPase that helps to preserve the ionic gradient across the cell membrane. The glial Na^+–K^+-ATPase differs significantly from its neuronal counterpart (Sweadner, 1979; Walz and Hertz, 1982; Grisar et al., 1979). The neuronal Na^+–K^+ pump is maximally active at physiological levels of $[K^+]_o$ and is highly dependent on extracellular Na^+. In astrocytes, on the other hand, the rate of activity of the glial Na^+–K^+-ATPase, and the affinity of the enzyme for ATP, are enhanced by raising the K^+ concentration from 5–20 mM—a level that would decrease the activity of the neuronal enzyme (Grisar et al., 1979).

Studies of turtle and rabbit Müller cells also revealed a glia-specific isoform of the Na^+–K^+-ATPase (Stirling and Sarthy, 1985; Reichenbach et al., 1985, 1986, 1992) with properties similar to those seen in astrocytes. In rabbit, pump activity is independent of $[Na^+]_o$, but is greatly enhanced by extracellular K^+, reaching a maximum when $[K^+]_o$ is about 10 mM (Reichenbach et al., 1985, 1986, 1992). Estimates of the distribution of pump sites, based on immunohistochemical localization and autoradiography of ^{3}H-ouabain binding to Müller cell membranes, indicate that they are present at high density in the region of the inner and outer plexiform layers, and that the maximal density is in the microvillar membranes at the apical region (Stirling and Sarthy, 1985; Reichenbach et al., 1992). These sites correspond to the locations at which there are significant K^+ effluxes at the onset or offset of illumination.

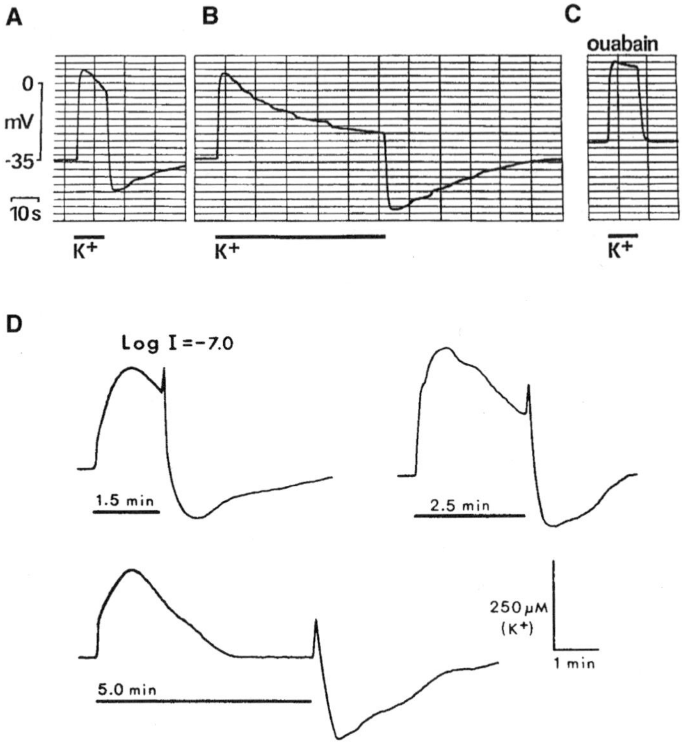

Figure 5.10. Membrane potential recordings from an isolated rabbit Müller cell in response to application of a test solution containing 100 mM K^+. A. With the cell bathed in a normal Ringer solution containing 4.5 mM K^+, a 10 sec exposure to the high K^+ produced a transient depolarization, followed by a brief undershoot of the resting potential when restored to normal Ringer. B. With a longer exposure to 100 mM K^+, the cell repolarized further, and the baseline undershoot (after hyperpolarization) was more pronounced. C. Both the repolarization during exposure to high K^+, and the subsequent undershoot, are essentially eliminated in cells treated with ouabain for 3 min prior to the high K^+ application. D. Light-induced potassium changes in the proximal retina of the intact skate eyecup show a similar phenomenon. The retinal irradiance of the illuminated area (corresponding to log I = −7) was 1.7×10^{-7} mW/cm^2, and was presented for the times indicated. Note that the longer the exposure duration, the greater the repolarization during the exposure, and the greater the baseline undershoot at light offset (Reichenbach et al., 1986; Kline et al., 1985). (A–C copyright 1986 Elsevier Science, D copyright 1985 Elsevier Science, reprinted with permission.)

The presence of a metabolically active pump mechanism with properties characteristic of the glial isoform indicates that coupled Na^+–K^+ exchange is increased at elevated levels of $[K^+]_o$. Membrane potential recordings from isolated rabbit Müller cells provide evidence of this activity (Reichenbach et al., 1986). When exposed briefly to high levels of K^+_o (100 mM), the membrane potential depolarizes rapidly, but begins to slowly repolarize; a return to a more physiological K^+_o (4.5 mM) causes the membrane to hyperpolarize below the prepulse baseline (Fig. 5.10A). The longer the duration of exposure to the high K^+_o, the more pronounced the repolarization and the greater the undershoot (Fig. 5.10B). Both aftereffects can be blocked by applying ouabain before introducing the high K^+_o solution (Fig. 5.10C), indicating that they are mediated by an active membrane pump. An analogous phenomenon was observed with intraretinal recordings of K^+ activity in the intact skate retina (Kline et al., 1985). The undershoot (below the resting K^+_o level) following an initial light-evoked K^+ increase in the IPL of skate retina (Fig. 5.10D) was thought to reflect the reuptake of K^+ by a similar process, but it was not possible to ascertain whether uptake was into neurons or glial cells.

Although the photically induced rise in $[K^+]_o$ may be considered too small to activate the glial Na^+–K^+ ATPase, measurements in the ECS with relatively large K^+-sensitive electrodes may grossly underestimate the changes in $[K^+]_o$ occurring within the confines of the synaptic clefts (cf. Ripps and Wikovsky, 1985). It is apparent that Müller cells are capable of accumulating K^+, i.e., exhibiting net uptake, in addition to K^+ siphoning. This uptake not only would contribute to the clearance of extracellular K^+ when $[K^+]_o$ is abnormally high, but also could provide a readily accessible source of K^+ to replenish that lost by active neurons. The Na^+–K^+ pump and several of the voltage-activated channels through which Müller cells regulate the neuronal environment are illustrated schematically in Fig. 5.11.

5.5. WHY VOLTAGE-DEPENDENT ION CHANNELS?

It is puzzling to find that the "passive" Müller cell membrane contains such an array of voltage-dependent channels, particularly when it appears many are without any apparent function under normal physiological conditions. It is often questioned whether the various ion channels found in cultured glial cells are expressed *in vivo*, but there is little evidence to suggest otherwise. In those instances in which studies have been conducted on glial cells that have been acutely isolated, or are *in situ* in slice preparations or intact tissues, ion channels with similar properties have been detected (cf. Barres et al., 1990). The more pressing questions are: (1) What are the circumstances that could induce a change in membrane potential sufficient to activate voltage-dependent channels on glial cells; and (2) assuming these ion channels participate to some extent in K^+ buffering, what other role might they serve?

To address the first question necessitates a consideration of pathological conditions such as Leão's spreading depression (Leão, 1944), a phe-

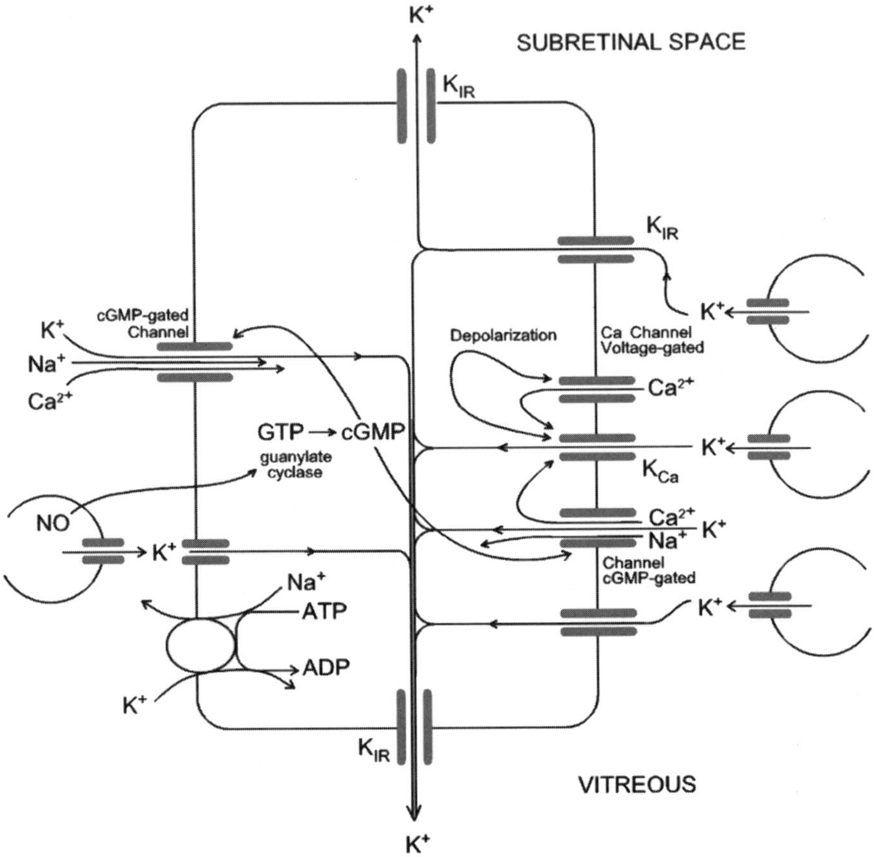

Figure 5.11. Schematic representation of several putative pathways by which the Müller cell serves to regulate the ionic environment of retinal neurons. K^+ released from active neurons in the plexiform layers enters primarily through inward rectifying (K_{IR}) channels, producing a local depolarization that drives K^+ to the distal and proximal ends of the Müller cell where it exits through K_{IR} channels to regions of low $[K^+]_o$. The redistribution of K^+ and other ions may be influenced also by the Na^+–K^+ ATPase and a depolarization-induced activation of voltage-gated calcium channels, as well as by the production of nitric oxide (NO). The latter may activate guanylate cyclase to form cGMP, and subsequently open nonspecific cationic cGMP-gated channels (see Chapter 4). The entry of Ca^{2+} via these routes can bring into play calcium-activated potassium channels (K_{CA}) to further enhance the influx of K^+.

nomenon associated with unusually high levels of $[K^+]_o$ that has been studied extensively in brain and retina. With regard to the second question, it has been speculated that Müller cells might serve as a storage and supply depot for neuronal channels.

5.5.1. Spreading Depression

The dependence of membrane voltage on ion concentration shown in Fig. 5.1 indicates that $[K^+]_o$ would have to increase by more than 10 mM to depolarize the glial cell membrane sufficiently to activate several of the voltage-dependent channels that could come to the aid of the "normal" pathways for K^+ clearance. One condition known to cause such an extreme change in $[K^+]_o$ is spreading depression (SD), described originally by Leão (1944) as the transient extinction of all spontaneous and evoked cortical activity for periods of tens to hundreds of seconds following its induction by focal electrical stimulation. The area of depressed electrical activity spreads slowly (~3 mm/min) across the brain to invade neighboring regions, and is accompanied by local increases in $[K^+]_o$ variably estimated to be between 20–60 mM (cf. Sugaya et al., 1975; Nicholson and Kraig, 1981), well above the ceiling of ~10 mM reached during intense neuronal activity (Heinemann and Lux, 1977; Somjen, 1979). This gross departure from normal brain activity can be detected extracellularly with a combination of ion-selective and voltage-sensing electrodes which record both the increase in $[K^+]_o$ and the large negative potential (>10 mV) that is evoked at the center of the wave of depression (Nicholson and Kraig, 1981).

Although the K^+ released during intense neuronal activity was postulated to be the primary event in SD (Grafstein, 1956), the underlying mechanisms are not fully understood. For example, glutamate elicits SD at significantly lower concentrations than K^+ (Van Harreveld and Fifkova, 1970), and the front of the spreading wave of SD is accompanied by acid–base shifts in the interstitial space as well as decreases in the extracellular concentrations of Ca^{2+}, Na^+, and Cl^- (Nicholson and Kraig, 1981; Kraig and Cooper, 1987). Despite this diversity of interrelated reactions, there is a good deal of support for the view that both a rise in $[K^+]_o$ and the release of excitatory amino acids are capable of triggering a sequence of events that lead to the generation and propagation of SD (Van Harreveld, 1978; Tuckwell and Hermansen, 1981; Somjen et al., 1992).

There was some uncertainty as to whether SD occurs *in vivo,* but numerous studies have shown that under pathological conditions neural tissues are particularly susceptible to the induction of SD. Indeed, $[K^+]_o$ levels as high as 100 mM have been recorded in the rat brain following 10 min of anoxia (Vyskocil et al., 1972), and comparable changes were reported in the anoxic cat cortex (Blank and Kirshner, 1977). Moreover, SD occurs in experimentally induced cortical seizures (Sypert and Ward, 1974), and it has been suggested that SD is responsible for the transient visual field defect (scintillating scotoma) that often signals the onset of migraine headache (Milner, 1958; Gardner-Medwin, 1981).

There are also many ocular conditions, e.g., diabetic retinopathy, arterial or venous infarct, trauma, etc. that probably induce SD and bring into play the spatial buffering mechanisms of Müller cells. The effectiveness of this system may account for the unusual resistance of the mammalian eye to prolonged periods of total occlusion of the retinal circulation. In the arterially perfused cat eye, for example, electroretinal activity recovers fully after more than one hour occlusion and subsequent reperfusion (Hoff and Gouras, 1973; Peachey et al., 1993).

5.5.2. Spreading Depression Propagation in Retina

The first evidence that SD occurs in retina was provided by Gouras (1958), who reported that waves of depressed electrical activity often arose spontaneously after excising and hemisecting the toad eye. He noted that the SD wave could be followed by observing the passage of a milky color traveling across the tissue, preceded by an intense epileptiform-like discharge of ganglion-cell action potentials. Studies confirming these observations revealed that SD could be triggered by a variety of mechanical, electrical, chemical, and photic stimuli (Gouras, 1958; Hanawa et al., 1968; Sheardown, 1993, 1997; Ripps et al., 1981). Both photometric and electrophysiological studies indicate SD originates in the inner plexiform layer (Martins-Ferreira and DeOliveira Castro, 1966; Mori et al., 1976a) owing to pathologically enhanced synaptic activity (Mori et al., 1976b).

There is good reason to believe that glial cells are involved in the propagation of SD. Using potassium-selective electrodes to record the movement of K^+ across the neocortical surface of anesthetized rats during SD, Gardner-Medwin (1983a, b) produced convincing evidence that the major K^+ flux passed through glial cells driven by voltage and concentration gradients—i.e., the transfer of K^+ by a "spatial buffering" mechanism was far more effective than extracellular diffusion in dispersing the excess K^+. In the retina, results consistent with this conclusion were obtained recently by Nedergaard et al. (1995), who demonstrated that the transcellular spread of SD is mediated by gap junctions, probably between astrocytes (and perhaps involving other cell types) of the inner retina.

Müller cells also make a significant contribution to the clearance of K^+ during generation of SD. Simultaneous measurements of the negative potential that signals the onset of SD, and the concomitant changes in

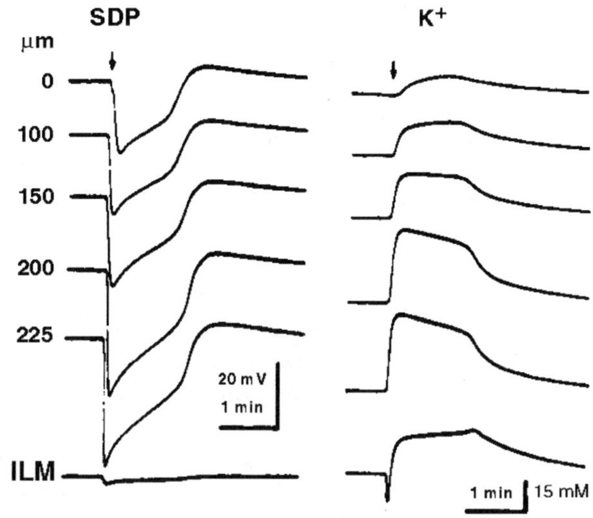

Figure 5.12. Spreading depression (SD) associated with intraretinal changes in K^+ in the intact frog retina. A chloride-free Ringer solution was used to predispose the retina to photically-induced SD. Simultaneous depth recordings of the negative extracellular potential (SDP) generated by SD and the concomitant rise in K^+ in response to the light flash (arrow) show that the largest changes in both responses occur in the inner retina (depth of 200–225 μm), where the K^+ rise typically exceeded 30 mM. Smaller responses are recorded near the photoreceptor layer (0 μm depth) or as the electrodes approach the inner limiting membrane (ILM) (Mori et al., 1976b). (Copyright 1976 W. Miller, reprinted with permission.)

$[K^+]_o$, showed that the peak amplitudes of both signals were maximal in the proximal retina, and decreased as the electrodes approached the distal layers of the retina (Mori et al., 1976a; Fig. 5.12). The rise in $[K^+]_o$ typically reached >30 mM. Intracellular recordings of $[K^+]$ with superfine ion-selective electrodes indicated that the excess $[K^+]_o$ was taken up by Müller cells, after which there was a large efflux of K^+ into the vitreous (Mori et al., 1976a, b). Although it was not known at the time, much of this process is probably attributable to K^+-induced activation of the voltage-sensitive channels on Müller cells.

5.5.3. Translocation of Ion Channels

One intriguing possibility for the many glial voltage-regulated ion channels is that the glial cell may act as a "channel factory" that expresses and stores ion channels for eventual transfer to neuronal membranes (Shrager et al., 1985; Bevan et al., 1985). This notion was put forth originally as a possible explanation for the presence of ion channels on Schwann cells (Shrager et al., 1985). It was postulated that the Schwann cells may provide a local source of Na^+ and K^+ channels for the axolemma of the neurons they ensheathe.

Although the transfer of macromolecules from glia to axon has been well documented in squid giant axons (Lasek et al., 1974; Gainer, 1978; Tytell et al., 1986), there is no unequivocal experimental evidence supporting the intercellular translocation of the protein constituents that form voltage-activated ion channels. Were translocation to occur, it could serve as a supplement to axonal transport, particularly for nerve cells with terminals far distant from the cell body. In this connection, it is important to recall that Müller cells ensheathe the perikarya and axon hillocks of retinal ganglion cells (cf. Chapter 1). However, the likelihood of channel interchange between the different cell types has been questioned based on differences in their channel properties (e.g., kinetics, TTX-sensitivity, etc.), and on evidence of cell-specific expression of glial and neuronal Na^+-channel mRNA (Gautron et al., 1992). Although these findings do not rule out the possibility that channel proteins can be transferred from glia to neurons, it will require some innovative experimental approaches to demonstrate that this does occur *in vivo.*

5.6. LIGHT-EVOKED CHANGES IN EXTRACELLULAR K^+

Having considered the membrane properties of Müller cells, it is important to identify the intraretinal loci at which light-evoked changes in extracellular $[K^+]_o$ arise. The issue has been studied extensively with ion-specific electrodes in the eyecup preparation (Karwoski and Proenza, 1978; Dick and Miller, 1978; Kline et al., 1978, 1985), in the intact eye (Steinberg et al., 1980), and in the retinal slice (Karwoski et al., 1985). There is now general agreement that the changes originate primarily at three principal sites (Fig. 5.13). Illumination produces a decrease in K^+ in the distal retina at the level of the photoreceptors, and induces an efflux of K^+ into the extracellular space at the depths of the outer and inner plexiform (synaptic) layers.

The neural events that lead to these ionic changes have also been well documented. The distal decrease recorded in the outer nuclear layer is almost entirely a consequence of the light responses of photoreceptors. In darkness, there is a sustained radial current mediated mainly by the entry of Na^+ through cyclic-nucleotide-gated (CNG) channels in the photoreceptor outer segments (Baylor et al., 1979; Fesenko et al., 1985) and its extrusion by a Na^+/K^+ pump located in the membranes of the receptor inner seg-

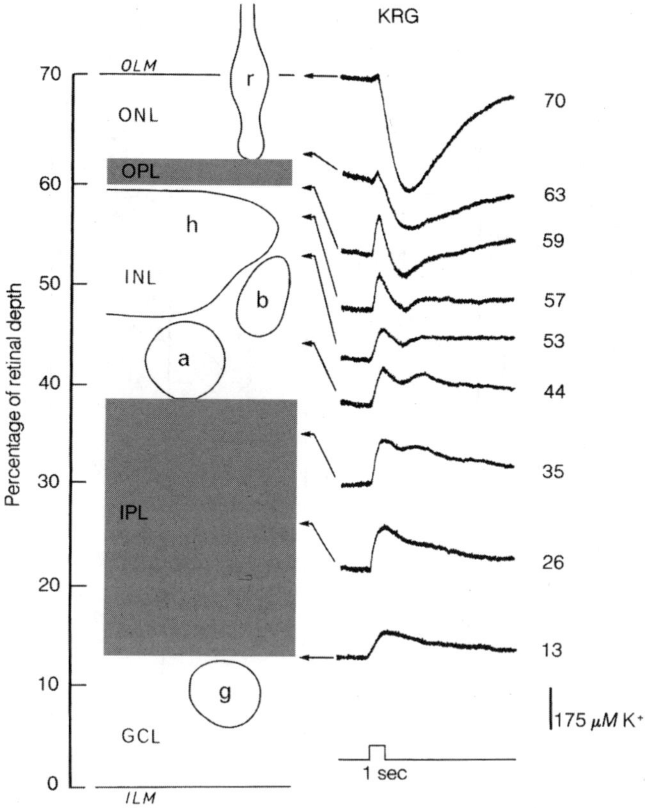

Figure 5.13. The light-evoked changes in K^+ recorded with an extracellular ion-selective electrode as it was advanced from the vitreal surface (ILM) to the outer limiting membrane (OLM). The retinal schematic shows the approximate locations of the photoreceptor (r), horizontal (h), bipolar (b), amacrine (a), and ganglion (g) cells. The large, prolonged efflux of K^+ seen in the IPL becomes more transient and smaller as the electrode approaches the ONL. In the photoreceptor region, the fall in K^+ dominates the KRG (Dowling, 1987). (Copyright 1987 Belknap Press, reprinted with permission.)

ments (Bok and Filerman, 1979). Although the pump drives K^+ into the cell as the excess Na^+ is removed, a passive efflux of K+ (Fig. 5.14) restores the ionic balance. However, light triggers an enzymatic cascade that leads to closure of the CNG channels and causes the photoreceptors to hyperpolarize, thereby reducing the passive K^+ efflux. Because the activity of the Na^+/K^+ pump is not immediately affected by these events, the result is a net influx of K^+, and a large transient decrease in $[K^+]_o$ in the region of the receptor inner segments (Matsuura et al., 1978; Oakley et al., 1979).

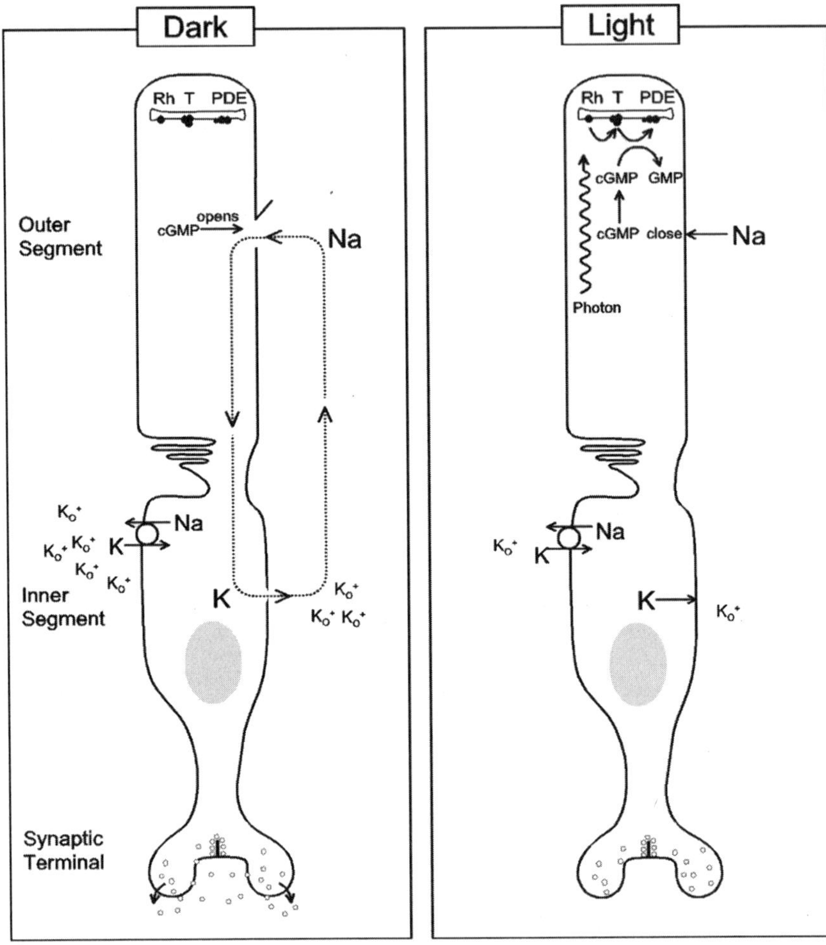

Figure 5.14. In darkness (left side of figure), a sustained inward current carried primarily by Na^+ enters the photoreceptor outer segment through cGMP-gated channels. A Na^+–K^+ exchanger in the inner segment extrudes the excess Na^+ and brings in K^+, which then exits the cell through passive ion channels to maintain the dark level of extracellular K^+. The influx of cations depolarizes the cell, and induces the release of glutamate from the photoreceptor terminal. The effect of light (shown on the right) is to trigger the sequential activation of rhodopsin (R), transducin (T), and phosphodiesterase (PDE). This enzymatic cascade leads to the hydrolysis of cGMP, channel closure, and suppression of the dark current. The resultant hypolarization reduces both the discharge of transmitter, and the passive efflux of K^+. However, the inward K^+ flux mediated by the Na^+–K^+ pump is not significantly affected, thus producing a fall in extracellular K^+.

The $[K^+]_o$ increases detected in the outer and inner plexiform layers are attributable to the activation of cells that depolarize in response to photic stimulation. In the distal retina, the principal source of the K^+ efflux in the OPL is the depolarizing (ON) bipolar cell (Dick and Miller, 1978; Kline et al., 1978). The light-evoked response recorded at this locus with an ISM appears in Fig. 5.13 as a small, transient rise in K^+. More proximally, in the region of the IPL, there is a larger, more sustained K^+ efflux that derives primarily from the depolarizing synaptic potentials of the spike-generating amacrine and ganglion cells. Both cell types respond to the onset and offset of illumination (Werblin and Dowling, 1969), and both exhibit center-surround antagonism (Burkhardt, 1974), features that are reflected in recordings of the K^+ changes induced by these stimulus parameters (Fig. 5.15; cf. Kline et al., 1985).

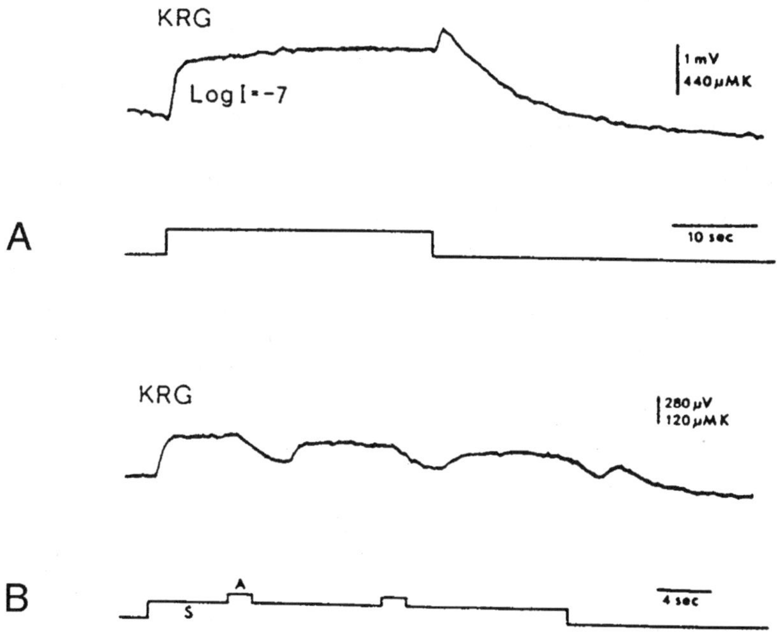

Figure 5.15. Recordings of extracellular K^+ (the KRG) showing that the changes in $[K^+]_o$ mirror the inhibitory effect of surround illumination on neuronal responses of the inner retina. A. The K^+ efflux is sustained in response to a 30 sec light exposure, and there is a transient K^+ increase at light offset. B. The K^+ level induced by the presence of a small spot of light (S) is reduced by the addition of annular illumination (A) (Kline et al., 1985). (Copyright 1985 Elsevier Science, reprinted with permission.)

5.7. CELLULAR ORIGINS OF THE ELECTRORETINOGRAM

The ionic fluxes described above give rise to radial currents that underlie various components of the electroretinogram (ERG), a light-evoked transretinal potential widely used as a noninvasive, objective test of retinal function in the differential diagnosis of visual disorders (cf. Goodman and Ripps, 1960; Ripps, 1982; Hood and Birch, 1990; Berson, 1993). Whether recorded across the retina, i.e., between electrodes in the vitreous and at the basal surface of the RPE, or with a contact lens electrode on the cornea of the intact eye, the ERG response evoked by the onset of illumination consists primarily of three readily detectable components: the *a*-, *b*-, and *c*-waves. The variation in response polarity and time course indicate that this complex waveform reflects the summation of potentials arising from different sources. Each component is itself a composite derived from more than one cellular source.

There has long been interest in the cellular origins of the ERG potentials and the mechanisms by which they are generated. Investigators have attempted to address these questions with intracellular recordings, pharmacological intervention, and current-source density (CSD) analysis. CSD provides a means for determining the "sources" and "sinks" of extracellular current flow based on depth recordings of tissue resistivity and intraretinal voltage in response to photic stimulation (cf. Nicholson and Freeman, 1975; Xu and Karwoski, 1994a). The results have shown that, for the most part, the potentials that underlie the ERG derive from the light-evoked ionic currents of radially oriented retinal neurons, Müller cells, and pigment epithelial cells. An oversimplified scheme depicting how several of these cell types can contribute to a response envelope that resembles the ERG is illustrated in Fig. 5.16, and described more fully in the following sections.

5.7.1. The *a*-wave

The initial response to photic stimulation is a rapid cornea-negative deflection referred to as the *a*-wave, or the "fast PIII" component of the ERG. By recording the light-induced voltage between two microelectrodes positioned near the distal and proximal ends of the photoreceptors in the rat retina, Penn and Hagins (1969) showed that the leading edge of the *a*-wave reflects the *change* in radial current flow that results from closure of the Na^+ channels in the outer segment membranes; i.e., the response is essentially a reduction of the "dark current" and a reversal of the source–sink current paths (Fig. 5.17). Although potassium ions are not a significant factor in the generation of the *a*-wave, it is important to recall that the ionic

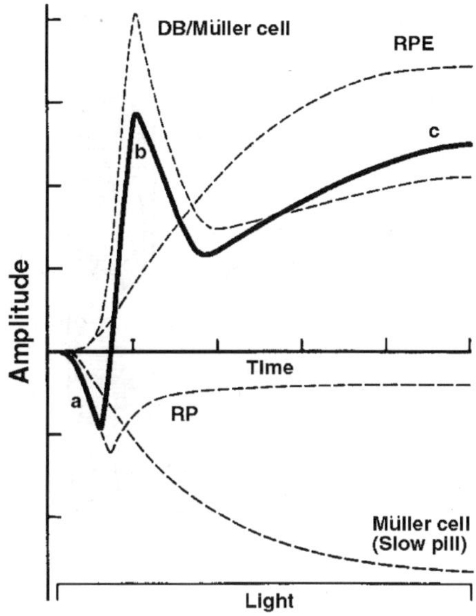

Figure 5.16. Schematic representation of several of the components that summate to give rise to the *a*, *b*, and *c*-waves of the transretinal ERG (bold continuous line). Light-evoked radial currents are generated by the photoreceptors, bipolar cells, the retinal pigment epithelium, and the Müller cells. The leading edge of the *a*-wave is dominated by the hyperpolarizing receptor potential (RP), whereas the rapidly rising transient phase of the *b*-wave reflects activation of depolarizing ON bipolar cells (DB) and the K^+-mediated response of Müller cells. The slowly rising *c*-wave receives contributions of opposite polarity from the retinal pigment epithelium (RPE) and the Müller cell (slow PIII). In this simplified drawing, contributions from cells of the inner retina (e.g., oscillatory potentials) and the response to light offset are omitted (Ripps and Witkovsky, 1985). (Copyright 1985 Elsevier Science, reprinted with permission.)

mechanisms underlying the photoreceptor response are responsible for the *distal decrease* in $[K^+]_o$ recorded in the region of the outer nuclear layer (cf. Fig. 5.14).

5.7.2. The *c*-wave

The slowly-developing cornea-positive *c*-wave is itself a composite potential consisting for the most part of contributions from the RPE and Müller cells (Fig. 5.17). Both signals arise as a consequence of the *distal decrease* in $[K^+]_o$, but owing to their cellular origins and extracellular current pathways, the two components are of opposite polarity. The RPE is largely responsible for the positivity of the waveform, although a sustained DC-potential, generated probably by depolarizing bipolar cells (Steinberg, 1969; Knapp and Schiller, 1984; Katz et al., 1992), contributes to the response. The negative component, usually of lesser amplitude (Fig. 5.17), derives primarily from the Müller cell with a small contribution from the plateau phase of the hyperpolarizing photoreceptor potential. The magnitude and time course of the various positive and negative potentials determine the amplitude of the *c*-wave and can lead to considerable response

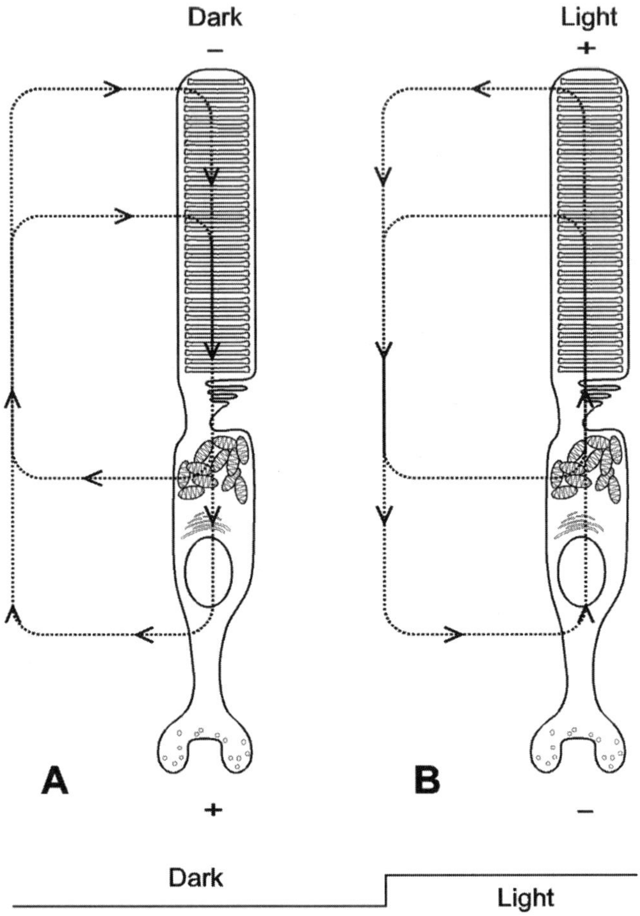

Figure 5.17. The radially oriented current of a dark-adapted photoreceptor (A) is suppressed by light. Electrodes at the distal and proximal ends of the photoreceptor record this change as a reversal of the current path (B), and the generation of the cornea negative *a*-wave.

variability. In some normal human subjects, for example, the ERG is lacking the *c*-wave, probably as a result of an atypical balance between these opposing potentials (Täumer et al., 1976; Hock and Marmor, 1983).

The RPE Component: The involvement of the RPE in the generation of the *c*-wave was first demonstrated by Noell (1953), who selectively poisoned the RPE with sodium acetate and showed that the *c*-wave of the rabbit

ERG disappeared. Subsequently, intracellular recordings from RPE cells in the intact cat eye revealed a hyperpolarizing response with the same temporal characteristics as the extracellularly recorded *c*-wave (Steinberg et al., 1970). Simultaneous recordings of K^+_o and *c*-wave responses provided convincing evidence that the hyperpolarization resulted from the light-evoked *distal decrease* of $[K^+]_o$ in the subretinal space (Oakley and Green, 1976; Oakley, 1977; Matsuura et al., 1978). Indeed, these studies showed that the *distal decrease* in $[K^+]_o$ and the *c*-wave exhibit similar kinetics (Fig. 5.18A), display virtually identical intensity-response functions (Fig. 5.18B), and that both responses persist after application of sodium aspartate, which suppresses the activity of postreceptoral neurons.

The decrease in $[K^+]_o$ within the subretinal space produces changes in a number of voltage-gated and transport mechanisms that affect the potential across the RPE apical and basal membranes (cf. Miller et al., 1978; Hughes and Steinberg, 1990; Gallemore et al., 1997) and the high resistance barrier formed by tight junctions between adjacent RPE cells (Cohen, 1965). However, the predominant effect is on the K^+ conductance of the apical membrane. The light-induced decrease in K^+_o hyperpolarizes the apical membrane relative to the basal membrane resulting in an increase in the transepithelial potential. This event establishes a current path, which gives rise to the positive component of the ERG *c*-wave (Fig. 5.19); the magnitude of the potential will depend in large measure on the intercellular junctional resistance and the resistances of the apical and basal membranes of the RPE.

The Müller Cell and Slow PIII: By acting also on the Müller cell, the *distal decrease* in $[K^+]_o$ generates a cornea-negative potential that follows a similar time course as the positive RPE component of the *c*-wave. This "slow PIII" response can be unmasked by recording across the isolated retina (free of the RPE) after blocking signal transmission to second-order neurons with aspartate or other chemical agents (Witkovsky et al., 1975; Xu and Karwoski, 1997). Under these conditions, only the leading edge of the receptor potential (*a*-wave) and slow PIII are seen in the electroretinographic traces (Fig. 5.20A). Slow PIII exhibits an incremental threshold curve similar to that of the *c*-wave, saturates at the same intensity level, and like the *c*-wave, displays a rod-dominated spectral sensitivity function (Witkovsky et al., 1975). However, CSD analysis of the spatial and temporal properties of slow PIII indicates that the current path is via the Müller cell (Fig. 5.20B), with the major "source" in the distal retina and the current "sink" at the vitreal surface (Faber, 1969; Xu and Karwoski, 1997). As a result, the transretinally recorded slow PIII is of opposite polarity and very nearly the mirror image of the RPE-derived component of the *c*-wave potential. Thus, the

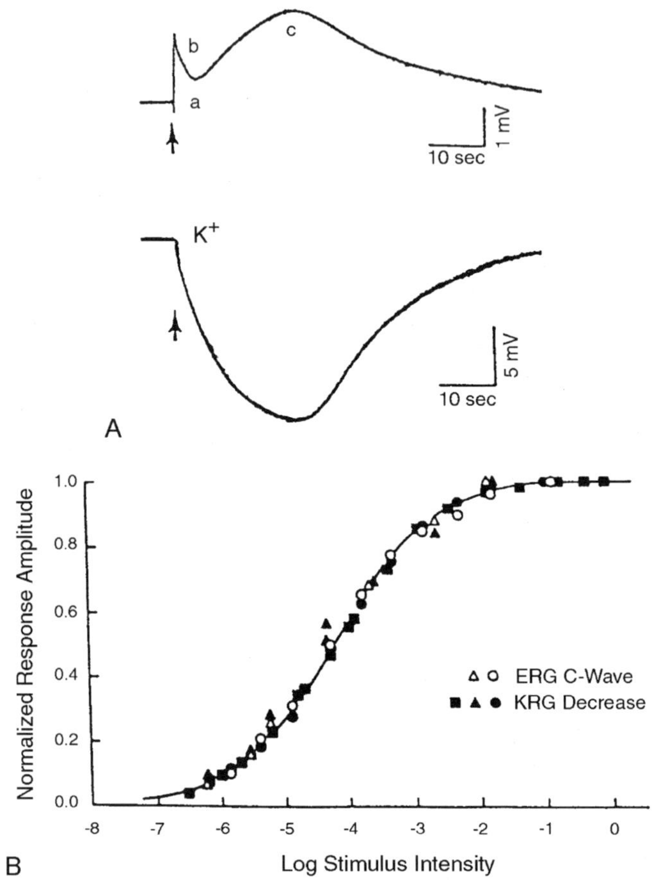

Figure 5.18. A. The transretinal ERG of a dark-adapted frog eyecup in response to a brief light flash is shown above the corresponding potassium change recorded with an ion-specific electrode (ISE) placed in the photoreceptor layer about 25 μm from the receptor surface. The voltage generated by the ISE represents the change in $[K^+]_o$, where 5 mV ≈ 1 mM K^+. Note the similar time course followed by the cornea-positive *c*-wave and the large fall in K^+ in the distal retina (Matsuura et al., 1978). (Copyright 1978 Elsevier Science, reprinted with permission.) B. Simultaneous recordings of the ERG *c*-wave and the K^+ decrease in the distal retina show that the magnitudes of the responses (normalized to their respective maxima) describe identical intensity-response functions (Oakley and Green, 1976). (Copyright 1976 The American Physiological Society, reprinted with permission.)

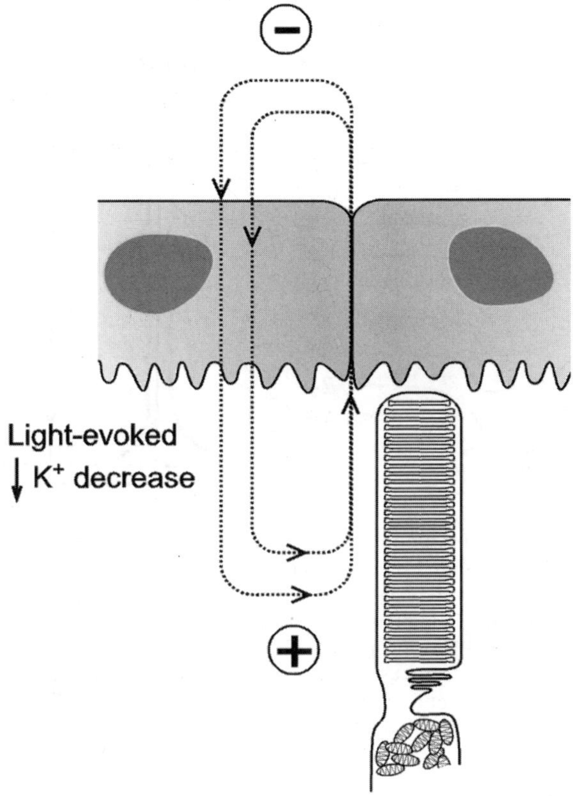

Figure 5.19. The *distal decrease* in K^+ establishes a current path through and along the retinal pigment epithelial cells that gives rise to the cornea-positive component of the *c*-wave.

amplitude of the *c*-wave is greatly enhanced when the slow PIII response is suppressed with DL-α-aminoadipic acid (Szamier et al., 1981; Welinder et al., 1982), a cytotoxin that affects primarily Müller cells (Olney et al., 1971; Pedersen and Lund Karlsen, 1979).

5.7.3. The *b*-wave

Unlike the ERG components described above, the large, transient cornca-positive *b*-wave arises postreceptorally, and thus the response can be suppressed by any substance that is capable of blocking synaptic transmission between the photoreceptors and second-order neurons (cf. Furakawa and Hanawa, 1955; Dowling and Ripps, 1972, 1973; Ripps et al., 1976). There

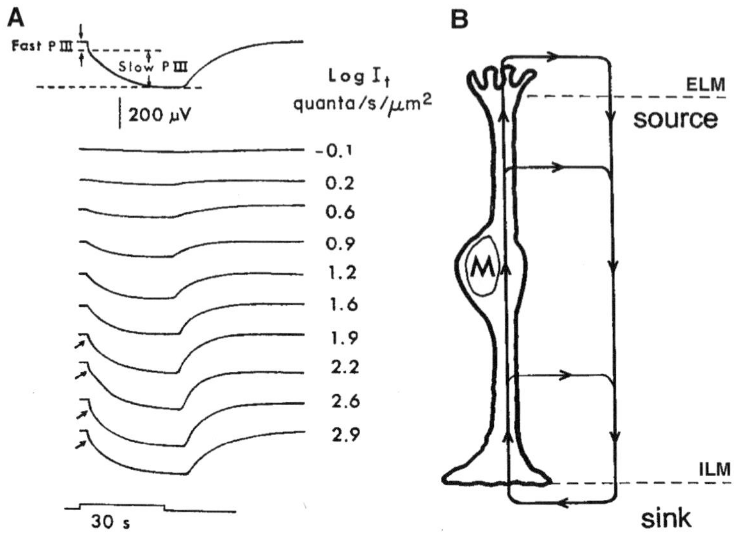

Figure 5.20. A. The slow PIII response is revealed by recording across the isolated retina (free of the RPE) after blocking signal transmission to second-order neurons with aspartate. The leading edge of the receptor potential (fast PIII) precedes a cornea-negative slow potential that grows in amplitude with increasing flash intensity (log I_t), and follows the same time course as the *c*-wave (Witkovsky et al., 1975). (Copyright 1975 The Rockefeller University Press, reprinted with permission.) B. Current source density analysis of the slow PIII potential shows that its primary source is at the distal end of the Müller cell, with a current sink at the ILM (Xu and Karwoski, 1997). (Copyright 1997 Cambridge University Press, reprinted with permission.)

is general agreement that the potential originates with the light-evoked activity of depolarizing "ON" bipolar cells (Dick and Miller, 1978; Stockton and Slaughter, 1989; Gurevich and Slaughter, 1993; Xu and Karwoski, 1994b; Robson and Frishman, 1995; Hanitzsch et al., 1996). However, the cellular origins of the current "sources" and "sinks" have long been a subject of controversy. The argument revolves essentially around the question of whether the *b*-wave reflects the K^+-mediated spatial-buffering currents of Müller cells, or whether it represents the voltage generated by the radial currents carried by the bipolar cells themselves. Results obtained from CSD analysis and a variety of other experimental approaches have yielded conflicting findings.

The Müller Cell Hypothesis: The view that the ERG *b*-wave is generated by the light-evoked radial currents of Müller cells was proposed initially by Faber (1969), whose CSD results in rabbit retina revealed a large *b*-wave sink in the region of the outer plexiform layer, with sources located distal to that site as well as extending proximally to the vitreal surface of the retina. These observations led Faber to conclude that "the only retinal element which has a spatial distribution consistent with the distribution of *b*-wave sources and sinks is the Müller cell." Soon thereafter, Miller and Dowling (1970) produced evidence that the intracellularly recorded Müller cell potential and the transretinal *b*-wave exhibited similar waveforms and intensity-response functions over a substantial range of stimulus intensities (Fig. 5.21A). They suggested that an increase in $[K^+]_o$ in the outer plexiform layer would depolarize the distal ends of the Müller cells and produce a current flow that agrees with the polarity of the *b*-wave. Simultaneous recordings of the ERG *b*-wave and the light-evoked K^+ increase in the distal layers of the skate retina gave results consistent with this prediction (Kline et al., 1985). It was evident that the two parameters displayed nearly equivalent intensity-response functions (Fig. 5.21B). With only minor variations, this concept remains the prevailing view of Müller cell involvement in *b*-wave generation. We know now that the only neurons that could produce a significant light-induced K^+ efflux at this distal site are the depolarizing "ON" bipolar cells.

Numerous studies have implicated light-evoked changes in $[K^+]_o$, and the radial current paths established by the spatial buffering activity of Müller cells in the generation of the *b*-wave (Miller, 1973; Dick and Miller, 1978; Kline et al., 1978; Newman, 1980; Newman and Odette, 1984; Wen and Oakley, 1990). Fig. 5.22 illustrates the main features of this process and their relationship to the current sources and sinks derived from CSD analysis (Newman, 1980). Light-evoked increases in $[K^+]_o$ occur mainly in the plexiform layers of the retina as the result of neural activity; the excess K^+_o enters the Müller cell at these sites and is transferred to the high conductance endfoot, where it is released into the vitreous (spatial buffering). CSD analysis, indicating current sinks in the outer and inner plexiform layers and a large current source at the vitreal surface of the retina, is consistent with this pattern of ionic fluxes (solid lines); the net effect is the creation of a current dipole that corresponds to the cornea-positive *b*-wave. It should be apparent that on this view, the *b*-wave is fundamentally an *epiphenomenon*, and not in the direct pathway of signal transmission, an important consideration in assessing some of the studies cited below.

Evidence favoring the Müller-cell hypothesis has come also from developmental studies showing the correspondence between Müller cell maturation and the development of the *b*-wave (Rager, 1979), as well as from experiments in which selective components of the K^+/Müller-cell pathway are interrupted pharmacologically (cf. Szamier et al., 1981; Laties, 1983; Wen and Oakley, 1990; Katz et al., 1992). For example, using multiple electrode combinations to record the ERG, the intraretinal changes in $[K^+]_o$, and

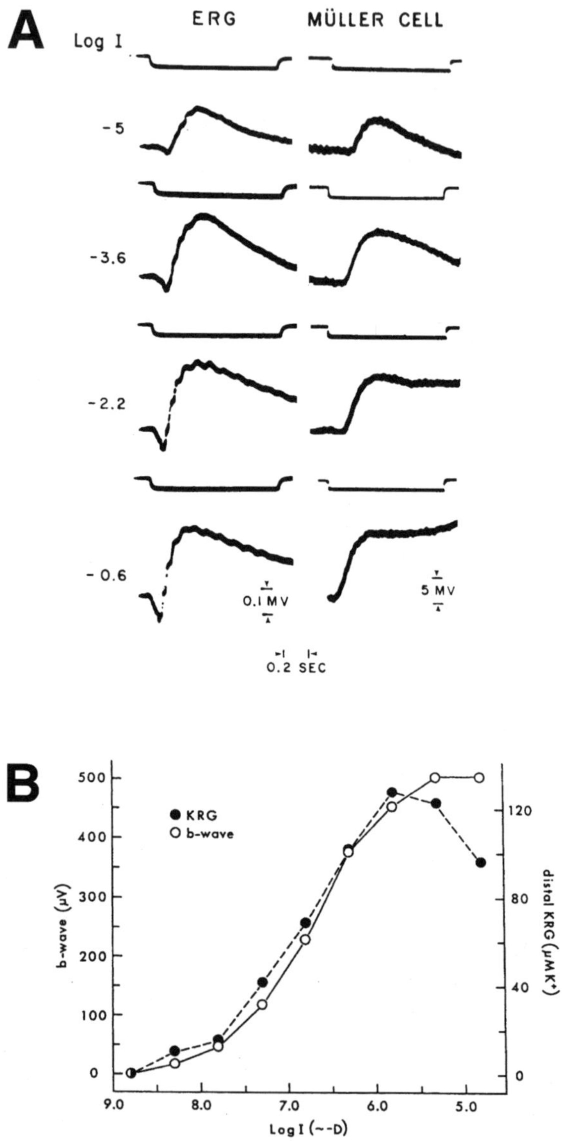
A
ERG
MÜLLER CELL
Log I
-5
-3.6
-2.2
-0.6
0.1 MV
5 MV
0.2 SEC
B
KRG
b-wave
b-wave (μV)
distal KRG (μM K+)
Log I (~-D)
500
400
300
200
100
0
120
80
40
0
9.0
8.0
7.0
6.0
5.0

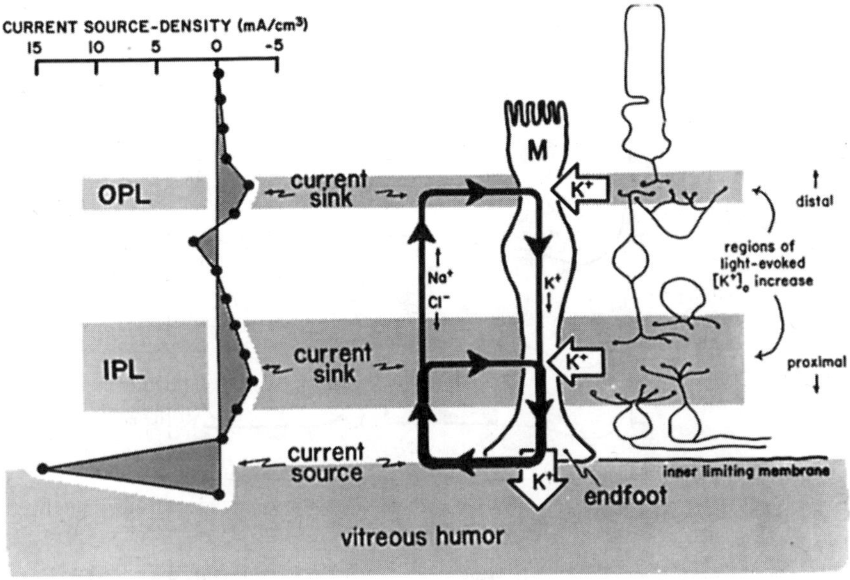

Figure 5.22. A light-evoked efflux of K^+ occurs in the inner and outer plexiform layers of the retina (right side of schematic). K^+ enters the Müller cell at these sites and exits via high conductance K^+ channels at the endfoot, thereby establishing a current source at the vitreous surface, and current sinks in the OPL and IPL. These spatial buffering currents (solid lines), carried by K^+ intracellularly and by Na^+ and Cl^- extracellularly, are thought to generate the ERG *b*-wave. Note that, on this view, the current pathway underlying the *b*-wave potential is independent of currents generated by neurons engaged in the transmission of signals through the retina (line drawings on the right) (Newman, 1985). (Copyright 1985, Elsevier Science, reprinted with permission.)

the intracellular responses of Müller cells, Wen and Oakley (1990) found that 200 μM Ba^{2+}, which blocks K^+ channels in Müller cells, reduced both the ERG *b*-wave and the Müller cell response. However, it did not suppress the K^+ increase within the OPL that is thought to reflect the activity of depolarizing bipolar cells (Fig. 5.23A). These findings and those of related

Figure 5.21. A. Intracellular potential recordings from mudpuppy (*Necturus*) Müller cells show nearly identical latencies to the ERG *b*-wave at all stimulus intensities, but as expected, lack the *a*-wave and oscillatory potentials that appear in the ERG traces. However, the Müller cell responses do not replicate precisely the kinetics of the *b*-wave; at the higher intensities, the rise time appears slower and the potentials are more sustained (Miller and Dowling, 1970). (Copyright 1970 American Physiological Society, reprinted with permission.) B. Intensity-response data obtained with simultaneous recording of the ERG *b*-wave and the distal increase in K^+. The two vertical axes have been adjusted arbitrarily, but scaling involved computations based on the relationship between the K^+ increase and the resultant *b*-wave (Kline et al., 1985). (Copyright 1985 Elsevier Science, reprinted with permission.)

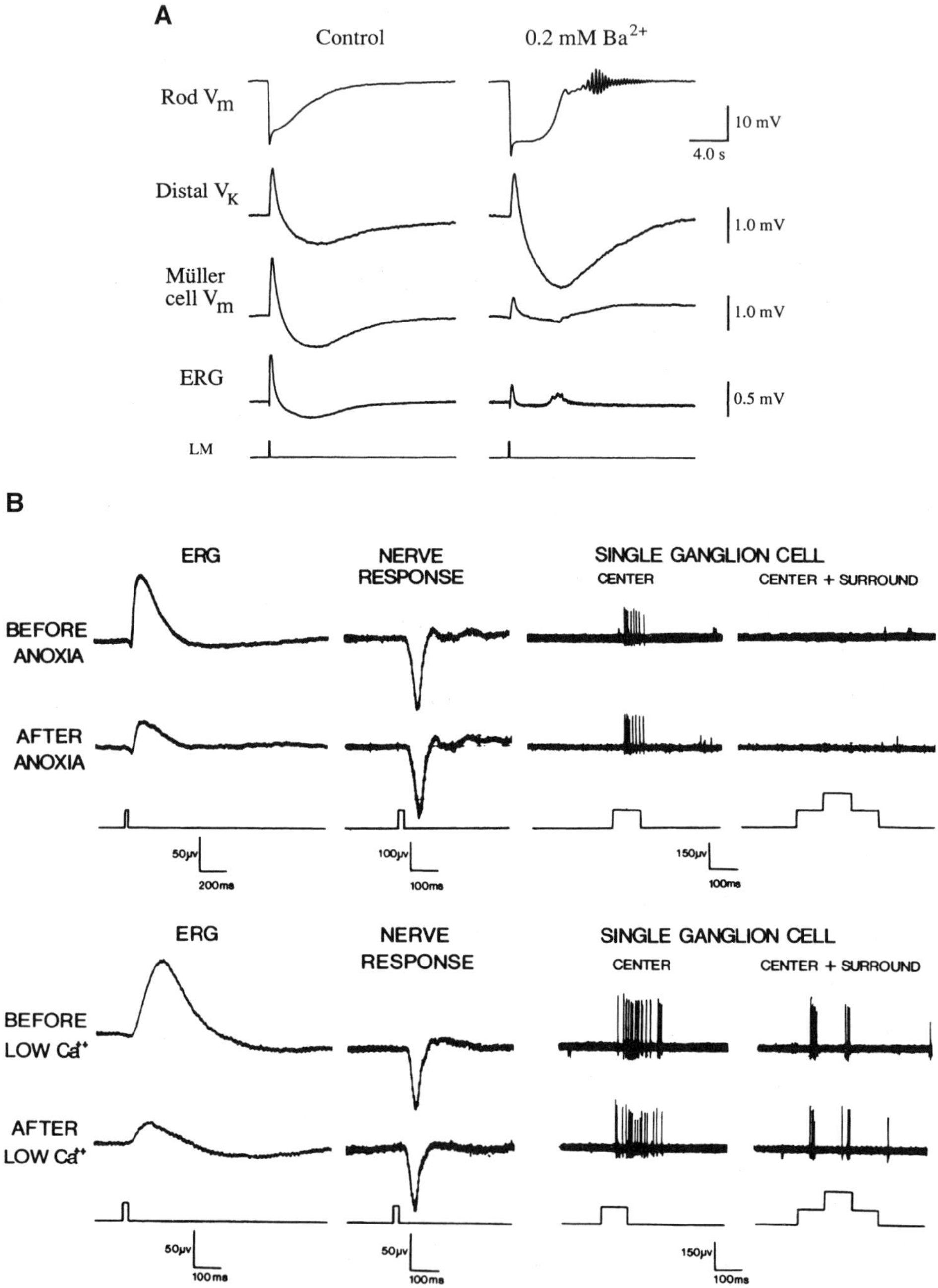
A
Control
0.2 mM Ba2+
Rod Vm
10 mV
4.0 s
Distal VK
1.0 mV
Müller cell Vm
1.0 mV
ERG
0.5 mV
LM
B
ERG
NERVE RESPONSE
SINGLE GANGLION CELL
CENTER
CENTER + SURROUND
BEFORE ANOXIA
AFTER ANOXIA
50μv
200ms
100μv
100ms
150μv
100ms
ERG
NERVE RESPONSE
SINGLE GANGLION CELL
CENTER
CENTER + SURROUND
BEFORE LOW Ca++
AFTER LOW Ca++
50μv
100ms
50μv
100ms
150μv
100ms

experiments on the pharmacology of the DC component of the ERG (Katz et al., 1992) suggest that both the *b*-wave and the DC component (cf. Fig. 5.17) are generated primarily by the distal K^+–Müller cell mechanism in response to the K^+ efflux from depolarizing bipolar cells. In addition, Masland and Ames (1975) demonstrated that following a short period of anoxia, or after incubation in a low calcium medium, there was marked diminution of the *b*-wave with no apparent reduction in ganglion cell response (Fig. 5.23B). Comparable results were obtained after selectively poisoning the Müller cell with D,L-α-aminoadipic acid (Bonaventure et al., 1981). If the reduced *b*-wave seen in these studies was the result of abnormal bipolar cell function, it is puzzling to find retention of seemingly normal signal transmission from photoreceptors to ganglion cells.

The Bipolar Cell Hypothesis: Despite the previously cited findings, there is reason to believe that the *b*-wave may not reflect radial current flow generated by a K^+-induced depolarization of Müller cells (cf. Tomita and Yanagida, 1981). For example, membrane potential measurements from amphibian Müller cells correlate poorly with the waveform and time course of the *b*-wave, but appear to mirror an extracellular field potential (the M-wave) recorded in the proximal retina (Karwoski and Proenza, 1977, 1978). Inconsistencies between the current sinks and sources for the *b*-wave (CSD analysis) and the kinetics and loci of K^+ fluxes contradict the Müller cell hypothesis (Vogel and Green, 1980). In addition, a question has been raised as to whether the magnitude of the light-evoked $[K^+]_o$ increase recorded in the distal retina is sufficient to generate the large transretinal *b*-wave, although it is generally recognized that electrode-induced damage, interference from the large distal K^+ decrease, as well as other factors may dilute considerably the K^+ measurements. (Karwoski et al., 1985).

←

Figure 5.23. A. Recordings of light-evoked responses from the toad retina superfused with a normal Ringer (control) solution or one containing 0.2 mM Ba^{2+}; the stimulus marker is shown on the bottom trace. The uppermost traces show the extracellular ERG and the distal K^+ increase; the lower pair are intracellular membrane potential recordings (V_m) from a Müller cell and rod photoreceptor. Note that the rod potential and the K^+ efflux are relatively unaffected by exposure to barium, but both the Müller cell response and the ERG *b*-wave are substantially reduced (Wen and Oakley, 1990). (Copyright 1990 National Academy of Sciences, U.S.A., reprinted with permission.) B. Effects of anoxia (upper pair of traces) and low Ca^{2+} (lower traces) on the transretinal ERG, the compound action potential recorded from the optic nerve, and the spike activity of a single ganglion cell. Both anoxia and the low calcium solution reduce significantly the ERG *b*-wave, but have no detectable effect on the responses recorded from ganglion cells or their optic nerve bundles (Masland and Ames, 1975). (Copyright 1975 John Wiley & Sons, Inc., reprinted with permission.)

Nevertheless, a large body of evidence supports the view that the *b*-wave potential is predominantly a direct expression of the activities of rod-driven depolarizing (ON) bipolar cells (cf. Xu and Karwoski, 1994b; Robson and Frishman, 1995; Hanitzsch et al., 1996; Green and Kapousta-Bruneau, 1999; Shiells and Falk, 1999; Lei and Perlman, 1999). The depolarizing responses of these cells reflect the opening of cGMP-activated cation channels in postsynaptic membranes, as a result of the light-induced suppression of glutamate release from photoreceptor terminals (Shiells and Falk, 1990; Nawy and Jahr, 1990, 1991). Membrane depolarization and the radial orientation of the bipolar cell would be effective in creating an extracellular current path consistent with the polarity of the *b*-wave.

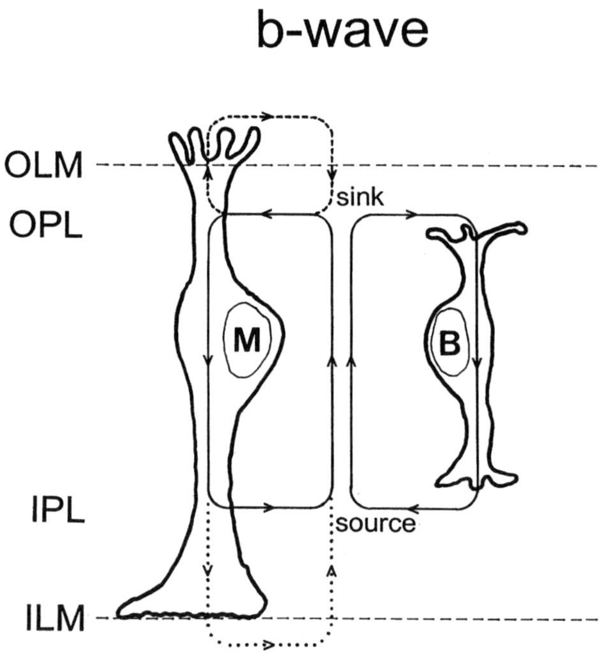

Figure 5.24. Current dipoles established by the sinks and sources of the ERG *b*-wave. CSD analysis of the intraretinal depth profiles of light-evoked field potentials and resistivity indicate that the major currents (bold lines with arrows) are established between the inner (source) and outer (sink) plexiform layers. These loci do not distinguish between bipolar and Müller cells in the generation of the *b*-wave. However, minor currents (dashed lines) in the region of the ILM are suppressed by barium (an effective blocker of glial K^+ channels), suggesting that the barium-resistant component of the *b*-wave originates with the depolarizing bipolar cells (Xu and Karwoski, 1994b). (Copyright 1994 American Physiological Society, reprinted with permission.)

More direct evidence that the *b*-wave potential arises from the depolarizing bipolar cell has come from extensive CSD analyses of current sinks and sources underlying the various components of the amphibian ERG (Xu and Karwoski, 1994a, 1994b, 1995; Karwoski et al., 1996). These data indicate that at light onset there is a large transient sink near the OPL, and a large source in the IPL that exhibits similar kinetics to the distal sink (Fig. 5.24). This current pattern closely approximates the extent of the bipolar cell, although the presence of a Ba^{2+}-sensitive slow current source at the internal limiting membrane (where K^+ exits the Müller cell), suggests a lesser current path representing a contribution from the Müller cell to the ERG *b*-wave.

A further indication of the link between the light-activated responses of depolarizing bipolar cells and the *b*-wave potential is provided by the strong correlation between the waveforms of the chemically isolated *b*-wave and the intracellulary recorded ON bipolar cell potential (Gurevich and Slaughter, 1993; Tian and Slaughter, 1995; Shiells and Falk, 1999), although this does not preclude an intervening stage mediated by the K^+–Müller cell spatial buffering mechanism. However, results showing enhancement of the *b*-wave response in the presence of barium (in concentrations sufficient to effectively block the K^+ channels on Müller cells) implicate the primacy of the bipolar cell in the generation of the ERG *b*-wave (Green and Kapousta-Bruneau, 1999; Lei and Perlman, 1999).

Given the conflicting evidence on the *b*-waves source, electrophysiological studies alone may not be able to establish unequivocally the relative contribution of bipolars and Müller cells to this transretinal potential. Recently, however, further support for the bipolar cell hypothesis was obtained from studies on $K_{IR}4.1$ knockout mice (Kofuji et al., 2000). Inactivation of the principal K^+ channel subunit expressed in mouse Müller cells produced a total loss of the slow PIII response but had little or no effect on the ERG.

Role in Retinal Pathophysiology

6

Müller cells undergo significant morphological, cellular, and molecular changes when retinal conditions become abnormal or when there is injury to retina. Some of the changes are related to wound healing and tissue repair, whereas others reflect Müller cell involvement in protecting the retina from further damage. For example, in atrophic macular lesions characterized by the loss of RPE and photoreceptors, Müller cells adhere to the denuded Bruch's membrane and may help to maintain the outer blood–retina barrier (Rentsch, 1977; Eagle, 1984; Birnbach et al., 1994). Similarly, a role for Müller cells in ELM repair is suggested by the observation that following vitrectomy, breaks in the ELM are sealed off by Müller cell processes (Madeperla et al., 1994; cf. Gass, 1999). In addition, Müller cells express growth factors as well as neurotransmitter transporters that have important functions in preventing excitotoxic damage to retinal neurons (Wen et al., 1995; Otori et al., 1994; Harada et al., 1998). This chapter will describe the putative role of Müller cells in retinal disease and the cellular and molecular changes in response to retinal injury.

6.1. RETINAL ISCHEMIA AND EXCITOTOXICITY

Retinal ischemia develops when blood supply is insufficient to fulfill the metabolic needs of the retina. The effects can vary from short-term impairment of function to infarction and death of the retina (Kohner, 1994). Neuronal damage and death following ischemia and hypoxia appear to result from an accumulation of excess glutamate in the tissue (Choi and Rothman, 1990). According to the "glutamate toxicity" hypothesis, conditions of hypoxia–ischemia lead to an increase in glutamate levels in the extracellular fluid. This could be due to excessive glutamate release from presynaptic terminals, leakage from cellular compartments, or impairment of glutamate transporters. The net result is prolonged stimulation of NMDA

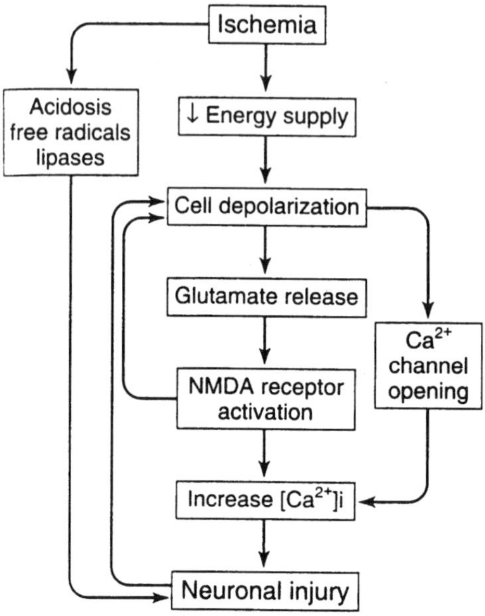

Figure. 6.1. Potential pathways leading to neuronal injury resulting from an episode of ischemia. Ischemia causes depletion of cellular energy stores, the release of free radicals, and subsequent neuronal injury. In addition, cellular depolarization due to the decreased energy supply induces the release of glutamate, which leads, in turn, to sustained activation of NMDA receptors and Ca^{2+} entry into the cell. The elevation of $[Ca^{2+}]i$ causes excessive activation of a variety of Ca^{2+}-dependent enzymes, such as proteases, leading to neuronal injury and possibly cell death. Recent evidence suggests that in addition to NMDA receptors, AMPA-type glutamate receptors may also be involved in this process (Dingledine and McBain, 1993). (Copyright 1993 Raven Press, reprinted with permission.)

receptors, followed by an influx of large amounts of extracellular Ca^{2+} and other cations into neurons, leading to neuronal death (Choi, 1988). The glutamate neurotoxicity model is supported by several lines of evidence and includes a role for glial cells in the cytotoxic process (Fig. 6.1) (Choi and Rothman, 1990; Lee et al., 1999). Although there is some disagreement regarding the role of Ca^{2+} ions in excitotoxic death of retinal neurons (Romano et al., 1998; Chen et al., 1998), it is clear that excessive levels of glutamate are toxic in the retina as well.

Because glial cells possess high affinity glutamate uptake systems and can convert glutamate to glutamine, a nonneurotoxic substance (see Chapter 3), glial cells are likely to be intimately involved in regulating extracellular glutamate levels in the CNS. It has been observed that neurons in mixed cultures containing neurons and astrocytes are more resistant to glutamate-induced cell death than neurons in glia-free cultures (Rosenberg and Aizenman, 1989). Conversely, incubation of neuronal cultures with either glutamate-uptake blockers or antisense oligonucleotides to glutamate transporters increases the excitotoxic effects of glutamate (Rosenberg et al., 1992; Rothstein et al., 1996).

Retinal neurons are also susceptible to glutamate-induced damage. As first shown by Lucas and Newhouse (1957), subcutaneous injections of

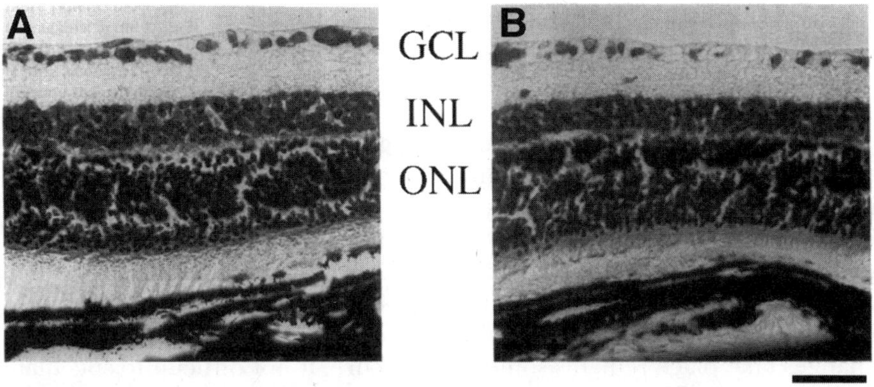

Figure. 6.2. The effect of chronic glutamate exposure leads to ganglion cell death. A. Transverse section of a normal retina. B. Retina from a rat that received intravitreal glutamate. Note the loss of a significant number of cell bodies in the ganglion cell layer (GCL) (Vorwerk et al., 1996). (Copyright 1996 Association for Research in Vision and Ophthalmology, reprinted with permission.)

sodium L-glutamate in adult and neonatal mice result in severe destruction of ganglion cells and partial loss of cells in the INL. Subsequent studies confirmed the retinotoxic effects of glutamate and its relative selectivity for neurons of the inner retina when administered subcutaneously (Cohen, 1967) or intravitreally (Sisk and Kuwabara, 1985). A surprisingly small elevation in the concentration of glutamate, when maintained for long periods of time, is toxic to retinal ganglion cells (Fig. 6.2). Repeated intravitreal injections of glutamate in the rat eye, titrated so as to induce a sustained rise in the vitreous concentration of glutamate from endogenous levels (5–12 mM) to the range of 26–34 mM were shown to cause the death of more than 40% of the retinal ganglion cells after three months exposure (Vorwerk et al., 1996). Memantine, an agent that blocks activation of NMDA receptors (Bormann, 1989; Chen and Lipton, 1997), gave partial protection against the effects of elevated glutamate (Vorwerk et al., 1996). Interestingly, the relatively small (micromolar) rise in glutamate that led to toxic changes in the rat eye, is typically seen in the vitreous of human and monkey eyes with glaucoma (Dreyer et al., 1996), a chronic disease that ultimately results in the loss of ganglion cell axons.

Considering the retinotoxic effects of glutamate, and its sustained release in darkness, there is clearly the need for an efficient means for its removal. Müller cell uptake is a vital part of this process in the normal retina. We have described already the carrier-mediated uptake of glutamate into Müller cells as a means to remove glutamate from the synaptic cleft and terminate its post-synaptic action (Chapter 4). Paradoxically, there is a remarkable number of ways in which factors affecting its normal activity can contribute to retinotoxicity.

The stoichiometry of the Müller cell glutamate transporter (see Chapter 4) and evidence that the carrier-mediated uptake of glutamate can be inhibited (Barbour et al., 1988) or even reversed (Szatkowski et al., 1990) have important implications for events that may be triggered because of pathological conditions. In retina subjected to trauma or anoxia, abnormal conditions can produce a precipitous rise in extracellular K^+ and elevation of glutamate to toxic levels. While the precise sequence in which these changes take place remains somewhat obscure, it not difficult to speculate on how they might come about (cf. Nicholls and Attwell, 1990).

Suppose, for example, that the extracellular $[K^+]$, normally in the range of 3–5 mM, rises to a level of 60 mM or greater, a situation encountered in the anoxic brain and during spreading depression (cf. Walz and Hertz, 1983). At this concentration, several interrelated processes are brought into play, all of which contribute to extracellular glutamate elevation and its potentially excitotoxic effects. First, there is the depolarization of glutamatergic nerve terminals and calcium-dependent release of glutamate. Second, both the rise in $[K^+]_o$ and its depolarizing effect on Müller cells (and neurons) not only inhibit glutamate uptake, but also induce calcium-independent release of glutamate (by reversed uptake) to increase further its extracellular concentration. In addition, glutamate activation of NMDA receptors on second order neurons causes the release of arachidonic acid, yet another source of uptake inhibition (Barbour et al., 1989). Last, the rise in glutamate will itself cause neurons to depolarize further, release more K^+, and stimulate, in turn, an even greater efflux of glutamate. Consequently, glutamate levels may exceed 100 mM, which is more than enough to induce neuronal death.

In addition to glutamate transporters, glutamine synthetase levels are also important for regulating glutamate toxicity in the retina (Gorovits et al., 1997). Chick retinas treated with cortisol to induce high levels of glu-

Figure. 6.3. Relationship between glutamine synthetase induction and lactate dehydrogenase (LDH) release, an indicator of cell damage. A. Cortisol was injected into E15, E16, or E17 eggs which were further incubated for 4, 24, or 48 hr respectively. Retinas were dissected out and organ-cultured for an additional 4 hr in the presence of glutamate. Levels of cellular GS and extracellular LDH in cortisol-treated chick retinas maintained *in vitro*. B. Concentration-dependence of GS and LDH levels on [Cortisol]. C. Inverse correlation between GS activity and LDH release. These data suggest that glutamine synthetase affords protection from glutamate-induced cellular injury (Gorovits et al., 1997). (Copyright 1999 National Academy of Sciences, U.S.A., reprinted with permission.)

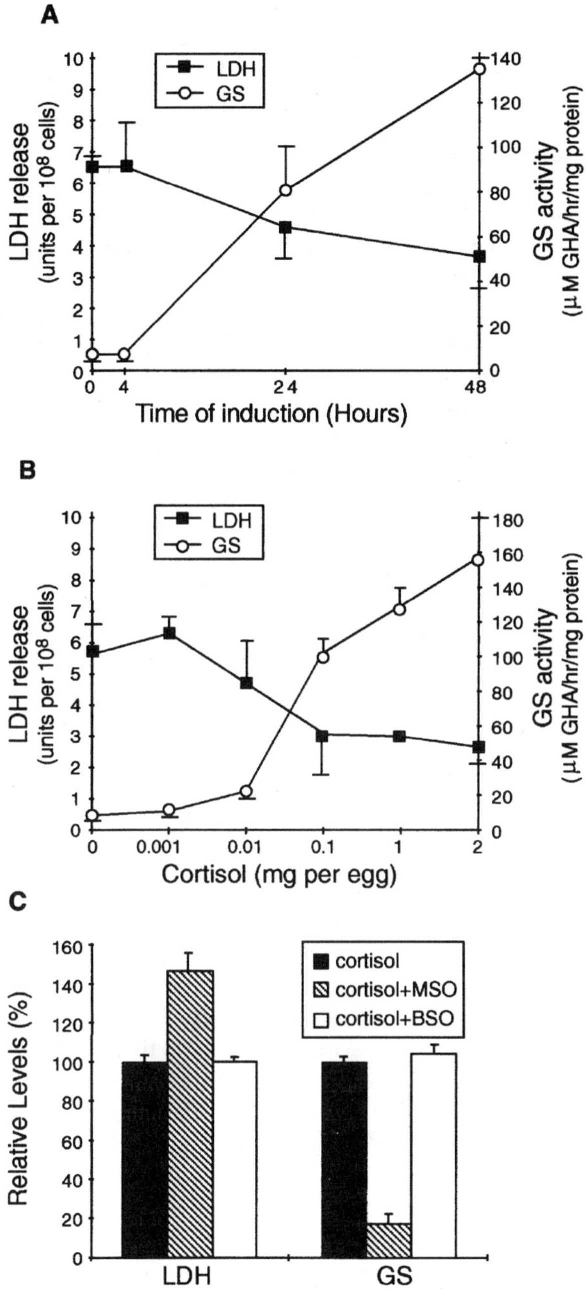
A
LDH
GS
LDH release
(units per 10^8 cells)
GS activity
(μM GHA/hr/mg protein)
Time of induction (Hours)
B
LDH
GS
LDH release
(units per 10^8 cells)
GS activity
(μM GHA/hr/mg protein)
Cortisol (mg per egg)
C
cortisol
cortisol+MSO
cortisol+BSO
Relative Levels (%)
LDH
GS

tamine synthetase are more resistant to glutamate-induced damage than retinas expressing low levels of the enzyme (Fig. 6.3). Conversely, treatment with the glutamate synthetase inhibitor, methionine sulfoxamine, renders chick retinas more susceptible to glutamate excitotoxicity (Gorovits et al., 1997). Finally, under conditions of retinal ischemia, when the intracellular ATP level drops in the retina and possibly in Müller cells as well, glutamine synthetase activity is expected to be inhibited because the enzyme requires ATP for conversion of glutamate to glutamine. Therefore, the intracellular glutamate level would increase in Müller cells, which may, in turn, trigger glutamate release.

In the retina, as in other regions of the nervous system, glutamate toxicity arises via a wide variety of intracellular pathways activated by the influx of Ca^{2+} through agonist- (mainly NMDA) and voltage-gated channels (Rothman and Olney, 1987; Meldrum and Garthwaite, 1990; Statkowski and Attwell, 1994; but cf. Izumi et al., 1999). This is supported by studies showing that blockers of either NMDA receptors (Hahn et al., 1988; Chen et al., 1992; Vorwerk et al., 1996) or calcium channels (Sucher et al., 1991) provide protection from the toxic effects of glutamate.

6.2. THE IMMUNE RESPONSE

The eye is generally regarded as an immunologically privileged organ (Niederkorn, 1990). There appear to be at least three reasons for this unique status. First, the eye lacks a lymphatic system that can capture potential ocular antigens released from cells. Second, the blood–retina barrier restricts entry into the retina of lymphoid and mononuclear cells as well as potential antigens. Last, ocular cells appear to express only low levels of major histocompatibility complex (MHC) antigens needed for induction and regulation of the immune response.

In the immune system, interactions between macrophages and lymphoid cells is essential for eliciting an immune response (Opremcak, 1994). Briefly, macrophages present foreign antigens to $CD4^+$ T lymphocytes in conjunction with class II MHC antigens. The "activated" T lymphocytes (T-helper cells) secrete cytokines that promote proliferation and differentiation of B cells into plasma cells which, in turn, secrete immunoglobulins directed against the antigens. In addition, the T-helper cells also regulate the function of other T lymphocytes involved in cell-mediated immunity (Fig. 6.4).

In organ-specific autoimmune diseases, "resident" cell types are involved in regulating tissue-specific autoimmunity (Unanue and Allen, 1987). It has been proposed that maintenance of the autoimmune state is achieved by presentation of autoantigens by resident cells aberrantly ex-

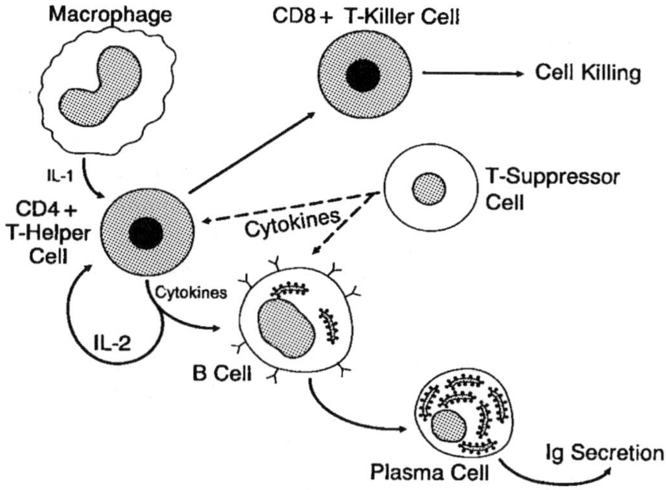

Figure. 6.4. Scheme showing steps in immune response. The macrophage (Antigen presenting cell, APC) presents foreign antigens in conjunction with class II MHC molecules to $CD4^+$ T-helper cells. Following activation, the T-helper cells secrete cytokines that initiate differentiation of B cells to plasma cells and also regulate T cell effector functions.

pressing class II MHC determinants (Unanue and Allen, 1987). Expression of MHC antigens on resident cells is of crucial importance since these molecules are essential for T-lymphocyte-mediated immune responses. Moreover, induction of MHC antigens has been correlated with the capacity to function as antigen-presenting cells (APC).

In the nervous system, astrocytes, oligodendrocytes, and microglia secrete cytokines, respond to cytokines, and present antigens to T lymphocytes (Benveniste, 1993). Antigen-presenting cells in the eye may be classified into bone marrow-derived cells such as uveal dendritic cells and retinal microglia, or nonbone marrow-derived cells represented by RPE, Müller cells, and vascular endothelial cells. There is good evidence both *in vivo* and *in vitro* that these cell types express MHC class II molecules (cf. Ishimoto et al., 1999).

The question as to whether Müller cells function as APC in experimental allergic uveitis (EAU) has been studied extensively by Robert Nussenblatt and his associates at the National Eye Institute. EAU is a $CD4^+$, T-lymphocyte-mediated autoimmune disease, which has served as a useful paradigm for human ocular inflammatory diseases such as sympathetic ophthalmia (Gery et al., 1986). EAU can be readily induced in animals by immunization with retinal S-antigen and other retinal proteins, and presents itself as a delayed, hypersensitive reaction, resulting in ocular inflammation and loss of photoreceptors (Gery et al., 1986). The identity of the resident APC and the factors that suppress inflammatory response in EAU are not well understood.

Müller cells from adult rat retina normally do not grow in cell culture. However, Roberge et al. (1985) found that the addition of conditioned media from concanavalin A-activated spleen cells, or from an S-antigen-activated T lymphocyte cell line, led to proliferation of Müller cells. Furthermore, about 30% of the cells in these cultures expressed class II MHC antigen, suggesting that Müller cells might be involved in retinal immune reactions. To further understand T cell and Müller cell interactions, S-antigen activated rat T helper (T_HS) cells were cocultured with syngeneic Müller cells expressing Ia determinants (class II MHC antigens) in the presence of the antigen. Surprisingly, little T cell proliferation was observed in the cultures (Table 6.1; Caspi et al., 1987). The effect appeared not to be cell type-specific since proliferation of a T helper control cell line (specific to a tuberculin-derivative) was also strongly suppressed. The involvement of a cytotoxic effect could be ruled out since the T_HS cells recovered from cocultures were fully viable.

To study whether the suppressive effect was mediated by a soluble inhibitory factor, experiments were carried out with cocultures of Müller cells and T helper cells in a two-chamber system in which there was free mixing of culture media although the cells themselves were not in contact (Caspi et al., 1987). The failure to inhibit T cell proliferation under these conditions showed that physical contact between Müller cells and T cells was essential to induce suppression.

In subsequent studies, trypsin treatment of Müller cells was also shown to abolish the suppressive effect, suggesting that a membrane-bound protein was probably involved in this process (Roberge et al., 1988). In a later study, it was shown that Müller cell-mediated inhibition of T cell proliferation could be overcome by treating Müller cells with glucocorticoids, presumably by preventing expression of the Müller cell inhibitory factor (Roberge et al., 1991). These studies provide support for a role for Müller cells as antigen-presenting cells in the retina.

Müller cell involvement in the immune response is further supported by the studies of Mano et al. (1991) who showed that glial cells derived from postmortem donor eyes express MHC class I and class II antigens when exposed to γ-interferon. In agreement with these findings, histopathologic studies of a patient with subretinal fibrosis and uveitis syndrome showed that Müller cells express class II MHC antigens (Kim at al., 1987). Finally, antibodies directed against Müller cells have been reported in other autoimmune diseases, e.g., Vogt-Koyanagi-Harada syndrome, Behcet's disease, and sympathetic ophthalmia (Chan et al., 1985).

Table 6.1. Inhibitory Effect of Müller Cells on Antigen-Driven Proliferation of SA_g-specific (T_HS) and PPD-specific (T_HP) Cell Lines[a]

Culture	[^{3}H] thymidine uptake (cpm × 10^{-3})	
	T_HS	T_HP
Mu-T_H-APC-Ag[b]	3.5 (97.6)	7.9 (90.1)
Mu (24hr)-T_H-APC-Ag[c]	18.3 (88.1)	ND
T_H-APC-Ag	153.5	80.1
Mu-T_H-APC	3.2	0.8
T_H-APC	2.7	0.5

[a]Longterm cultures of T-helper lymphocytes (T_H) and Müller cells (Mu) were generated from Lewis rats and maintained. The assays were carried out in a mixture of the culture media from each of the two cell types [minimal essential medium with 10% fetal bovine serum and RPMI 1640, with 1.5% rat serum; each culture contained additional supplements]. Rested T_H cells (2 × 10^4) and APCs (syngeneic thymocytes irradiated with 3000R) (4 × 10^5) were incubated with antigen in the presence or absence of 1 × 10^4 Müller cells (irradiated with 3000R) in 96-well tissue culture trays, in triplicate 0.2 ml cultures. In the wells where coculture with Müller cells was to be dealyed by 24 hours, the Müller and the T_HS cultures were plated separately at time 0, and the lymphocyte cultures were quantitatively transferred into the Müller cell well at 24 hours of incubation. All cultures were incubated for 60 hours. One microcurie of [^{3}H] thymidine was added to each well during the last 14 hours. The background radioactivity of Müller cells alone was 0.4 × 10^{-3} cpm. Standard errors were less than 10% of the mean. Numbers in parentheses represent percent of inhibition.

[b]SA_g or PPD (10 μg/ml) in T_HS and T_HP cultures, respectively.

[c]Coculture with Müller cells started after 24 hours of activation with antigen and APCs; ND, not done.

Müller cells appear to be an attractive candidate for preventing autoimmune responses in both the uvea and retina. These cells are involved in formation of the outer limiting membrane, which can potentially act as an anatomical barrier and thus prevent entry of circulating lymphocytes into the retina when the blood–retina barrier is compromised. If antigen-specific T cells escape into the retina, Müller cells may become activated and suppress T helper cell proliferation by direct cell–cell contact (Caspi and

Roberge, 1989). In addition to serving as immunoregulatory cells, Müller cells may participate in healing and scar formation following EAU (Chan et al., 1985; Rao et al., 1986).

It is important to remember that the potential role of Müller cells in the immune response is based on experimental evidence obtained using *in vitro* cell culture systems. Whether Müller cells have a similar function *in vivo* remains to be established. Indeed, the studies of Ishimoto and associates (1999) using bone marrow-transplanted chimeric rats suggest that bone marrow-derived antigen-presenting cells are the ones likely to be involved in initiating uveoretinitis.

6.3. PHAGOCYTOSIS

If Müller cells are to function as antigen-presenting cells, they must be able to take up and process antigens. Indeed, Müller cells have been found to phagocytize a variety of substances. The earliest documented evidence of phagocytic activity by Müller cells was reported more than six decades ago by Friedenwald and Chan (1932). After injecting a suspension of melanin granules into the vitreous of albino rabbits, they noted that the pigment accumulated in monocytic phagocytes and Müller cells. Subsequently, it was shown that a variety of materials, such as carbon and copper particles, erythrocyte debris, and subretinal hemorrhage, are phagocytized by Müller cells (Algvere and Kock, 1983; Rosenthal and Appleton, 1975; Miller et al., 1986; Koshibu, 1978; Mano and Puro, 1990). In human eyes with chalcosis, copper particles are found in retinal glial cells (Rao et al., 1976).

The phagocytic activity of Müller cells has been examined in Müller cell cultures and also in experimental animals. Mano and Puro (1990) found that Müller cell cultures from postmortem eyes were able to phagocytize retinal cell fragments as well as latex beads (Fig. 6.5). Furthermore, they reported that uptake was partially dependent on extracellular [Ca^{2+}] and was blocked by the calcium channel blocker, nifedipine. Phagocytic activity was also reduced by the addition of 8-bromo-cAMP while vitamin D3 stimulated uptake suggesting that the phagocytic process used by Müller cells is similar to that associated with macrophages.

Freshly dissociated Müller cells, Müller cell cultures, and isolated rabbit retina have been shown to take up latex beads (Stolzenburg et al., 1992). In addition, Müller cells have been reported to phagocytize egg-lecithin-coated silicone particles following intraocular injection (Nishizona et al., 1993). Similarly, Müller cells in explant cultures of goldfish retina also accumulate latex beads, although no uptake was observed *in vivo* (Li et al., 1987; Wagner and Raymond, 1991). Phagocytosis of exogenous particles, cell debris, and hemorrhagic products may be an important scavenging func-

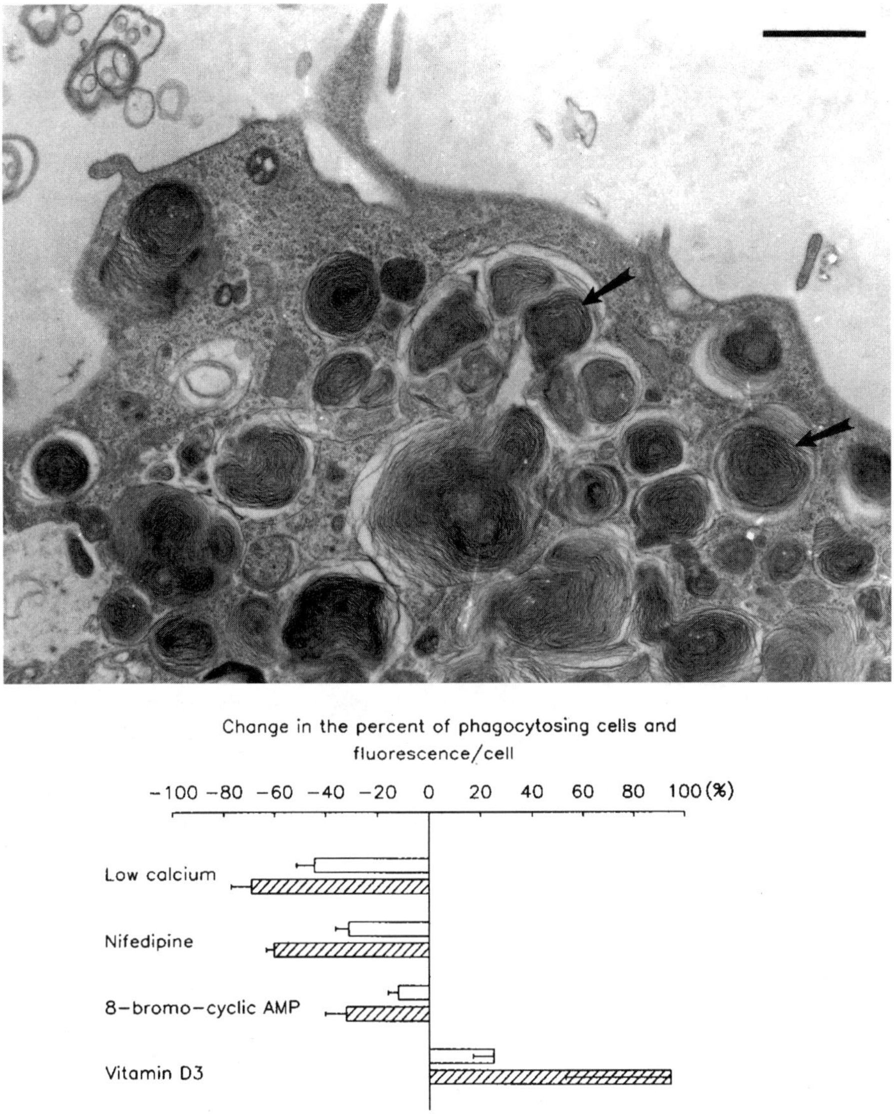

Figure. 6.5. Phagocytosis by Müller cell cultures. Electron micrograph of a Müller cell exposed to retinal debris. Arrows point to cellular debris inside the Müller cell. The histogram shows the effect of various conditions on the percentage of cells exhibiting phagocytic activity (open bar) and the mean fluorescence per cell (hatched bar). Mean changes from the appropriate control values are shown. Glial cell cultures were preincubated in the presence of 1 mM EGTA, 10 mM nifedipine, 1 mM 8-bromo-cyclic AMP, or 10 mM vitamin D3, and exposed to microspheres and subsequently analyzed by flow cytometry (Mano and Puro, 1990). (Copyright 1990 Association for Research in Vision and Ophthalmology, reprinted with permission.)

tion of Müller cells. Indeed, it is conceivable that Müller cells phagocytose outer segment membranes in degenerative retinal diseases in which the RPE is unable to perform this function.

6.4. RETINAL CONDITIONS WITH POTENTIAL MÜLLER CELL INVOLVEMENT

There is no compelling evidence that the Müller cell is the primary site affected in any retinal disease. However, there are many retinal conditions in which Müller cells appear to play a prominent role. For instance, in non-proliferative diabetic retinopathy, Müller cells are a major site for the synthesis and secretion of VEGF, an angiogenic agent that contributes to vascular proliferation (Amin et al., 1997). Gass (1999) postulated that the Müller cell cone, a cellular component of fovea centralis, is involved in the pathogenesis of some macular diseases such as idiopathic macular holes and foveomacular schisis. Retinal disorders in which there is potential for direct Müller cell involvement are considered in the next section.

6.4.1. X-linked Juvenile Retinoschisis

Juvenile retinoschisis (also termed congenital hereditary retinoschisis) is a rare, X-linked recessive disease that develops early in life and leads to parafoveal intraretinal cysts and marked vision loss in affected males (Yanoff et al., 1968; Harris and Yeung, 1976; Condon et al., 1986). The disease is characterized by bilateral splitting of the retina, often involving the macula and especially the inferior temporal quadrant of the peripheral retina. Retinal splitting at the level of the nerve fiber layer has been observed in histopathology studies (Condon et al., 1986) (Fig. 6.6). A major clinical feature of the disease is that the *b*-wave of the ERG is drastically reduced although the *a*-wave is normal (Harris and Yeung, 1976). Since Müller cells participate in the generation of the *b*-wave (Chapter 5), the disorder has been linked to Müller cell dysfunction (Harris and Yeung, 1976; Condon et al., 1986), and as expected, ultrastructural changes are seen in the Müller cells near the internal limiting membrane (Condon et al., 1986).

The Mizuo phenomenon (also known as Mizuo–Nakamura phenomenon) refers to a rapid change in the color of the fundus from red in the dark-adapted state to golden after light onset (Mizuo, 1913). A similar fundus appearance has been reported in X-linked retinoschisis (de Jong et al., 1991). Interestingly, a transient whitish-yellow sheen is also seen in "spreading depression" (see Chapter 5), a phenomenon triggered by a rise in extracellular K^+ in the retina. Accordingly, de Jong et al. (1991) suggested

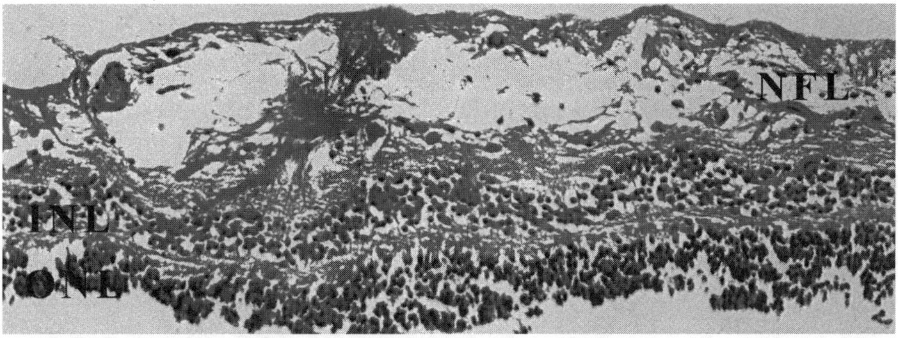

Figure. 6.6. Histopathology of juvenile X-linked retinoschisis. A histological section showing retinal split in the nerve fiber layer (NFL). INL, inner nuclear layer, and ONl, outer nuclear layer (Eagle, 1999). (Copyright 1999 W.B. Saunders, reprinted with permission.)

that the color changes associated with retinoschisis result from spreading depression due to the inability of the damaged Müller cell endfeet to clear K^+ from the extracellular space.

The gene associated with X-linked juvenile retinoschisis (*XLRS1*) has been cloned (Sauer et al. 1997). It encodes a 224-amino acid protein containing a "discoidin" domain at the C-terminus, which suggests that the protein is involved in cell–cell interactions such as cell adhesion or intercellular signaling. Surprisingly, the gene is expressed in photoreceptors but not in Müller cells (Reid et al., 1999). It remains to be learned whether the retinal splitting results from a direct action of the mutant protein on the Müller cell or is a secondary response to neuronal dysfunction.

6.4.2. Cystoid Macular Edema

Macular edema is yet another ocular disease in which Müller cells have been implicated. Cystoid macular edema (CME) is a condition in which there is an accumulation of fluid in the macula (Gass and Norton, 1966; Jampol, 1994). It is associated with a variety of ocular conditions that include retinovascular diseases, retinal degenerations, intraocular inflammation, tumors, and most frequently, as a complication of postcataract surgery (Gass and Norton, 1966; Jampol, 1994). A common feature of the disorder is the presence of clear cysts in Henle's layer in the foveal region and occasionally in other retinal layers (Fig. 6.7). CME involves breakdown of the blood–retina barrier with accumulation of serous transudates in the retina, but no single etiologic factor is likely to be responsible for the occurrence of CME in such a wide variety of conditions. The question as to whether the fluid

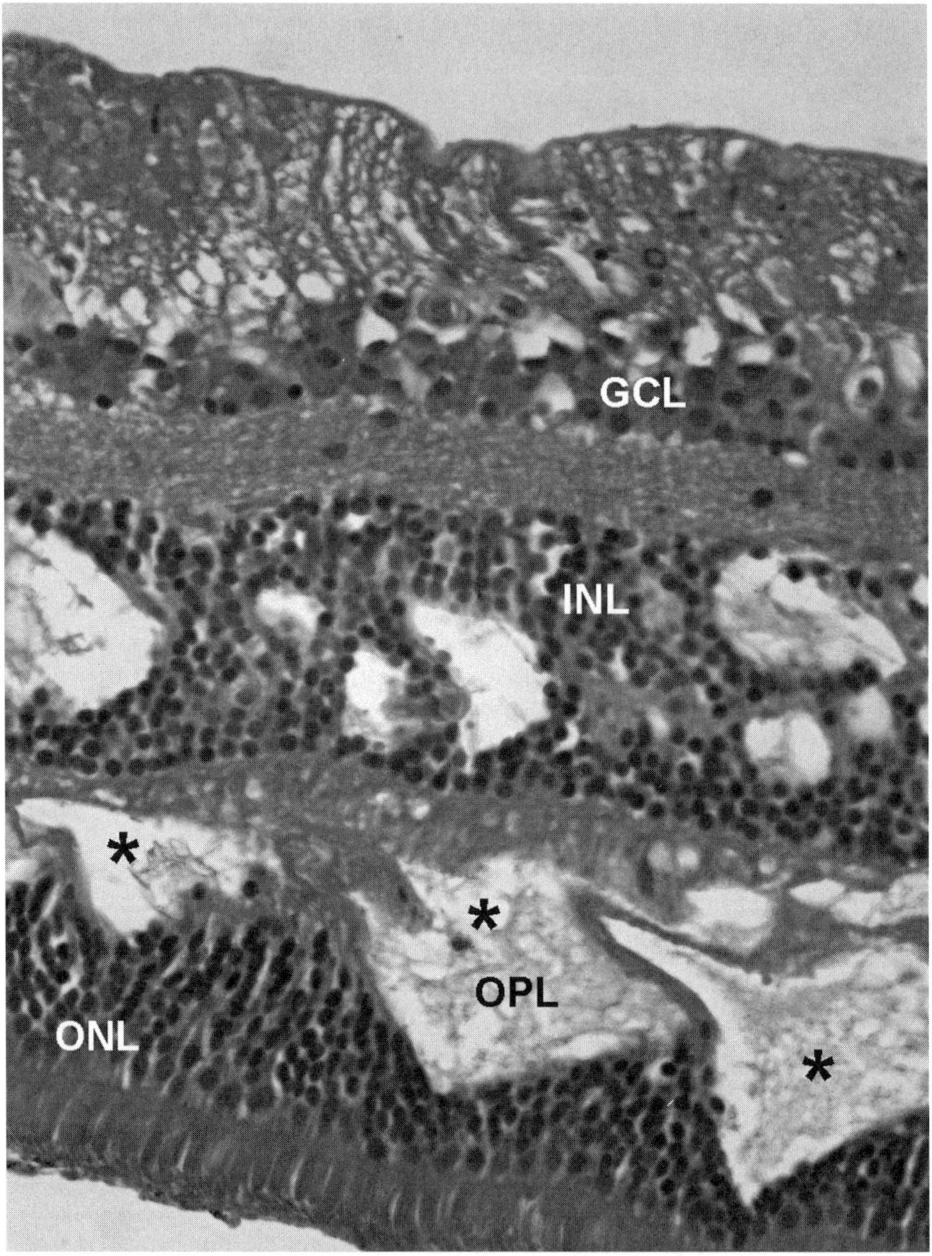

Figure. 6.7. Histology of cystoid macular edema. Retina from a patient with pseudophakic cystoid macular edema. Note the presence of large cystic spaces in the outer (asterisk) and inner retinas. Courtesy of Dr. Deepak Edward, University of Illinois Medical School.

accumulates in extracellular spaces or within Müller cells remains controversial owing to limited access to clinical specimens at early stages of CME development.

Early histopathological studies showing swelling and degeneration of Müller cells led to the suggestion that the cystoid spaces represented swollen Müller cell processes (Fine and Brucker, 1981; Yanoff et al., 1984). Either the edema, subsequent necrosis of Müller cells, or both, may cause further degenerative changes and the formation of larger cystic spaces. At the ultrastructural level, Müller cell enlargement occurs without much accumulation of extracellular fluid (Fine and Brucker, 1981). This suggests that the primary site of edema is the Müller cell itself, and that the buildup of extracellular fluid is a secondary event (Yanoff et al., 1984). A similar conclusion was reached based on a histopathologic study of dominantly inherited cystoid macular edema, a rare clinical entity (Loeffler et al., 1992).

The results of other studies, however, suggest that the fluid is accumulated primarily in the extracellular space, and that neuronal loss and Müller cell changes are secondary events (Gass et al., 1985; Tso, 1982 Wolter, 1981). In an ultrastructural study of an eye with CME, Gass et al. (1985) reported finding no cellular components in the enlarged spaces of the INL and the OPL. Surprisingly, they found no evidence for Müller cell degeneration or necrosis. In an extensive histopathological study of eyes with CME, Tso (1982) reported finding cystic spaces in the outer and inner retina, but was unable to resolve the question of whether the spaces were due to extracellular fluid accumulation or Müller cell swelling.

Although results from the various CME studies appear to be at odds, the observed differences might also be due to a number of complicating factors: stage of disease; processing of the tissue; prior systemic disease or treatment; and previous surgical or medical history. In this context, development of a representative animal model for CME would be of great value (Tso, 1982; Bellhorn, 1984), but at present no suitable model exists.

6.4.3. Müller Cell Sheen Dystrophy

In addition to retinoschisis and CME, Müller cell dysfunction has been implicated in an autosomal dominant retinal dystrophy termed "Müller cell sheen dystrophy" (Polk et al., 1997; Kellner et al., 1998). In this condition, the fundus shows multiple folds at the level of the internal limiting membrane at the posterior pole. In some cases, this is accompanied by macular edema. Histological studies indicated a diffuse thickening and undulation of the internal limiting membrane (Polk et al., 1997), and the ERG shows a reduction in the amplitude of the *b*-wave and flicker response (Kellner et al., 1998). It has been proposed that membrane thickening and subsequent

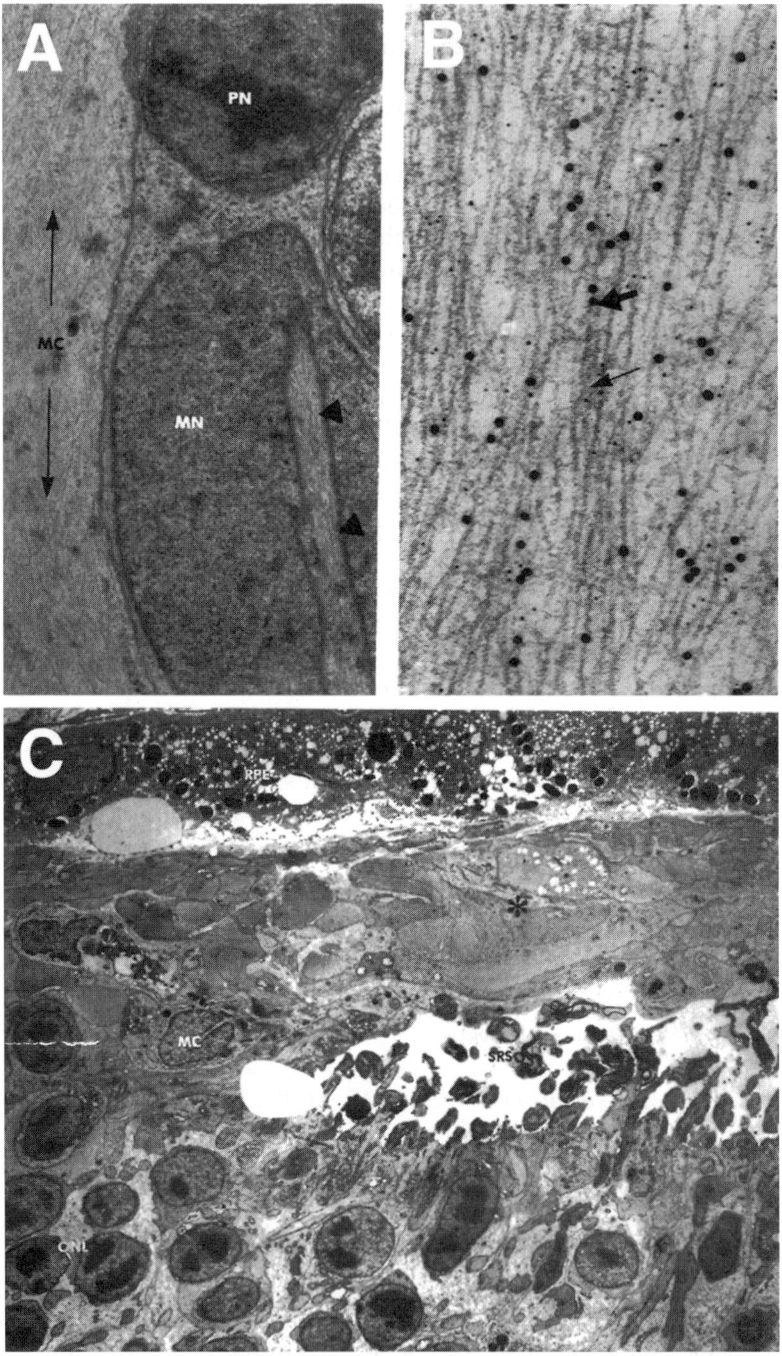
A
PN
MC
MN
B
C
RPE
MC
SRS
ONL

folding result from accumulation of a protein synthesized and secreted by Müller cells (Kellner et al., 1998).

6.4.4. Retinal Detachment

Retinal detachment results not only in physical separation of the neural retina from the RPE, but also leads to rapid cellular changes in the retina and RPE (Fisher and Anderson, 1994). The condition can have a potentially devastating effect on visual function. Experimental studies of retinal detachment have been made possible by the development of a superb animal model in cats where the cellular changes accompanying retinal detachment have been extensively documented (Fisher and Anderson, 1994). In cats with surgically detached retinas, detachment evokes striking cytological changes in photoreceptors and Müller cells (Fig. 6.8; Fisher and Anderson, 1994; Okada et al., 1990). In the early stages of detachment, Müller cells undergo hypertrophy and proliferation. At more advanced stages, Müller cell processes extend into subretinal space to cover the entire area of detachment, resulting in a scar-like structure in the distal retina (Fisher and Anderson, 1994; Erickson et al., 1987; Fisher et al., 1991; Lewis et al., 1994). These morphological changes are accompanied by increases in RNA and protein synthesis in Müller cells (Erickson et al., 1990; Fisher and Anderson, 1994).

The consequences of the Müller cell changes can be profound. Formation of abnormal membranes in the subretinal space results in subretinal fibrosis which can prevent regeneration of photoreceptor outer segments after reattachment surgery and thus can lead to subsequent photoreceptor degeneration (Fisher and Anderson, 1994; Korte et al., 1992).

6.4.5. Müller Cells in Retinal Membranes

In many ocular conditions, such as proliferative diabetic retinopathy, a series of pathological processes result in cells invading the vitreous, adhering to the ILM, proliferating, and secreting extracellular matrix molecules. The net effect of these changes is the formation of epiretinal membranes

Figure 6.8. Cytological changes in Müller cells following retinal detachment. A. The electron micrograph shows a Müller cell nucleus (MN) that has migrated to the outer plexiform layer after detachment of cat retina. PN, photoreceptor nucleus. B. Micrograph showing numerous 10 nm filaments in the Müller cell cytoplasm; some of the filaments show immunolabeling for GFAP (gold particles). C. Micrograph showing extensive Müller cell (MC) hypertrophy in the subretinal space (SRS) following detachment (Fisher and Anderson, 1994). (Copyright 1994 W. B. Saunders Co./Mosby, reprinted with permission.)

and development of traction forces on the retina (McDonald et al., 1994). The presence of epiretinal membranes leads to adverse effects on vision and is frequently associated with ocular morbidity (McDonald et al., 1994). Epiretinal membranes appear to contain multiple cell types, and it was generally assumed that most are derived from astrocytes, endothelial cells, fibroblasts, pericytes, macrophages, and RPE cells. Recently, however, it was shown that Müller cells participate in the formation of epiretinal membranes (Guerin et al., 1990b; Kono et al., 1995).

The availability of Müller cell-specific antibodies has been crucial in this regard. From previous work it was known that GFAP is expressed by fibrous astrocytes and reactive Müller cells but not by RPE, fibroblasts, or macrophages (Sarthy, 1991). CRALBP, on the other hand, is expressed by Müller cells and RPE but not by astrocytes (Bunt-Milam and Saari, 1983). Based on these antigenic differences, Guerin et al. (1990) used double labeling methods to distinguish Müller cells from fibrous astrocytes and RPE in epiretinal membranes. They found cells that were $GFAP^{+}/CRALBP^{-}$ (astrocytes), $GFAP^{-}/CRALBP^{+}$ (RPE cells), and $GFAP^{+}/CRALBP^{+}$ (Müller cells). These data suggest that in addition to astrocytes and RPE cells, Müller cells are an important component of epiretinal membranes (Fig. 6.9).

6.5. REACTIVE GLIOSIS

The vertebrate CNS is remarkable for its morphological complexity and functional sophistication. Therefore, it is surprising that such a highly evolved system lacks efficient repair mechanisms for neuronal regeneration and functional recovery following trauma. Many factors are responsible for the limited regenerative capacity of the mammalian CNS. For example, there may be an inability to reinitiate critical steps that occur during normal development, which are dependent upon expression of specific trophic and guidance molecules. Alternatively, there may be active inhibitory molecules in the adult tissue. In addition, reactive gliosis followed by scar formation may create physical barriers for axonal growth.

Reactive gliosis refers to cytological changes observed in glial cells at the site of nerve damage (Landis, 1994; Ridet et al., 1997). The major changes exhibited by glial cells include hypertrophy, enlargement of the cell body, and hyperplasia (an increase in cell number). The nature of the glial response appears to depend on the type of injury and its location (Reier, 1986). In stab wounds, for example, proliferation appears to be a major response, while in axonal degeneration of the optic nerve, hypertrophy is more prevalent (reviewed in Lindsey, 1986). Finally, reactive gliosis does not require signals that come from breakdown of the blood–brain

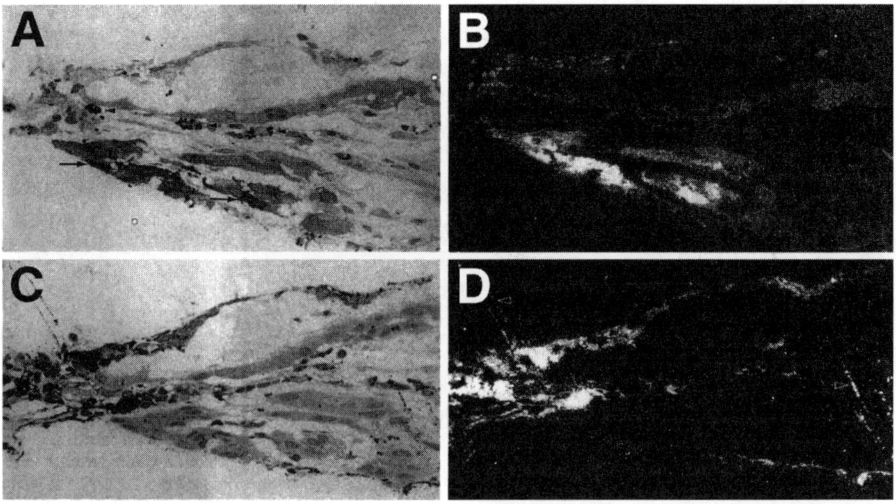

Epiretinal Cell Type Identification by Immunohistochemical Labeling Pattern

Anti-GFAP	Anti-CRALBP	Anti-vimentin	Cell type[a]
+	+	+	Müller's glia
−	+	±	Retinal pigment epithelium
+	−	+	Astrocyte
−	−	±	Monocyte/ macrophage

[a]Based upon *in situ* immunolabeling pattern.
+ = labeling; − = no labeling.

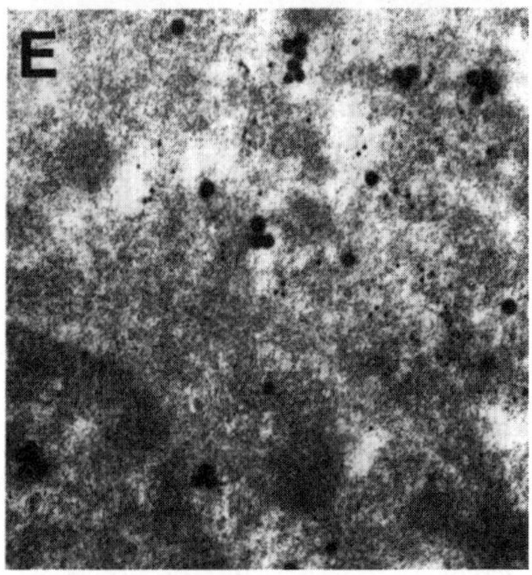

Figure. 6.9. Immunocytochemical demonstration of Müller cells in epiretinal membranes. Micrographs of sequential sections stained with anti-CRALBP (A,B), and GFAP (C,D). A,C are bright field and B,D are polarized epi-illumination images of the fields, in which only the silver particles signify concentrations of enhanced antibody staining. E. Lower right-hand panel shows an electron micrograph of a section double-labeled for GFAP (10 nm gold) and CRALBP (30 nm gold). The table provides a list of markers used for cell identification (Guerin et al., 1990b). (Copyright 1990 Association for Research in Vision and Ophthalmology, reprinted with permission.)

barrier; indeed, gliosis can occur even when there is no breach of the blood–brain barrier (Murabe et al., 1992).

Although the specific function of reactive gliosis is not known, it is presumed to be involved in phagocytosis of neuronal debris and in restoring breaches in the blood–brain barrier by scar formation (Reier, 1986). In addition, glial cells undergoing reactive gliosis have been found to upregulate production of cytokines and neurotrophic factors which may be crucial for the viability of injured neurons. A prevailing idea is that reactive microglia produce inflammatory molecules such as interleukin-1, whereas reactive astrocytes secrete neuroprotective agents such as ciliary neurotrophic factor (CNTF).

6.5.1. Morphological Changes

Reactive gliosis in Müller cells has been reported in response to a wide variety of ocular conditions such as mechanical injury, retinal detachment, chemical toxicity, light damage, experimental allergic uveitis, and hereditary rod–cone degeneration (cf. Sarthy, 1991). Although the morphological changes that occur in reactive astrocytes have been investigated extensively (e.g., Reier, 1996), gliosis in Müller cells has not been studied as thoroughly. The hypertrophic changes seen in Müller cells involve swelling of the cellular processes and enlargement of the cell body. In addition to the pronounced hypertrophy associated with gliosis, electron microscopic study shows that the fine processes of Müller cells are either retracted or lost; the distal processes become more tubular in appearance; the cell body transforms from an oval to a more rounded shape; and the Müller cell endfeet become enlarged and more electron dense (Fisher and Anderson, 1994). In the case of retinal detachment, displacement of Müller cell nuclei from the middle of the INL to the OLM has been reported. Other changes observed include an increase in the number of glycogen granules, Golgi cisternae and rough endoplasmic reticulum.

At early stages of retinal degeneration, little change is observed in the apical processes at the OLM, but as the disease progresses, the apical microvilli become shorter and are eventually lost. This does not occur until the photoreceptor layer is lost and the pigment epithelium is in contact with the apical processes of Müller cells (Fig. 6.10; Korte et al., 1992).

Korte and associates (1992) studied the cytological changes in the Müller cell membrane that take place during subretinal scar formation in rabbits with sodium iodate-induced retinopathy (Fig. 6.10). Following iodate injection, large regions of retina lose photoreceptors, and the underlying pigment epithelium also degenerates. In these areas, the apical processes of Müller cells can be seen to form scars. Müller cells undergo many cytological alterations during this process. The highly polarized apical mem-

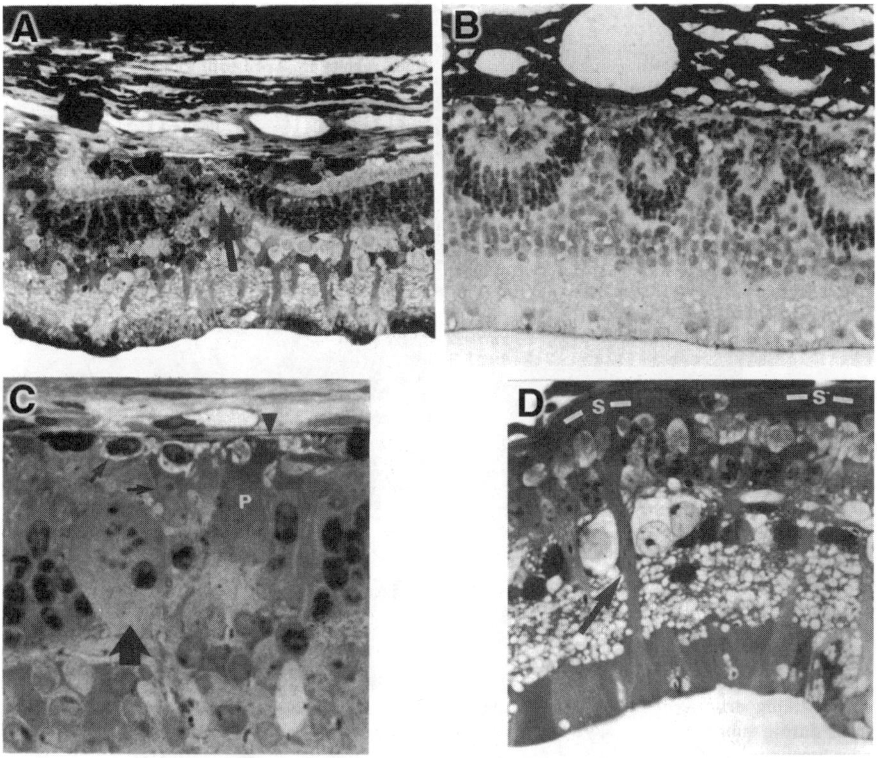

Figure. 6.10. Histological features of reactive Müller cells. Light micrographs of subretinal scars in rabbits after intravitreal injection of sodium iodate, which leads to retinopathy. A. Site of subretinal scar formation (arrow) flanked by zones of RPE loss and photoreceptor atrophy. B. Photoreceptor rosettes derived from the scalloped photoreceptor layer. C. In formative subretinal scars, the external limiting membrane breaks down and Müller cell processes (p) approach the RPE basement membrane. Large arrow shows a putative mitotic Müller cell, and small arrows show neighboring cells. D. A mature scar(s) is formed mainly by hypertrophic ascending processes of Müller cells (arrow) that extend up to the RPE basement membrane. The endfeet of Müller cells also show hypertrophy (Korte et al., 1992). (Copyright 1992 Academic Press, reprinted with permission.)

branes lose microvilli and the intercellular junctions disappear. In addition, the slender apical processes are transformed into thick processes that remodel to form a glial membrane at the subretinal space.

6.5.2. Müller Cell Proliferation

Glial cell hypertrophy is often followed by cell proliferation and scar formation around the sites of injury (Reier, 1996). Generally, the extent to

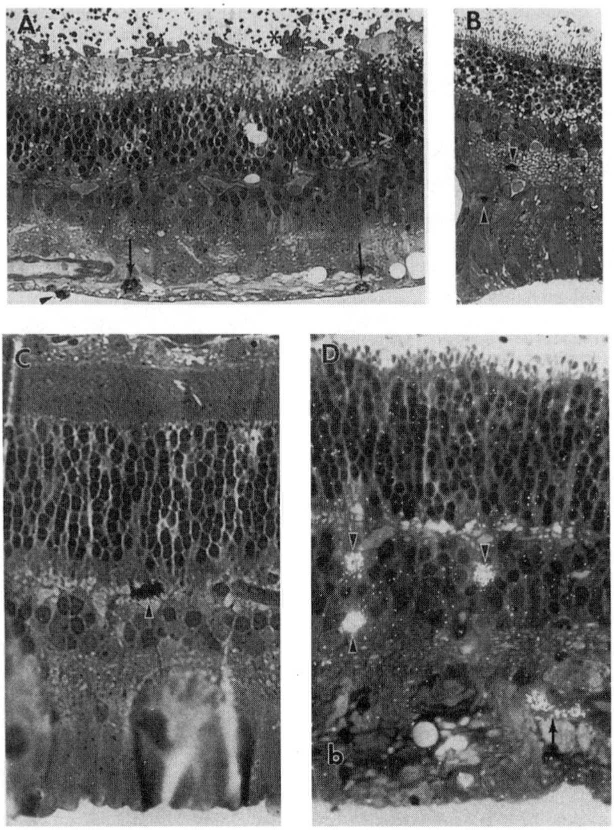
A
B
C
D
b

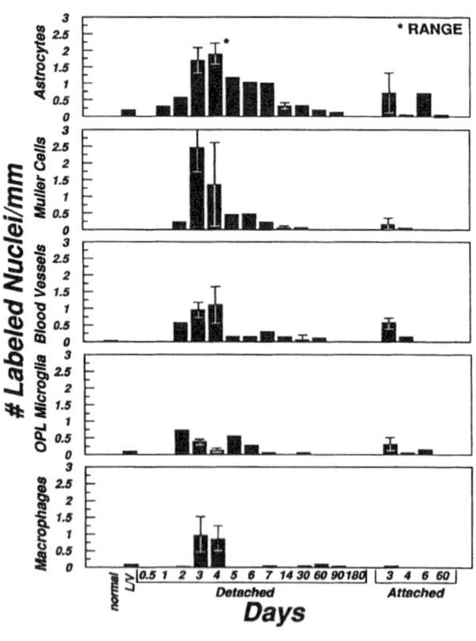
* RANGE
Astrocytes
Muller Cells
Blood Vessels
OPL Microglia
Macrophages
Labeled Nuclei/mm
normal
L/V
0.5 1 2 3 4 5 6 7 14 30 60 90 180
Detached
3 4 6 60
Attached
Days

which glial cells proliferate depends on the severity and type of trauma. Whereas microglia proliferate more extensively in highly inflammatory wounds, astrocytes make up the major population of mitotic cells in other situations; oligodendrocytes show the least response to injury (cf. Lindsey, 1986).

In the human retina, glial cell proliferation has been reported in a variety of pathological conditions including proliferative vitreoretinopathy, macular pucker, proliferative diabetic retinopathy, and idiopathic preretinal macular gliosis (Hjelmland and Harvey, 1988). In these cases, retinal astrocytes undergo hypertrophy, proliferation, and fibrosis, but Müller cells exhibit only hypertrophy. However, Müller cell proliferation has been reported in proliferative diabetic retinopathy and massive retinal gliosis (Nork et al., 1986, 1987, 1990).

Glial cell proliferation has been best examined in studies where retinal detachment was induced experimentally. Fisher et al. (1991) studied ^{3}H-thymidine incorporation into RPE and retinal cells in cats with experimentally-induced retinal detachment. They provide conclusive evidence that in addition to RPE, all non-neuronal cells in the retina (Müller cells, astrocytes, pericytes, and endothelial cells) incorporate ^{3}H-thymidine following retinal detachment. The proliferative response occurred about 48 hours after detachment and reached maximal levels by 3–4 days. After about two weeks, the mitotic activity returned to baseline even though the retina remained detached (Fig. 6.11).

6.5.3. Müller Cell Proliferation: Putative Mitogens

Neither the mitogens that trigger glial cell proliferation in response to specific pathological conditions are known, nor have the mechanisms that suppress glial cell mitotic activity in normal tissue been identified. Moreover, we have yet to determine whether the mitogenic agents are derived from degenerating neurons, retinal storage compartments, or from activated macrophages. Three different classes of mitogens have been reported in the retina: growth factors, neurotransmitters, and cations. Many growth factors and cytokines known to be glial mitogens, e.g., NGF, bFGF, EGF, and

←

Figure. 6.11. Müller cells undergo mitosis in detached retina. ^{3}H-thymidine was injected intravitreally into cats with retinal detachment and labeled cells were localized by autoradiography. A–D show examples of ^{3}H-labeled cells: astrocytes, Müller cells, vascular cells, microglia, and macrophages. In D, three labeled Müller cells (arrowheads) are prominent. Labeled astrocytes are indicated by arrows in A and D. The histogram shows the number of ^{3}H-labeled nuclei for each cell type as a function of days following detachment. The peak of mitotic activity appears to be around 3–4 days (Fisher et al., 1991). (Copyright 1991 Association for Research in Vision and Ophthalmology, reprinted with permission.)

interleukin-1 (IL-1), are also active on Müller cells (Puro, 1994b). In addition, small molecules such as glutamate and the cations, Ca^{2+} and K^{+}, are also mitogenic to Müller cells (Reichelt et al., 1989; Sahel et al., 1990, 1991; Uchihori and Puro, 1993b).

6.5.4. Cytokines–Growth Factors

The mitogenic activity of a large number of known growth factors has been examined in Müller cell cultures and *in vivo* (Macarelli et al., 1991; Small et al., 1991; Lewis et al., 1992; Puro et al., 1990; Puro and Mano, 1991a,b; Uchihori and Puro, 1991, 1993a; Ikeda and Puro, 1994; Scherer and Schnitzer, 1994). Although most known growth factors are found to be mitogenic, a few others such as TGFβ turn out to have an antiproliferative effect (Ikeda and Puro, 1995).

One of the first growth factors to be examined for mitogenic activity was bFGF. In Müller cell cultures derived from postmortem retinas, Puro and Mano (1991a) found bFGF was able to increase mitotic activity by 40% (Fig. 6.12). To determine how the effect was mediated, they examined concomitant changes in Ca^{2+} current by means of whole cell, patch clamp recording. They found that bFGF increased the currents only through L-type Ca^{2+} channels, suggesting that Ca^{2+}-influx may trigger mitotic activity in Müller cells (Puro and Mano, 1991a). Consistent with these findings is the observation that there are high affinity-binding sites for bFGF in bovine Müller cell cultures (Mascarelli et al., 1991).

Experiments in which bFGF was intravitreally injected into cat and rabbit eyes showed that bFGF is also an effective mitogen *in vivo* (Fig. 6.12; Lewis et al., 1992). Significantly greater mitotic activity was seen in Müller cells, astrocytes, microglia, and vascular cells of bFGF-injected eyes compared to sham-injected controls, although Müller cells were the least responsive. The proliferative response was observed as early as 4 days after injection but not after 4 weeks. These data show bFGF can induce mitotic activity in Müller cells.

Many other growth factors (Table 6.2), including EGF, PDGF, and NGF are also potential mitogens for Müller cells *in vitro* (Puro, 1994b). Whether these substances are mitogenic *in vivo* remains to be established. An age-dependent difference in proliferative response to growth factors has also been reported in Müller cell cultures from guinea pig retina (Small et al., 1991). First-passage cultures showed a strong response to bFGF but not to interleukin-2 (IL-2). Older cultures showed minimal response to bFGF but were stimulated by IL-2. These observations suggest that Müller cells might respond to different growth factors during development, maturation, and

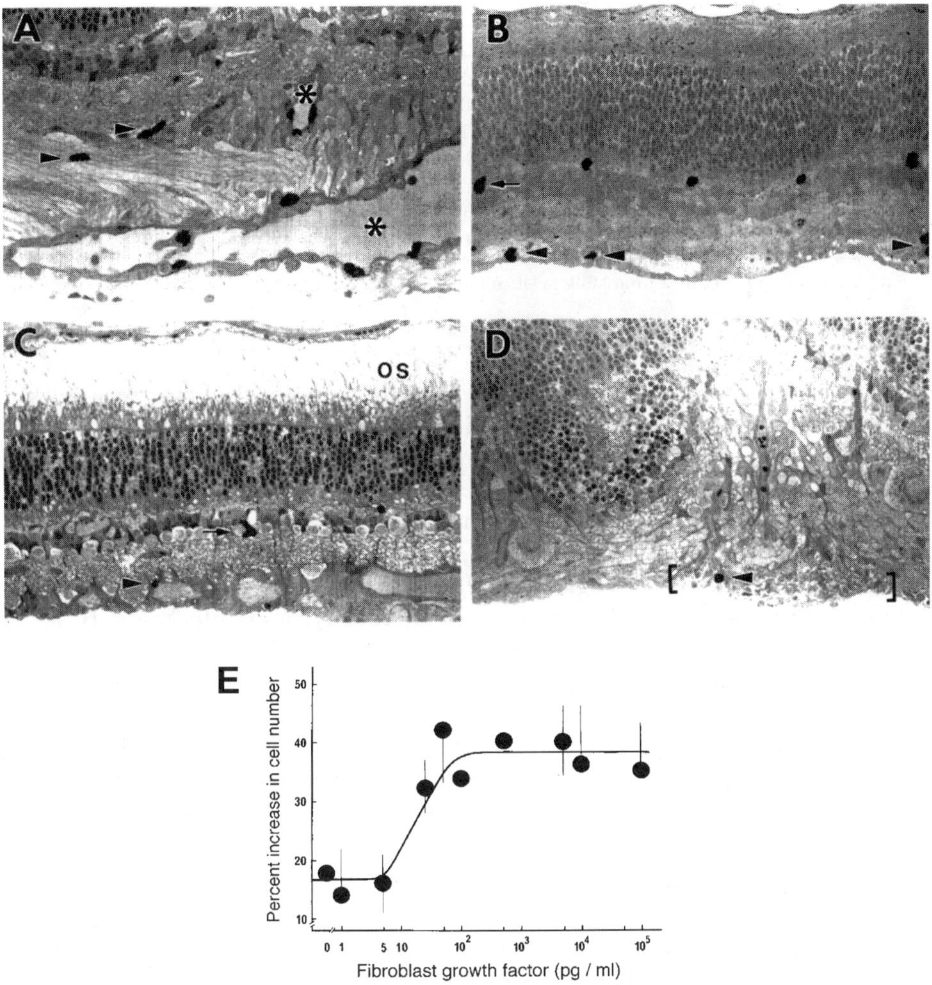

Figure. 6.12. Basic fibroblast growth factor (bFGF) is mitogenic to Müller cells. A–D show micrographs of retinas from cats injected with bFGF and ^{3}H-thymidine. ^{3}H-labeled Müller cells (arrows) are seen in the sections. Astrocytes (arrowhead) are also labeled. Asterisks (A) indicate blood vessels and a mitotic cell at retinal surface is seen in D (Lewis et al., 1992). (Copyright 1992 *The Journal of Neuroscience*, reprinted with permission.) E. The proliferation of human retinal glial cells as a function of bFGF concentration (Puro and Mano, 1991a). (Copyright 1991 *The Journal of Neuroscience*, reprinted with permission.)

Table 6.2. Potential Mitotic and Antimitotic Agents for Müller Cells

Agent	Müller cell	Reference
b-FGF	+ R/C	Puro and Mano, 1991a; Lewis et al., 1992
EGF	+ C	Scherer and Schnitzer, 1994; Roque et al., 1992
Glutamate	+ R/C	Sahel et al., 1990, 1991; Uchihiro and Puro, 1993
IL-2	+ C	Small et al., 1991
K^+	+ C	Puro et al., 1989; Reichelt et al., 1989; Mascarelli et al., 1991
NGF	+ C	Ikeda and Puro, 1994
PDGF	+ C	Uchihori and Puro, 1991
Thrombin	+ C	Puro et al., 1990

R, data from retina; C, data from Müller cell cultures.

aging. In this context, a comparative study of the growth factor receptors expressed by Müller cells in developing and aged retinas is needed.

In addition to growth factors, other proteins also appear to be effective Müller cell mitogens. Puro et al. (1990) found that thrombin, a plasma component involved in wound healing, stimulated proliferation in a dose-dependent manner. In other systems, thrombin-induced mitotic activity is blocked by the pharmacological agents pertussis toxin (inactivator of G-proteins), amiloride (Na^+/H^+ exchanger inhibitor), and nifedipine (Ca^{2+} channel blocker). Thus, when there is a breach in the blood–retina barrier, plasma thrombin that leaks in might trigger the mitogenic response (Puro et al., 1990).

In contrast to the stimulatory agents discussed above, transforming growth factor β2 and a metabotrophic glutamate agonist, t-ACPD, have been reported to block Müller cell proliferation (Ikeda and Puro, 1995).

6.5.5. Other Mitogens

In addition to the macromolecules discussed above, small molecules such as glutamate also appear to be effective mitogens for Müller cells. For example, intravitreal injection of kainic acid (a glutamate analog) in adult rats was found to induce mitotic activity in cells located in the middle of the INL, presumably Müller cells (Sahel et al., 1990, 1991). In addition to mitotic activity, the other changes observed were swelling, vacuolization, and pyknosis in OPL, INL, and IPL, all of which occurred within 24–72 hr following injection (Sahel et al.,1991).

To understand whether kainic acid acts directly on Müller cells or if its *in vivo* effect is mediated through retinal neurons, Uchihori and Puro (1993b) examined kainate effects on Müller cell cultures derived from postmortem retina. They found kainate-induced cell proliferation occurred in a dose-dependent manner, and that the addition of 1 mM glutamate led to a 25% increase (maximal) in cell number. To identify the glutamate receptor subtypes involved in the mitogenic response, they tested several pharmacological agents. Both D-AP-5 (a competitive NMDA receptor blocker) and MK-801 (a noncompetitive NMDA receptor blocker) inhibited glutamate- induced mitotic activity, whereas CNQX (an inhibitor of non-NMDA receptors) had no effect.

NMDA itself was mitogenic and was more effective than glutamate (Uchihori and Puro, 1993b). However, NMDA-induced mitotic activity was blocked when the incubation medium contained low Ca^{2+} suggesting that the proliferative response was induced by Ca^{2+} influx (Fig. 6.13). In agreement with these observations, NMDA has been shown to elicit membrane currents in isolated human Müller cells, and biochemical studies show that Müller cell cultures express the NMDAR1 subunit (Puro, 1994b).

These data are, however, not in agreement with other observations on the action of NMDA. In the experiments of Sahel et al. (1991) NMDA did not elicit a mitotic response in rat Müller cells. Intracellular recordings from dissociated, salamander Müller cells, and rabbit retinal astrocytes failed to reveal the presence of functional NMDA receptors (Brew and Attwell, 1987; Schwartz and Tachibana, 1990; Clark and Mobbs, 1992). The observed discrepancies may be due to species differences, length of NMDA treatment, or the cell recording conditions (Uchihiro and Puro, 1993b).

Several experiments suggest that K^+ may also be mitogenic to Müller cells *in vitro.* Patch clamp studies of rat Müller cells showed cells exposed to a mitogen-conditioned medium had an increase in tetraethyl ammonium (TEA)-sensitive K^+ channel activity. Furthermore, TEA was reported to block the mitotic activity of Müller cells in the conditioned medium (Puro et al., 1989). Müller cell cultures from rabbit retina exposed to 20 mM K^+ for 48 hr had an increase in ^{3}H-thymidine incorporation (Reichelt et al., 1989).

The extensive literature in this area might convey the impression that Müller cell proliferation is an important phenomenon. This is certainly not the case *in vivo* where Müller cells exhibit limited capacity for proliferation as compared to retinal astrocytes. This feature of Müller cells may be a biological necessity for retinal survival and function because extensive Müller cell proliferation is likely to distort retinal architecture and disrupt neuronal connectivity.

6.5.6. Re-entry into Mitosis

Although retinal neurons and Müller cells are postmitotic in the mature retina, it appears that the Müller cells alone retain the capability to

A

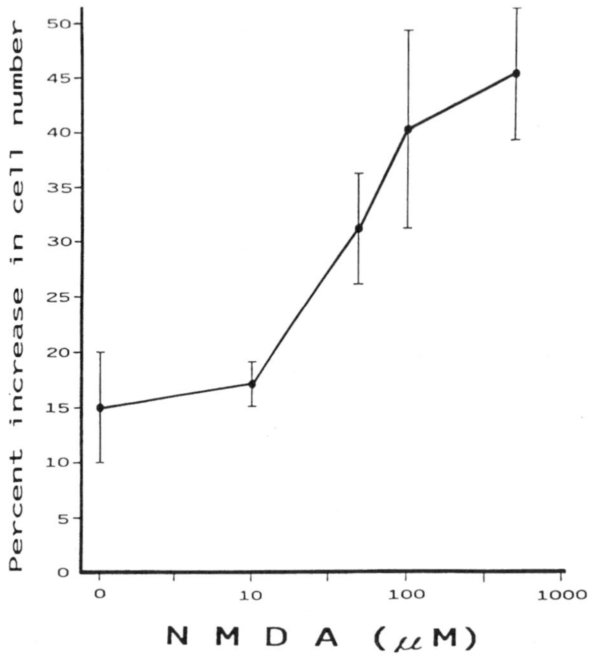

B

Conditions	Increase in cell number (%)
No NMDA, 1 mM Mg^{2+}	9±4
NMDA, 1 mM Mg^{2+}	28±7
NMDA, 10 mM Mg^{2+}	8±4
NMDA, low Na^{+}	23±3
NMDA, low Ca^{2+}	6±6

Figure. 6.13. Effect of NMDA on retinal glial cell proliferation. A. Dose-response curve for NMDA effect on proliferation. B. Effect of extracellular concentrations of magnesium, sodium, and calcium on the NMDA-induced proliferation of human retinal glial cells in culture. While sodium removal had minimal effect on proliferation, low Ca^{2+} prevented NMDA-induced proliferation (Uchihori and Puro, 1993b). (Copyright 1993 Elsevier Science, reprinted with permission.)

undergo mitotic activity in an appropriate environment. It is possible that as the retina matures, either the supply of endogenous mitogens becomes limiting or that mitogen-receptors are down-regulated in Müller cells. Under pathological conditions, the availability of mitotic agents and the expression of their receptors are probably responsible for initiating mitotic activity. The inherent capacity of Müller cells to divide makes them excellent candidates as mediators of ocular pathology in conditions such as massive retinal gliosis (Nork et al., 1986; Rodrigues et al., 1987), retinal tumors (Jakobiec et al., 1983; Craft et al., 1985), and proliferative diabetic retinopathy (Nork et al., 1987).

6.5.7. Molecular Changes

The cytological changes observed in Müller cells in response to injury are accompanied by significant alterations in gene expression. Whereas proteins such as GFAP, Glutamate/aspartate transporter (GLAST), CNTF, and target of the antiproliferative antibody (TAPA) are upregulated under pathological conditions, other proteins appear to be down-regulated (Table 6.3) (Sagar et al., 1991; Chen et al., 1994; Grosche et al., 1995; Yoshida et al., 1995; Otori et al., 1995; Wen et al., 1995; Clarke and Geisert, 1998; Lewis et al., 1989; Sarthy, 1982). The significance of these changes and the underlying gene regulatory mechanisms, however, remain to be elucidated.

A prominent feature of reactive gliosis is that the Müller cell cytoplasm becomes crowded with organelles and filaments (Fig. 6.8). In particular,

Table 6.3. Proteins Highly Expressed or Induced in Müller Cells

Protein	Normal condition	Abnormal condition	Reference
APP	N	Y	Grosche et al., 1995
Bcl2	N	Y	Grosche et al., 1995
bFGF	ND	Y	Wen et al., 1995; Cao et al., 1997a,b
Carbonic anhydrase C	Y	Y	Linzer and Moscona, 1981a
CD44	Y	Y	Chaitin et al., 1994, 1998
C-fos	N	Y	Yamada et al., 1993
CNTF	N	Y	Wen et al., 1995; Ju et al., 1999
CRALBP	Y	Y	Sarthy, unpublished
F11	Y	ND	Schlosshauer, 1994
GFAP	N	Y	Sarthy, 1991
GLAST	Y	Y	Otori et al., 1995
Glutamine synthetase	Y	N	Linzer & Moscona, 1979
Jun	N	Y	Ohki et al., 1995
LIF	Y	ND	Neophytou et al., 1997
NO	N	Y	Goreau et al., 1994
S-100 protein	Y	Y	Linser and Moscona, 1981b
TAPA	N	Y	Geisert, 1998
Thy-1	N	Y	Dabin and Barnstable, 1995
Vimentin	Y	Y	Drager and Edwards, 1983

Y, high expression observed; N, little or no expression found; ND, not determined.

there is a copious increase in the content of intermediate filaments. This increase is seen first in the endfoot region, and later involves both the distal and apical processes (Fisher and Anderson, 1994). The major intermediate filament expressed by reactive Müller cells appears to be GFAP (see Chapter 1). Because GFAP is the best known marker for reactive gliosis throughout the nervous system, there is substantial interest in the mechanisms regulating GFAP gene expression in glial cells.

GFAP expression in Müller cells displays a unique feature. Normally, GFAP is expressed at a low level or is not detectable in mammalian Müller cells. However, loss of retinal integrity as a result of mechanical injury, detachment, or photoreceptor degeneration results in the appearance of intense GFAP-immunoreactivity in Müller cells, and an increase in GFAP content of retina (Fig. 6.14; Table 6.4.) (Erickson et al., 1987; Guerin et al., 1990a; Eisenfeld et al., 1984; Sarthy and Fu, 1989).

Neuronal loss does not appear to be a prerequisite for GFAP induction because increased GFAP levels have been reported in Müller cells in ischemic retinas (Osborne et al., 1991; Fitzgerald et al., 1990) and in a rat model for retinopathy of prematurity (Penn et al., 1988) where histological sections showed no detectable neuronal loss. In spite of the diverse and seemingly unrelated conditions that evoke GFAP formation (Table 6.4), a shared signaling pathway may be responsible for GFAP induction in Müller cells.

Regulation of GFAP expression in Müller cells has been extensively studied by Sarthy and his associates (1998). First, it is clear than GFAP

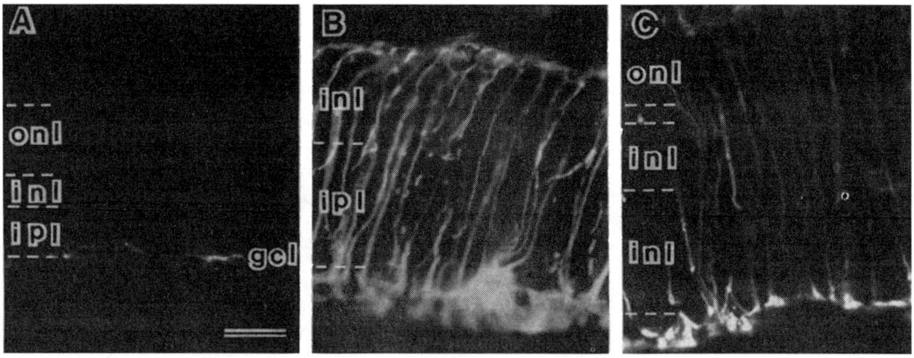

Figure. 6.14. GFAP immunostaining in the mouse retina. Immunocytochemical localization of GFAP in normal and photoreceptor-deficient retinas. A. BALB/C retina shows labeling of astrocytic processes in the ganglion cell layer (gcl). B. Retinal degeneration (*rd/rd*) retina shows strong immunostaining of Müller cell processes. C. BALB/c retina with light damage also exhibits GFAP-immunostaining in the radial fibers of Müller cells (Sarthy and Fu, 1989). (Copyright 1989 Mary Ann Liebert Publishers, reprinted with permission.)

expression in Müller cells is not due to a "de-differentiation" event resulting in re-expression of an "embryonic" marker in Müller cells. Immunocytochemical and *in situ* hybridization studies have shown GFAP is not expressed by Müller cells in embryonic or adult mouse retina (Sarthy et al., 1991; Huxlin et al., 1992). Second, following synthesis, GFAP is integrated into Müller cell cytoskeleton and turns over extremely slowly or not at all. In contrast, GFAP mRNA is transcribed for a limited time and the gene is subsequently turned off (Sarthy and Egal, 1995).

The increase in GFAP expression has been shown to be due to transcriptional activation of the *GFAP* gene in Müller cells (Sarthy and Fu, 1989). However, the *cis* and *trans*-activating factors that regulate *GFAP* gene expression in Müller cells have not been identified so far. Cell transfection and *GFAP-lacZ* transgenic mice studies indicate that *cis* elements that stimulate GFAP transcription in astrocytes and Müller cells are different (Brenner, 1994; Verderber et al., 1995; Johnson et al., 1995).

6.5.8. GFAP-inductive Signal

As we have seen, a hallmark of reactive gliosis in the retina is the induction of GFAP in Müller cells. What is the nature of the extracellular signal that induces GFAP? One possible factor is the loss of cell–cell contact between retinal neurons and Müller cells. This seems unlikely since in dissociated retinal cultures, GFAP is not expressed by Müller cells although cell–cell contacts are disrupted (Hicks and Courtois, 1990; Lewis et al., 1988).

It is more likely that GFAP induction is mediated by the action of extracellular, diffusible substances. Several studies show that focal damage to retina induced by mechanical injury, high-intensity light damage, or laser photocoagulation results in GFAP accumulation in Müller cells remote from the site of injury (Bignami and Dahl, 1979; Verderber et al., 1995; Humphrey et al., 1993). In rodent retinas, a penetrating needle wound to the retina results in GFAP accumulation not only in Müller cells in the damaged area, but also in Müller cells throughout the retina (Verderber et al., 1995).

There is some evidence that growth factors and cytokines are the signaling molecules involved in GFAP induction (Kahn et al., 1995; Winter et al., 1995; Ridet et al., 1997). For example, intravitreal injection of bFGF has been reported to induce GFAP in Müller cells in cat, rabbit, and mouse retinas (Lewis et al., 1992; Sarthy et al., 1997). Other potential candidates for the signaling molecule are members of the cytokine family of intercellular mediators. Indeed, intravitreal injection of the cytokine, CNTF results in rapid GFAP induction in Müller cells in the mouse retina (Fig. 6.15; Sarthy et al., 1997).

Table 6.4. Ocular Conditions in Which GFAP is Accumulated in Müller Cells

Mouse	
Retinal degeneration (rd)	Sarthy and Fu, 1989
Retinal degeneration slow (rds)	Sarthy and Fu, 1989; Ekstrom et al., 1988
Vitiligo (mi^{Vit})	Smith et al., 1997
Nervous (nr)	Sarthy, unpublished
Purkinje cell degeneration (pcd)	Sarthy, unpublished
Light-induced damage	Sarthy and Fu, 1989
Trauma	Verderber et al., 1995
GM-CSF, γ-IFN transgenic mice	Cuthbertson et al., 1990; Geiger et al., 1994
bFGF injection	Sarthy et al., 1997
CNTF, LIF injection	Sarthy et al., 1997
Rat	
Royal College of Surgeons (RCS)	Eisenfeld et al., 1984; Roque et al., 1990; Li et al., 1993
Light-induced damage	Eisenfeld et al., 1984; Burns and Roble, 1990
Experimental allergic uveitis	Chan et al., 1991
Trauma	Bignami and Dahl, 1979
Ischemia	Osborne et al., 1991
Urethane retinopathy	Tyler and Burns, 1991
ROP model	Penn et al., 1988
Cat	
Retinal detachment	Erickson et al., 1987
Abyssinian	Nofstrom and Nilsson, 1986
bFGF injection	Lewis et al., 1992
Dog	
Proliferative diabetic retinopathy	Nork et al., 1987
Gliotic	Davidson et al., 1990
Progressive cone-rod degeneration (pcrd)	Sarthy, unpublished
RD	Sarthy, unpublished
Rabbit	
IRBP-induced uveitis	Eisenfeld et al., 1987
Laser damage	Humphrey et al., 1993; Ishigooka et al., 1989; Tassignon et al., 1991
Lens/vitrectomy	Durlu et al., 1990; Yoshida et al., 1993
bFGF injection	Lewis et al., 1992
Pigeon	
Ischemia	Fitzgerald et al., 1990
Monkey	
Retinal detachment	Guerin et al., 1990
Glaucoma model	Tanihara et al., 1997a

Table 6.4. (*Continued*)

Human	
Retinitis pigmentosa	Rodriguez et al., 1987; Milam et al., 1996
Coat's disease	Mondon et al., 1970
Proliferative diabetic retinopathy	Ohira and de Juan, 1990
Age-related macular degeneration	Madigan et al., 1994
Fundus flavimaculatus (Stargardt)	Birnbach et al., 1994
AIDS	Hofman and Hinton, 1992
Proliferative vitreoretinopathy	Nork et al., 1990; Okada et al., 1987
Gliosis	Molnaar et al., 1984
Photic retinopathy	Green et al., 1991
Retinal detachment	Okada et al., 1990
Massive retinal gliosis	Nork et al., 1986

Because molecules such as CNTF do not diffuse over long distances in the tissue, one might wonder as to how a localized injury can result in GFAP induction in Müller cells across the entire retina. A plausible mechanism is that signal spreading occurs through a paracrine mechanism. For example, at the site of injury, a Müller cell might start to produce and secrete CNTF which could diffuse away and act on a neighboring Müller cell which in turn would respond by synthesizing and secreting CNTF that would activate a Müller cell farther away. Such a process could be expected to result in lateral spread of CNTF effect over a large distance, possibly the entire retina.

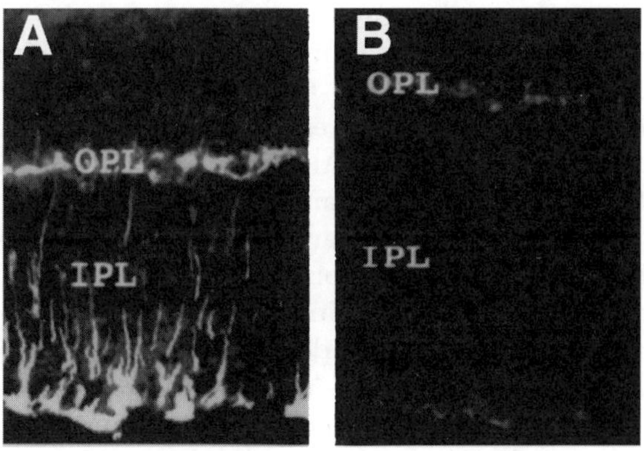

Figure 6.15. Ciliary neurotrophic factor (CNTF) can induce GFAP in Müller cells. A. GFAP immunostaining in mouse retina three days after intravitreal injection of CNTF. B. Mouse eye injected with physiological saline. The result suggests that reactive gliosis and GFAP induction might be mediated by endogenous CNTF.

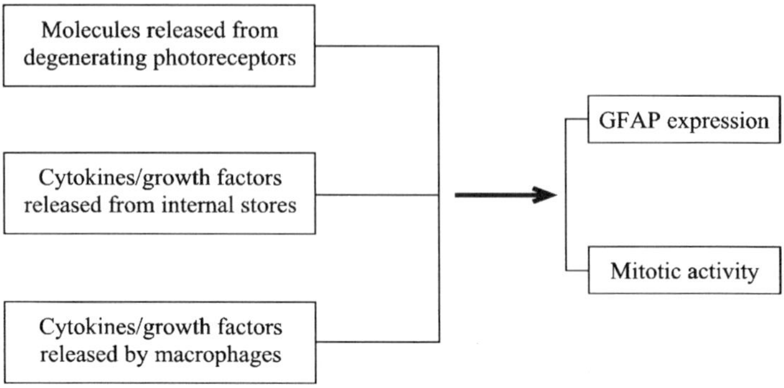

Figure 6.16. Schematic representation of the cellular mechanisms responsible for GFAP expression or mitotic activity in Müller cells. Molecules emanating from degenerating photoreceptors, and cytokines released from retinal compartments or secreted by activated macrophages, might act on Müller cells to induce GFAP expression or mitotic activity.

The cellular source of the extracellular signaling molecules is not known. There are at least three possible sources (Fig. 6.16): (1) neurons undergoing degeneration; (2) retinal compartments (Hageman et al., 1991); and (3) activated macrophages that invade the retina (Cuthbertson et al., 1990). In the case of bFGF and CNTF, cell damage appears to be a prerequisite since the substances lack signal sequences required for secretion.

6.5.9. GFAP Function

In spite of its robust upregulation in a multitude of retinal disorders, the cellular functions of GFAP, and the reason for its induction remain obscure. The high level of GFAP expression by reactive Müller cells is rather intriguing because the need for GFAP is certainly not due to a paucity of intermediate filaments; Müller cells are indeed well populated with vimentin filaments (see Chapter 2). It is clearly possible that GFAP expression is in some way tied to the structural reorganization of neurons and reactive Müller cells in the degenerating retina.

It is well known that in severe degenerative diseases that result ultimately in the loss of photoreceptors, Müller cells extend their processes to fill the spaces created by photoreceptor loss. In advanced cases of retinitis pigmentosa, for example, the distal ends of Müller cells are seen to abut the apical processes of the RPE (Gartner and Henkind, 1982). Similarly, in atrophic macular lesions or sodium-iodate-induced retinopathy, loss of RPE and photoreceptors leads to remodeling of the apical surface to form a glial membrane at the subretinal space (Korte et al., 1992). The membrane may serve as an outer blood–retina barrier and protect inner retina. It is possible GFAP filaments play a role in the membrane remodeling events at the OLM, or they may be involved in stabilizing Müller cell contacts with RPE or Bruch's membrane.

Another possibility is that GFAP expression and the resulting formation of vimentin–GFAP heteropolymers might function to limit Müller cell proliferation. There is some experimental support for this idea (Murakami et al., 1996). Although GFAP is upregulated in reactive astrocytes, the absence of vimentin may allow them to proliferate. Conversely, GFAP expression in nonglial cells, which contain vimentin, was found to inhibit cell proliferation (Murakami et al., 1996). Although GFAP-knockout mice have been generated and analyzed (Gomi et al., 1995; Pekny et al., 1995; McCall et al., 1996; Liedtke et al., 1997), it has not been possible to elucidate GFAP function in astrocytes from these studies.

Evidence is accumulating from animal models and human diseases that shows intermediate filaments play important roles in cell and tissue adhesion to the extracellular matrix (Fuchs and Cleveland, 1998). Recent molecular genetic studies revealed mutations in keratin genes result in epithelial cell detachment and lead to blistering skin diseases such as epidermolysis bullosa simplex (Fuchs, 1995). Thus, by analogy with other intermediate filaments, GFAP might function to provide added mechanical stability to the retina.

Heinrich Müller (1820–1864). Born in Castell, he became professor of topographical and comparative anatomy at Würzburg (Duke–Elder, 1961). (Copyright 1961 Butterworth–Heinemann, reprinted with permission.)

Appendix

On the Histology of the Retina

Dr. H. Müller
(Zeitschrift für wissenschaftlicke Zoologie III. Band, pp. 234–237, 1851)

The original account, reproduced at the end of the translation, is provided for readers versed in the terminology and units of measure used in mid-19th century Germany.

The examination of eyes that had been lying for some time in a chromic acid solution allows recognition of the individual elementary parts of which the retina consists, as well as their relative positions, which would otherwise be very difficult to investigate. Here I want to give a preliminary communication about only a few points, while I reserve further observations for a detailed description of the anatomy of the retina in various animals.

1) In all classes of vertebrate, numerous cylinders penetrate the entire thickness of the retina, standing perpendicular to the nerve extension,[a] and thus radial to the eyeball. Sometimes they are thin fibers, which, when hardened in chromic acid, bear some similarity to elastic fibers, and sometimes they are thicker striated bundles.

Their inner ends closely abut the nerve fibers; in some animals, this end is swollen to a bulbous, granular mass that looks like a fragment of a cell, in others, the fiber ends in a triangular membrane-like base that is sharply cut off. After penetrating through the inner, fine-grained layer of the retina that is absolutely similar to the gray brain matter, the radial fibers in many animals consistently display a swelling, which sometimes clearly contains a nucleus along with nucleoli, and which often has jagged processes going sideways and appearing to anastomose with the adjacent ones. Toward the

[a]Probably refers here to the nerve fiber layer.

outside, the perpendicular fiber enters the so-called granular layer, whereupon it often breaks up into several fibrils. In any case, it is in such close connection with the outwardly situated parts next to it that, not infrequently, when a retina is torn, a fiber will be isolated on whose outer part a number of the so-called granules along with rods or twin cones are adhering like currants on a stem. In other words, a thin cylinder splits off from the entire thickness of the retina; in a frog, for example, its length[b] amounted to 305 μm. One discerns the same vertical striation through the complete thickness in thin perpendicular sections that offer a cross-sectional view.

2) The well-known small filaments that frequently sit on the conically pointed ends of the rods are not turned toward the choroid, but rather are turned inward. They do not all begin at exactly the same height; for example, in most fish they go in between the twin cones and are connected with the adjacent inward, so-called granular layer. The name of this layer comes from the fact that it consists of nuclei, which are often vesicular and sometimes more and sometimes less elongated depending on the thickness of the retina, and which are connected with the rods in the same direction by small filaments of various length. Since at times one sees a second contour along a section of the periphery that merges into the small filaments, these "granules" should probably be regarded as very small cells.

In those fishes and birds in which the pigment epithelium forms inward-going processes, the rods themselves, rather than the small filaments, are inserted into the pigment. When one removes the pigmented layer from the outside up to the twin cones, one has removed at least the largest part of the rods also and only allowed the inner end with the small filaments to remain.

In the case of elasmobranchs, where there is no pigment between the rods, one sees the rods going uniformly outward up to a layer of polygonal cells, which resemble those in the tapetum of the ruminants. Behind it lies a structureless vascular membrane, which bears the scales that produce the silvery luster with the well-known fine needles, and only then comes the pigmented choroid. Also in the case of some other fishes the pigment extends inward between the rods for only a short distance.

3) The twin cones[1] of most fishes and of mammals likewise turn into a process at their inner, blunt end that extends into a filament; often a distinct nucleus forms the beginning of this filament. This filament is thicker than that found on the rods and, in any case, goes through the whole thickness

[b]In mid-19th century Germany, the inch was equivalent to 2.615 cm, and the symbol ‴ (referred to as Linien) was equal to 1/12 of an inch or 2.18 mm. Hence the length Müller designates as 0,14‴ = 305 μm.

[1]Since simple cones are found not only in tortoises (Hannover) but also in fishes and other animals, the term "twin" will probably have to be removed from the general description.

of the so-called granular layer, at the end of which it displays a swelling. In the case of twin cones, they have two filaments with two nuclei.

In birds there is a layer inward from the rods which corresponds in fishes to the filaments of the rods and the cones, namely, a layer of cylindrical bodies which do not have the same thickness as rods do, but rather are partially like a filament and in part thicker. Every rod extends continuously into one of these intertwined cylinders, and the well-known colored spherule[c] are at the place where the rods merge into these cones; these spherules are therefore to be found at the inner end of the actual rods, however, not all at completely the same height. Most are real spherules, not cone-shaped (Hannover), although single cones with larger dark-red spherules are colored red further inward. The ratio between these colored spherules also undergoes some modification depending on retinal location. The rods of frogs, where they lie on top of each other in a certain thickness, appear themselves to be somewhat reddish and an individual rod can be seen as alternately colored or colorless, according to whether it is lying or standing upright.

In frogs, too, the rods are connected on the inner side with a paler cylinder. It is not only sometimes filiform on various rods of various thicknesses, but also on each cylinder the diameters at various heights are not the same, so that thinner and thicker parts are interlocking. On the inner end of the cylinder sits a swelling; it is usually clearly formed by a nucleus. Moreover, pyramidal small bodies lie between these cylinders inward from the actual rods; Bowman already declared these as analogous to the cones of fishes. With a length of approximately 21.8 μm they have a bright tip going outward, and a thicker somewhat granular part going inward from which a filament extends. Internally they contain a yellowish spherule.[c]

Similarly in sharks, for example, the rods, which have a length of approximately 12 μm, and a width of 2.2 μm or slightly greater, are directly met by a second layer of cylinders whose length is 26.2 μm. These cylinders are distinguished from the shiny rods by their somewhat granular appearance. They are often separated from the rods along wide distances but can also be seen in isolation in connection with them. A small filament goes from their inner end more or less deeply into the granular layer in order to fasten itself to one of the particles therein.

Therefore, one finds everywhere, internal to the actual rods, a layer that sometimes consists of rather uniform cylinders and sometimes of large, thicker cones and very fine filaments next to each other. Often, at least, the sizes of the cones and rods, along with the nuclei attached to them, are in

[c]Throughout this section, Müller uses the term Kügelchen (spherule), but he is apparently referring to the structure we now refer to as the "oil droplet."

inverse proportion to each other. A sharp border line appears everywhere on the inner boundary of this cone layer, which at least in those specimens somewhat shrunken in chromic acid results from small projections that also sit here on the filiform parts. This is especially striking in birds, where at the same time a lanceolate extension against the granular layer is very clear; it is connected with the small bodies in this layer by a thinner filament.

4) A layer of cells with all the characteristic features of nerve cells is present next to the nerve extension[a] in all vertebrate classes. Bowman, Kölliker, and Corti have described processes of these cells in turtles and mammals that are also found in fishes and birds; indeed, it is scarcely to be doubted that these processes merge into nerve fibers although absolute certainty is difficult to achieve owing to the lack of contrast. Still, the processes are often very long, are sometimes clearly varicose, and also have otherwise the appearance of nerve fibers from the same eyes. However, there are not merely 2, but rather very frequently 3–4 divided processes present on the peculiarly shaped cells.

Cells that are more indeterminate are found further in the fine-grained substance of the retina in different numbers and clarities. An exquisite layer of cells, however, is also found inward from the so-called granular layer. In some cartilaginous fish and teleosts, a layer of flat, jagged, granulated cells is especially clear in the outermost region. In the whole cross-sectional view, they stand out because of their large oval nuclei whose longitudinal axis is parallel to the retina. Even if it cannot be doubted that an anastomosis occurs between these cells through their processes, this is even more apparent in the case of exceedingly beautiful cells that form a layer inside of the previously described layer; they appear striated in cross-section since these thin cells and their surfaces are parallel to the retina.

One can sometimes clearly distinguish two layers; one consists of irregular polygonal, somewhat granular cells, most having a diameter of 26–87 μm. They are connected with each other by short and in part very broad bridges in such a way that in many sections the only gaps remaining are smaller than the cells. The second layer consists of cells whose numerous processes are highly developed in proportion to the cell body in that the size of the cell body scarcely exceeds the caliber of the larger branches, and the lengths of these branches from the nucleus outward amounts to nearly 218 μm. Moreover, the processes are multiply branched and thickened at the places where they branch. These cells and their processes are somewhat yellowish, rather smooth, or more striated than granular, and their nucleus is not exquisitely vesicular and only of medium size. The most distant branchlets of these cells clearly likewise merge into each other so that one

[a]Probably refers here to the nerve fiber layer.

cell anastomoses with several adjacent ones at 2–3 points on each. In this way they form a meshwork through which pass the radial fibers, often by several fibers inclining together to the same interstice. In this way a latticework arises, made up of multiply crossing cords that is especially tightly meshed where the swellings on the perpendicular fibers are studded with jagged processes. These swellings are by the way consistently at the inner border of the cell layer, right where it meets the fine-grained mass.

If these cells were all to be regarded as nerve cells, as they have perhaps sometimes been so designated, their anastomoses would be highly peculiar. However, this designation is questionable not only because of these cells' flat and deeply-inscribed form and because of the nature of their matter and of their nucleus, but also because a network of striped cords is found at an analogous location in other fishes, which show scarcely a trace of cellular nature and appear more as fibrous tissue.

Continued comparative research will hopefully also permit physiological conclusions about the significance of elementary parts for the retina and for the nervous system in general, "but such conjectures can at present lead to nothing" (Bowman).[d]

Würzburg, May 15, 1851

Zur Histologie der Netzhaut von Dr. H. Müller

Die Untersuchung von Augen, welche einige Zeit in Chromsäurelösung gelegen waren, lässt sowohl in Betreff einzelner Elementartheile, aus denen die Netzhaut besteht, als auch der relativen Lage derselben Vieles erkennen, das ausserdem sehr schwierig zu eruiren ist. Ich will hier nur über einige Punkte eine vorläufige Mittheilung geben, indem ich Weiteres einer ausführlichen Darstellung des Baues der Netzhaut bie den verschiedenen Thieren vorbehalte.

1) Bei allen Wirbelthierklassen kommen in der Retina zahlreiche Cylinder vor, welche dieselbe der Dicke nach durchsetzen, indem sie senkrecht gegen die Nervenausbreitung, also radial zum Augapfel stehen. Es sind bald dünne Fasern haben, bald dickere, streifige Stränge.

Ihr inneres Ende stösst dicht an die Nervenfassern; bei manchen Thieren ist es zu einer kolbigen, körnigen Masse angeschwollen, die sich wie ein Bruchstück einer Zelle ausnimmt, bei andern gebt die Faser in eine membranartige dreiseitige Basis aus, die scharf abgeschnitten ist. Nach dem Durchtritt durch die innere, feinkörnige, der grauen Hirnsubstanz vollkommen ähnliche Schichte der Netzhaut zeigen die Radialfasern bei vielen Thieren constant eine Anschwellung, die manchmal deutlich einen Kern sammt Kernkörperchen anhält, such wohl zackige Fortsätze nach den Seiten hat, welche mit den benachbarten zu anastomosiren scheinen. Nach aussen geht die senkrechte Faser in die sogenannte Körnerschichte hinein, wobei sie sich öfters in mehreren Fäserchen auflöst. Jedenfalls steht sie mit den zunächst nach aussen liegenden Theilen in so enger Verbindung, dass nicht selten beim Zerreissen der Retina sich eine Faser

[d]The quote is in English in the original paper.

vollkommen isoliert, an deren äusserem Theil eine Anzahl der sogenannten Körner sammt Stäbchen oder Zwillingszapfen wie die Johannisbeeren an ihrem Stiel haften. Es spaltet sich also durch die ganze Dicke der Netzhaut einer schmaler Cylinder heraus, dessen Länge bei einem Frosch z. B. 0,14?" betrug. Dieselbe senkrechte Streifung durch die ganze Dicke erkennt man an dünnen senkrechten Schnitten, welche eine Profilansicht geben.

2) Die bekannten feinen Fädchen, welche häufig an den konisch zugespitzten Enden der Stäbchen sitzen, sind nicht gegen die Choroidea, sondern nach innen gekehrt. Sie beginnen nicht alle genau auf derselben Höhe, gehn z. B. bei den meisten Fischen zwischen die Zwillingszapfen hinein und stehn mit der nächsten innern, sogenannten Körnerschichte in Verbindung. Diese besteht nämlich aus Kernen, welche oft bläschenförmig, nach der Dicke der Netzhaut bald mehr bald weniger verlängert sind und in derselben Richtung durch längere oder kürzere Fädchen mit den Stäbchen zusammenhängen. Da man mitunter an einer Strecke des Umfangs eine zweite Contur sieht, die in das Fädchen übergeht, so sind diese "Körner" wohl für sehr kleine Zellen zu halten.

Bei denjenigen Fischen und Vögeln, wo das Pigment Fortsätze nach innen bildet, stecken nicht die Fädchen sondern die Stäbchen selbst im Pigment und wenn man die pigmentirte Schichte bis an die Zwillingszapfen von aussen wegnimmt, hat man die Stäbchen mindestens grösstentheils mitgenommen und nur das innere Ende mit den Fädchen stehen gelassen.

Bei Plagiostomen, wo kein Pigment zwischen den Stäbchen liegt, sieht man dieselben gleichmässig nach aussen gehen bis zu einer Schichte polygonaler Zellen, welchen denen des Tapetum der Wiederkäuer gleichen. Dahinter liegt dann eine strukturlose gefässreiche Membran, welche hier die Schuppen trägt, die durch die bekannten feinen Nadeln den Silberglanz erzeugen und dann erst kommt die pigmentirte Choroidea. Auch bei einigen andern Fischen erstreckt sich das Pigment nur eine kürzere Strecke zwischen den Stäbchen nach innen.

3) Die Zwillingszapfen[1] gehen bei den meisten Fischen und bei Säugethieren ebenfalls an ihrem innern stumpfen Ende in einen Fortsatz über, der sich in einen Faden auszieht; häufig bildet den Anfang des letztern ein deutlicher (footnote at bottom of page 235) Kern. Dieser Faden ist stärker als der an den Stäbchen befindliche hindurch anderen Ende er eine Anschwellung zeigt. Wo die Zapfen Zwillinge sind, haben sie zwei Fäden, mit zwei Kernen.

Bei Vögeln ist nach innen von den Stäbchen eine Schichte, welche den Fäden der Stäbchen und den Zapfen bei den Fischen entspricht, nämlich cylindrische Körper, die nicht von gleicher Dicke, wie die Stäbchen, sondern theils fadenformig, theils dicker sind. Jedes Stäbchen setzt sich in einen dieser zwischeneinandergeschobenen Cylinder continuirlich fort und wo die Stäbchen in diese Zapfen übergehen, sitzen die bekannten farbigen Kügelchen, die also im inneren Ende der eigentlichen Stäbchen zu finden sind, allerdings nicht alle ganz in gleicher Höhe. Die meisten sind wirkliche Kügelchen, nicht Kegel (Hannover) einzelne Zapfen mit grössern dunkelrothen Kügelchen aber sind ausserdem weiterhinein roth gefärbt. Die Verhältnisse dieser farbigen Kügelchen erleiden auch einige Modification nach den verschiedenen Stellen der Netzhaut. Die Stäbchen der Frösche erscheinen an sich selbst, wo sie in einer gewissen

[1]Da nicht blos bei Schildkröten (Hannover) sondern auch bei Fischen und sonst einfache Zapfen vorkommen, wird man wohl das „Zwilling" bei der allgemeinen Bezeichung streichen müssen.

Dicke übereinander liegen, etwas röthlich und man kann ein einzelnes Stäbchen abwechselnd farblos und gefärbt sehen, je nachdem es sich legt oder aufrichtet.

Auch bei den Fröschen stehen die Stäbchen nach innen (?) mit einen blasseren Cylinder in Verbindung, der nicht blos an verschiedenen Stäbchen von verschiedener Dicke manchmal fadenartig ist, sondern auch an jedem einzelnen sind die Stellen in verschiedener Höhe nicht gleich, so dass dickere und dünnere Theile in einandergeschoben sind. Am inneren Ende sitzt eine Anschwellung, die meist sehr deutlich durch einen Kern gebildet wird. Ausserdem liegen zwischen diesen Cylindern, innerhalb der eigentlichen Stäbchen pyramidale Körperchen, die schon *Bowman* für analog den Zapfen der Fische erklärte. Sie haben bei einer Länge von etwa $0{,}01^{m}$ eine hellere Spitze nach aussen, einen dickeren etwas körnigen Theil nach innen, von dem ein Faden ausgeht. Im innern liegt ein gelbliches Kügelchen.

Aehnlich stösst z. B. bei Haien innen unmittelbar an die Stäbchen, welchen etwa $0{,}025^{m}$ Länge haben, auf eine Breite von $0{,}004^{m}$ oder etwas mehr, eine zweite Schichte von Cylindern, deren Länge $0{,}042^{m}$ ist. Diese sind durch ein etwas granuliertes Ansehen von den glänzendern Stäbchen unterschieden, oft auf weiten Strecken von ihnen losgetrennt, oft aber auch mit solchen in Verbindung isoliert zu sehen. Von innern Ende geht ein Fädchen mehr oder weniger tief in die Körnerschichte, um sich an eines von deren Körperchen zu heften.

Man findet also überall innerhalb der eigentlichen Stäbchen eine Schichte welche bald aus ziemlich gleichmässigen Cylindern, bald aus grossen, dicken Zapfen und sehr feinen Fäden nebeneinander besteht. Häufig wenigstens steht die Grösse der Zapfen und der Stäbchen sammt die daran gehefteten Kernen in umgekehrtem Verhältnis. An der innern Grenze dieser Zapfenschichte zeigte sich überall eine scharfe Grenzlinie, welche wenigstens bei den in Chromsäure etwas geschrumpften Präparaten dadurch entsteht, dass auch an den fadenförmigen Theilen hier kleine Vorsprünge sitzen. Besonders auffallend ist dies bei Vögeln, wo zugleich eine lanzettförmige Verlängerung gegen die Körnerschichte sehr deutlich ist, mit deren Körperchen sie durch einen dünneren Faden in Verbindung steht.

4) Eine Schicht von Zellen mit allen Charakteren der Nervenzellen ist bei allen Wirbelthierklassen zunächst der Nervenausbreitung vorhanden. *Bowman, Kolliker, Corti* haben Fortsätze dieser Zellen bei Schilkröten und Sägethieren beschrieben; solche finden sich auch bei Fischen und Vögeln und zwar ist kaum zu sweifeln, dass sie in Nervenfasern übergehn, obwohl eine vollkommene Sicherheit hier wegen des mangelnden Criteriums der dunkeln Conturn schwerer zu erreichen ist. Dafür sind die Fortsätze oft sehr lang, manchmal deutlich varikos und haben auch sonst das ansehen von Nervenfasern aus denselben Augen. Es sind jedoch nicht blos 2, sondern sehr häufig 3–4 auch getheilte Forsätze an den eigenthümlich gestalteten Zellen vorhanden.

Unbestimmtere Zellen finden sich ferner in der feinkörnigen Substanz der Retina in verschiedener Zahl und Deutlichkeit. Eine exquisite Schichte von Zellen kommt aber auch nach innen von der sog. Körnerschichte vor. Bei einigen Knorpel- und Knochenfischen besonders deutlich ist hier zu äusserst eine Schichte platter, zackiger, granulirter Zellen, die in der ganzen Profilansicht durch ihre grossen, ovalen Kerne auffallen, deren Längsaxe der Retina parallel liegt. Wenn schon an diesen Zellen ein Anastamosiren durch ihre Fortsätze nicht zu bezweifeln ist, so ist dies doch viel mehr in die Augen fallend bei überaus schönen Zellen, welchen innerhalb der vorigen eine Schicht bilden, die im Profil streifig erscheint, da die dünnen Zellen mit ihrer Fläche der Retina parallel liegen.

Man kann bisweilen zwei Lagen deutlich unterscheiden; die eine besteht aus unregelmässig polygonalen, etwas körnigen Zellen, meist von 0,042–0,04^{m} Durchmesser, die durch kurze und zum Theil sehr breite Brücken mit einander so in Verbindung stehen, dass an manchen Strecken bloss Lücken bleiben, die kleiner sind als die Zellen. Die zweite Lage besteht aus Zellen, deren zahlreiche Fortsätze verhältnissmässig zum Körper sehr entwickelt sind, indem dieser die Breite der stärkeren Aeste manchmal kaum übertrifft und die Länge der letztern bis nahezu 0,4^{m} vom Kern aus beträgt. Dabei sind sie vielfach ästig, und an den Theilungsstellen verdickt. Diese Zellen mit den Fortsätzen sind etwas gelblich, ziemlich glatt, oder mehr streifig als körnig, ihr Kern nicht exquisit bläschenformig und nur mittlerer Grösse. Die äussersten Zweige dieser Zellen nun gehen ebenfals deutlich in einander über, so das eine Zelle mit mehreren benachbarten an je 2–3 Punkten anastomosirt. Sie bilden so ein Netz, durch dessen Maschen die radialen Fasern hindurchtreten, indem öfters mehere sich zu einer Lücke zusammenneigen. Dadurch entsteht ein Gitterwerk aus vielfach gekreuzten Strängen, das besonders dicht ist, wo die Anschwellungen an den senkrechten Fasern mit zackigen Fortsätzen besetzt sind. Diese Anschwellungen liegen übrigens constant an der inneren Grenze jener Zellenschichte, da wo sie an die feinkörnige Masse anstösst.

Wenn man diese Zellen alle für Nervenzellen halten dürfte, bei denen sie vielleicht schon manchmal mitgezählt worden sind, wurden ihre Anastomosen höchst merkwürdig sein. Es muss jedoch ausser ihrer platten and tief eingeschnittenen Form, der Beschaffenheit ihrer Substanz und ihres Kerns auch der Umstand bedenklich machen, dass bei andern Fischen an analoger Stelle ein Netz von streifigen Strängen vorkommt, die kaum eine Spur zelliger Natur zeigen und sich mehr wie ein Fasergewebe ausnehman.

Fortgesetzte vergleichende Untersuchungen werden hoffentlich auch physiologische Folgerungen uber die Bedeutung der Elementartheile für die Netzhaut und das Nervensystem überhaupt erlauben, "but such conjectures can at present lead to nothing" (*Bowman*).

Würzburg den 15 Mai 1851.

From: *Zeitschrift für wissenschaftliche Zoologie* III. Band 1851, pp. 234–237.

References

Adler, R., 1983, Taurine uptake of chick embryo retinal neurons and glial cells in purified culture. *J. Neurosci. Res.* **10**:369–379.

Adler, R., 1993, Determination of cellular types in the retina. *Invest. Ophthalmol. Vis. Sci.* **34**: 1677–1682.

Adler, R., Magistretti, P. J., Hyndman, A. G., and Shoemaker, W. J., 1982, Purification and cytochemical identification of neuronal and non-neuronal cells in chick embryo retina cultures. *Dev. Neurosci.* **5**:27–39.

Agulló, L., Baltrons, M. A., and García, A., 1995, Calcium-dependent nitric oxide formation in glial cells. *Brain Res.* **686**:160–168.

Ahmad, I., Leinders-Zufall, T., Kocsis, J. D., Sheppard, G. M., Zufall, F., and Barnstable, C. J., 1994, Retinal ganglion cells express a cGMP-gated conductance activatable by nitric oxide donors. *Neuron* **12**:155–165.

Akagawa, K., Hicks, D., and Barnstable, C. J., 1987, Histotypic organization and cell differentiation in rat retinal reaggregate cultures. *Brain Res.* **437**:298–308.

Akaike, A., Maruyama, T., Sikdar, S. K., and Yasui, S., 1987, Sodium-dependent suppression of γ-aminobutyric-acid-gated chloride currents in internally perfused frog sensory neurones. *J. Physiol.* **392**:543–562.

Algvere, P., and Kock, E., 1983, Experimental epiretinal membranes induced by intravitreal carbon particles. *Am. J. Ophthalmol.* **96**:345–353.

Amara, S. G., and Kuhar, M. J., 1993, Neurotransmitter transporters: Recent progress. *Annu. Rev. Neurosci.* **16**:73–93.

Amédée, T., Robert, A., and Coles, J. A., 1997, Potassium homeostasis and glial energy metabolism. *Glia* **21**:46–55.

Ames, III, A., Li, Y.-Y., Hehner, E. C., and Kimble, C. R., 1992, Energy metabolism of rabbit retina as related to function: High cost of Na^+ transport. *J. Neurosci.* **12**:840–853.

Amin, R. H., Frank, R. N., Kennedy, A., Eliott, A. D., Puklin, J. E., and Abrams, G. W., 1997, Vascular endothelial growth factor is present in glial cells of the retina and optic nerve of human subjects with nonproliferative diabetic retinopathy. *Invest. Ophthalmol. Vis. Sci.* **38**: 36–47.

Anderson, D. H., Guerin, C. J., Hageman, G. S., Pfeffer, B. A., and Flanders, K. C., 1995, Distribution of transforming growth-factor-beta isoforms in the mammalian retina. *J. Neurosci. Res.*.**42**:63–79.

Anderson, D. J., 1992, Molecular control of neural development, in: *An Introduction to Molecular Neurobiology* (Z. Hall, ed.), Sinauer Associates, Sunderland, MA, pp. 355–387.

Arber, S., and Caroni, P., 1995, Thrombospondin-4, an extracellular-matrix protein expressed in the developing and adult nervous system promotes neurite outgrowth. *J. Cell Biol.* **131**:1083–1094.

Armson, P. F., Bennett, M. R., and Raju, T. R., 1987, Retinal ganglion-cell survival and neurite regeneration requirements—the change from Müller cell-dependence to superior colliculi dependence during development. *Dev. Brain Res.* **32**:207–216.

Arriza, J. L., Eliasof, S., Kavanaugh, M. P., and Amara, S. G., 1997, Excitatory amino acid transporter 5, a retinal glutamate transporter coupled to a chloride conductance. *Proc. Natl. Acad. Sci. USA* **94**:4155–4160.

Artavanis-Tsakonas, S., Rand, M. W., and Lake, R. J., 1999, Notch signaling: Cell fate control and signal integration in development. *Science* **284**:770–776.

Arthur, F. E., Shivers, R. R., and Bowman, P. D., 1987, Astrocyte-mediated induction of tight junctions in brain capillary endothelium: An efficient *in vitro* model. *Brain Res.* **433**: 155–159.

Ashton, N., 1965, The blood-retinal barrier and vaso-glial relationships in retinal disease. *Trans. Ophthal. Soc. UK* **85**:199–230.

Ashton, N., and Cunha-Vaz, J. G., 1965, Effect of histamine on the permeability of the ocular vessels. *Arch. Ophthalmol.* **73**:211–223.

Ashwell, K. W. S., Höllander, H., Streit, W., and Stone, J., 1989, The appearance and distribution of microglia in the developing retina of the rat. *Visual Neurosci.* **2**:437–448.

Attwell, D., Barbour, B., Sarantis, M., and Szatkowski, M., 1989, The use of whole-cell patch clamping to monitor glutamate uptake electrically in glial cells isolated from the salamander and rabbit retina. *J. Physiol.* **415**:5P.

Attwell, D., Barbour, B., Brew, H., Sarantis, M., and Szatkowski, M., 1991, Electrogenic glutamate uptake: Basic properties, modulation and failure in anoxia. *Ann. N.Y. Acad. Sci.* **633**:610.

Attwell, D., Barbour, B., and Szatkowski, M., 1993, Nonvesicular release of neurotransmitter. *Neuron* **11**:401–407.

Aumailley, M., and Rousselle, P., 1999, Laminins of the dermo-epidermal junction. *Matrix Biol.* **18**:19–28.

Avisar, N., Shiftan, L., Ben-Dror, I., Havazelet, N., and Vardimon, L., 1999, A silencer element in the regulatory region of glutamine synthetase controls cell type-specific repression of gene induction by glucocorticoids. *J. Biol. Chem.* **274**:11399–11407.

Ayoub, G. S., and Lam, D. M.-K., 1984, The release of γ-aminobutyric acid from horizontal cells of the goldfish (*Carassius auratus*) retina. *J. Physiol.* **355**:191–214.

Azuma, N., Hida, T., Akiya, S., Uemura, Y., Kohsaka, S., and Tsukada, Y., 1990, Histochemical studies of hyaluronic acid in the developing human retina. *Graefés Arch. Clin. Exp. Ophthalmol.* **228**:158–160.

Babel, J., and Stangos, N., 1973, Essai de correlation entrel'ERG et les ultrastructures de la retinae. I. Action du monoiodoacetate. *Arch. Ophthalmol.* **33**:297–312.

Backus, K. H., Kettenmann, H., and Schachner, M., 1989, Pharmacological characterization of the glutamate receptor in cultured astrocytes. *J. Neurosci. Res.* **22**:274–282.

Barbour, B., Brew, H., and Attwell, D., 1988, Electrogenic glutamate uptake in glial cells is activated by intracellular potassium. *Nature* **335**:433–435.

Barbour, B., Szatkowski, M., Ingledew, N., and Attwell, D., 1989, Arachidonic acid induces a prolonged inhibition of glutamate uptake into glial cells. *Nature* **342**:918–920.

Barbour, B., Brew, H., and Attwell, D., 1991, Electrogenic uptake of glutamate and aspartate into glial cells isolated from the salamander (Ambystoma) retina. *J. Physiol.* **436**:169–193.

Barbour, B., Magnus, C., Szatkowski, M., Gray, P. T. A., and Attwell, D., 1993, Changes in NAD (P) H fluorescence and membrane current produced by glutamate uptake into salamander Müller cells. *J. Physiol.* **466**:573–597.

Barnes, S., and Bui, Q., 1991, Modulation of calcium-activated chloride current via pH-induced changes of calcium channel properties in cone photoreceptors. *J. Neurosci.* **11**:4015–4023.

Barres, B. A., 1991, New roles for glia. *J. Neurosci.* **11**:3685–3694.

Barres, B. A., Chun, L. L. Y., and Corey, D. P., 1989, Calcium current in cortical astrocytes: Induction by cAMP and neurotransmitters and permissive effect of serum factors. *J. Neurosci.* **9**:3169–3175.

Barres, B. A., Chun, L. L. Y., and Corey, D. P., 1990, Ion channels in vertebrate glia. *Annu. Rev. Neurosci.* **13**:441–474.

Bartsch, U., Kirchhoff, F., and Schachner, M., 1990, Highly sialylated N-CAM is expressed in adult mouse optic nerve and retina. *J. Neurocytol.* **19**:550–565.

Bartsch, U., Bartsch, S., Dorries, U., and Schachner, M., 1992, Immunohistological localization of tenascin in the developing and lesioned adult mouse optic nerve. *Eur. J. Neurosci.* **4**: 338–352.

Basinger, S. F., Gordon, W. C., and Lam, D. M. K., 1979, Differential labeling of retinal neurons by 3H-deoxyglucose. *Nature* **280**:682–684.

Batter, D. K., Corpina, R. A., Roy, C., Spray, D. C., Hertzberg, E. L., and Kessler, J. A., 1992, Heterogeneity in gap junction expression in astrocytes cultured from different brain regions. *Glia* **6**:213–221.

Baylor, D. A., 1987, Photoreceptor signals and vision. *Invest. Ophthalmol. Vis. Sci.* **28**:34–49.

Baylor, D. A., Fuortes, M. G. F., and O'Bryan, P. M., 1971, Receptive fields of cones in the retina of the turtle. *J. Physiol.* **214**:265–294.

Baylor, D. A., Lamb, T. D., and Yau, K.-Y., 1979, The membrane current of single rod outer segments. *J. Physiol.* **288**:589–611.

Beaulieu, C., Dyck, R., and Cynader, M., 1992, Enrichment of glutamate in zinc-containing terminals of the cat visual cortex. *NeuroReport* **3**:861–864.

Becquet, F., Courtois, Y., and Goureau, O., 1994, Nitric oxide decreases *in vitro* phagocytosis of photoreceptor outer segments by bovine retinal pigment epithelial cells. *J. Cell. Physiol.* **159**:256–262.

Bellhorn, R. W., 1984, Analysis of animal models of macular edema. *Surv. Ophthalmol.* **108**:520–524.

Benjamin, A. M., and Quastel, J. H., 1974, Fate of L-glutamate in the brain. *J. Neurochem.* **23**: 457–464.

Benveniste, E. N., 1993, Astrocyte-microglia interactions, in: *Astrocytes, Pharmacology and Function* (S. Murphy, ed.), Academic Press, San Diego, CA, pp. 355–382.

Berg-von der Emde, K., and Wolburg, H., 1989, Müller (glial) cells but not astrocytes in the retina of the goldfish possess orthogonal arrays of particles. *Glia* **2**:458–469.

Berka, J. L., Stubbs, A. J., Wang, D. M. Z., Dinicolantonio, R., Alcorn, D., Campbell, D. J., and Skinner, S. L., 1995, Renin-containing Müller cells of the retina display endocrine features. *Invest. Ophthalmol. Vis. Sci.* **36**:1450–1458.

Berl, S., Clarke, D. D., and Nickals, W. J., 1970, Compartmentation of citric acid cycle metabolism in brain: Effect of aminooxyacetic acid, ouabain and Ca^{2+} on the labelling of glutamate, glutamine, aspartate and GABA by [1–^{14}C]acetate, [U-^{14}C]glutamate and [U-^{14}C]aspartate. *J. Neurochem.* **15**:131–140.

Berson, E. L., 1993, Retinitis pigmentosa: The Friedenwald lecture. *Invest. Ophthalmol. Vis. Sci.* **34**:1659–1676.

Bevan, S., and Raff, M., 1985, Voltage-dependent potassium currents in cultured astrocytes. *Nature* **315**:229–232.

Bevan, S., Chiu, S. Y., Gray, P. T. A., and Ritchie, J. M., 1985, The presence of voltage-gated sodium, potassium and chloride channels in rat cultured astrocytes. *Proc. Roy. Soc. Lond. B* **225**:299–313.

Bevan, S., Lindsay, R. M., Perkins, M. N., and Raff, M. C., 1987, Voltage gated ionic channels in rat cultured astrocytes, reactive astrocytes and an astrocyte-oligodendrocyte progenitor cell. *J. Physiol.* **82**:327–335.

Bhattacharjee, J., and Sanyal, S., 1975, Developmental origin and early differentiation of retinal Müller cells in mice. *J. Anat.* **120**:367–372.

Biedermann, B., Eberhardt, W., and Reichelt, W., 1994, GABA uptake into isolated retinal Müller glial cells of the guinea-pig detected electrophysiologically. *NeuroReport* **5**:438–440.

Biedermann, B., Fröhlich, E., Grosche, J., Wagner, H.-J., and Reichenbach, A., 1995, Mammalian Müller (glial) cells express functional D_2 dopamine receptors. *NeuroReport* **6**: 609–612.

Bignami, A., 1984, Glial fibrillary acidic (GFA) protein in Müller glia: Immunofluorescence study of the goldfish retina. *Brain Res.* **300**:175–178.

Bignami, A., and Dahl, D., 1979, The radial glia of Müller in the rat retina and their response to injury: An immunofluorescence study with antibodies to the glial fibrillary acidic (GFA) protein. *Exp. Eye Res.* **28**:63–69.

Bignami, A., Eng, L. F., Dahl, D., and Uyeda, C. T., 1972, Localization of the glial fibrillary acidic protein in astrocytes by immunofluorescence. *Brain Res.* **43**:429–435.

Billups, B., Rossi, D., and Attwell, D., 1996, Anion conductance behavior of the glutamate uptake carrier in salamander retinal glial cells. *J. Neurosci.* **16**:6722–6731.

Birnbach, C. D., Jarvelainen, M., Possin, D., and Milam, A. H., 1994, Histopathology and immunocytochemistry of the neurosensory retina in fundus flavimaculatus. *Ophthalmology* **101**:1211–1219.

Biscardi, J. S., Cooper, N. F., and Maness, P. F., 1993, Phosphotyrosine-modified proteins are localized in Müller cells of the chick neural retina. *Exp. Eye Res.* **56**:281–289.

Bittar, P. G., Charnay, Y., Pellerin, L., Bouras, C., and Magistretti, P. J., 1996, Selective distribution of lactate dehydrogenase isoenzymes in neurons and astrocytes of human brain. *J. Cereb. Blood Flow Metab.* **16**:1079–1089.

Black, J. A., and Waxman, S. G., 1996, Sodium channel expression: A dynamic process in neurons and non-neuronal cells. *Dev. Neurosci.* **18**:139–152.

Blancher, C., Omri, B., Bidou, L., Pessac, B., and Crisanti, P., 1996, Nectinepsin: A new extracellular matrix protein of the pexin family: Characterization of a novel cDNA encoding a protein with an RGD cell binding motif. *J. Biol. Chem.* **271**:26220–26226.

Blank, W. F., Jr., and Kirshner, H. S., 1977, The kinetics of extracellular potassium changes during hypoxia and anoxia in the cat cerebral cortex. *Brain Res.* **123**:113–124.

Blanks, J. C., and Roffler-Tarlov, S., 1982, Differential localization of radioactive gamma-aminobutyric acid and muscimol in isolated and in vivo mouse retina. *Exp. Eye Res.* **35**: 573–584.

Bok, D., and Filerman, B., 1979, Autoradiographic localization of Na^+-K^+-ATPase in retinal photoreceptors and RPE with ^3H-ouabain. *Invest. Ophthalmol. Vis. Sci.* **18**:S224.

Bok, D., Ong, D. E., and Chytil, F., 1984, Immunocytochemical localization of cellular retinol binding protein in the rat retina. *Invest. Ophthalmol. Vis. Sci.* **25**:877–883.

Bolz, S., and Wolburg, H., 1992, First appearance of orthogonal arrays of particles in Müller cell membranes depends on retina maturation. *Dev. Neurosci.* **14**:203–209.

Bonaventure, N., Roussel, G., and Wioland, N., 1981, Effects of DL-alpha-amino-adipic acid on Müller cells in frog and chicken retinae in vivo: Relation to ERG b-wave, ganglion cell discharge and tectal evoked potentials. *Neurosci. Lett.* **27**:81–87.

Borden, L. A., Smith, K. E., Hartig, P. R., Branchek, T. A., and Weinshank, R. L., 1992, Molecular heterogeneity of the gamma-aminobutyric acid transport system. Cloning of two novel high affinity GABA transporters from rat brain. *J. Biol. Chem.* **267**:21098–21104.

Borgula, G. A., Karwoski, C. J., and Steinberg, R. H., 1989, Light-evoked changes in the extracellular pH in frog retina. *Vis. Res.* **29**:1069–1077.

Bormann, J., 1989, Memantine is a potent blocker of N-methyl D-aspartate (NMDA) receptor channels. *Eur. J. Pharmacol.* **166**:591–592.

Bormann, J., and Feigenspan, A., 1995, $GABA_C$ receptors. *Trends Neurosci.* **18**:515–519.

Bormann, J., and Kettenmann, H., 1988, Patch-clamp study of γ-aminobutyric acid receptor Cl^- channels in cultured astrocytes. *Proc. Natl. Acad. Sci. USA* **85**:9336–9340.

Bouvier, M., Szatkowski, M., Amato, A., and Attwell, D., 1992, The glial cell glutamate uptake carrier counter transports pH-changing anions. *Nature* **360**:471–474.

Bowery, N., 1989, $GABA_B$ receptors and their significance in mammalian pharmacology. *Trends Pharmacol. Sci.* **10**:401–407.

Boycott, B. B., and Hopkins, J. M., 1981, Microglia in the retina of monkey and other mammals: Its distinction from other types of glia and horizontal cells. *Neurosci.* **6**:679–688.

Boyle, P. J., and Conway, E. J., 1941, Potassium accumulation in muscle and associated changes. *J. Physiol. (Lond.)* **100**:1–63.

Brecha, N. C., and Weigmann, C., 1994, Expression of GAT-1, a high-affinity γ-aminobutyric acid plasma membrane trasnporter in the rat retina. *J. Comp. Neurol.* **345**:602–611.

Bredt, D. S., and Snyder, S. H., 1994, Nitric oxide: A physiologic messenger molecule. *Annu. Rev. Biochem.* **63**:175–195.

Bredt, D. S., Hwang, P. M., Glatt, C. E., Lowenstein, C., Reed, R. R., and Snyder, S. H., 1991, Cloned and expressed nitric oxide synthase structurally resembles cytochrome P-450 reductase. *Nature* **351**:714–718.

Brenneman, D. E., Schultzberg, M., Bartfei, T., and Gozes, I., 1992, Cytokine regulation of neuronal survival. *J. Neurochem.* **58**:454–460.

Brenner, M., 1994, Structure and transcriptional regulation of the GFAP gene. *Brain Pathology* **4**:245–257.

Brew, H., and Attwell, D., 1985, Is the potassium channel distribution in glial cells optimal for spatial buffering of potassium? *Biophys. J.* **48**:843–847.

Brew, H., and Attwell, D., 1987, Electrogenic glutamate uptake is a major current carrier in the membranes of axolotl retinal glial cells. *Nature* **327**:707–709.

Brew, H., Gray, P. T. A., Mobbs, P., and Attwell, D., 1986, Endfeet of retinal glial cells have higher densities of ion channels that mediate K^+ buffering. *Nature* **324**:466–468.

Brightman, M. W., and Reese, T. S., 1969, Junctions between intimately apposed cell membranes in vertebrate brain. *J. Cell Biol.* **19**:415–439.

Brittis, P. A., Canning, D. R., and Silver, J., 1992 Chondroitin sulfate as a regulator of neuronal patterning in the retina. *Science* **255**:733–736.

Brookes, N., 1997, Intracellular pH as a regulatory signal in astrocyte metabolism. *Glia* **21**:64–73.

Bunt-Milam, A. H., and Saari, J. C., 1983, Immunocytochemical localization of two retinoid-binding proteins in vertebrate retina. *J. Cell Biol.* **97**: 703–712.

Bunt-Milam, A. H., Saari, J. C., Klock, I. B., and Garwin, G. G., 1985, *Zonulae adherentes* pore size in the external limiting membrane of the rabbit retina. *Invest. Ophthalmol. Vis. Sci.* **26**:1377–1380.

Burke, J. M., and Kower, H. S., 1980, Collagen synthesis by rabbit neural retina *in vitro* and *in vivo*. *Exp. Eye Res.* **31**:213–226.

Burkhardt, D. A., 1974, Sensitization and centre-surround antagonism in *Necturus* retina. *J. Physiol. (Lond.)* **236**:593–610.

Burnashev, N., Khodorova, A., Jonas, P., Helm, P. J., Wisden, W., Monyer, H., Seeburg, P. H., and Sakmann, B., 1992, Calcium-permeable AMPA-kainate receptors in fusiform cerebellar glial cells. *Science* **256**:1566–1570.

Burns, M., and Robles, M. 1990, Müller cell GFAP expression exhibits gradient from focus of photoreceptor light damage. *Curr. Eye Res.* **9**:479–486.

Burnside, B., and Nagle, W., 1983, Retinomotor movements of photoreceptor and retinal pigment epithelium: Mechanisms and regulation, in: *Progress in Retinal Research*, Volume 2 (N. Osborne and G. Chader, eds.), Pergamon Press, Oxford, pp. 67–109.

Buskirk, D. R., Thiery, J. P., Rutishauser, U., and Edelman, G., 1980, Antibodies to a neural cell adhesion molecule disrupt histogenesis in cultured chick retina. *Nature* **285**:488–489.

Büssow, H., 1980, The astrocytes in the retina and optic nerve head of mammals: A special glia for the ganglion cell axons. *Cell Tissue Res.* **206**:367–368.

Cajal, S. R., 1972, *The Structure of the Retina* (translated by S. A. Thorpe and M. Glickstein), Charles C. Thomas, Springfield, IL.

Cammack, N. J., and Schwartz, E. A., 1993, Ions required for the electrogenic transport of GABA by horizontal cells of the catfish retina. *J. Physiol.* **472**:81–102.

Cao, W., Wen, R., Li, F., Cheng, T., and Steinberg, R. H., 1997a, Induction of basic fibroblast growth factor mRNA by basic fibroblast growth factor in Müller cells. *Invest. Ophthalmol. Vis. Sci.* **38**:1358–1366.

Cao, W., Wen, R., Li, F., LaVail, M. M., and Steinberg, R. H., 1997b, Mechanical injury increases bFGF and CNTF mRNA expression in the mouse retina. *Exp. Eye Res.* **65**:241–248.

Cao, Q.-L., Murphy, H. A., and Petry, H. M., 1999, Localization of nitric oxide synthase in the tree shrew retina. *Vis. Neurosci.* **16**:399–409.

Carmignoto, G., Comelli, M. C., Candeo, P., Cavicchioli, L., Yan, Q., Meright, A., and Maffei. I., 1991, Expression of NGF receptors and NGF receptor mRNA in the developing and adult rat retina. *Exp. Neurol.* **111**:302–311.

Carmignoto, G., Pasti, L., and Pozzan, T., 1998, On the role of voltage-dependent calcium channels in calcium signaling of astrocytes in situ. *J. Neurosci.* **18**:4637–4645.

Casper, D. S., and Reif-Lehrer, L., 1983, Effects of alpha-aminoadipate isomers on the morphology of the isolated chick embryo retina. *Invest. Ophthalmol. Vis. Sci.* **24**:1480–1488.

Caspi, R. R., and Roberge, F. C., 1989, Glial cells as suppressor cells: Characterization of the inhibitory function. *J. Autoimmun.* **2**:709–722.

Caspi, R. R., Roberge, R. G., and Nussenblatt, R. B., 1987, Organ-resident, non-lymphoid cells suppress proliferation of autoimmune T-helper lymphocytes. *Science* **237**:1029–1032.

Cepko, C., 1988, Retrovirus vectors and their applications in neurobiology. *Neuron* **1**:345–353.

Cepko, C. L., 1993, Retinal cell fate determination, in: *Progress in Retinal Research*, Volume 12 (N. Osborne and G. Chader, eds.), Pergamon Press, Oxford, pp. 1–12.

Cepko, C. L., Austin, C. P., Yang, X., Alexiades, M., and Ezzeddine, D., 1996, Cell fate determination in the vertebrate retina. *Proc. Natl. Acad. Sci. USA* **93**:589–595.

Chai, L., and Morris, J. E., 1994, Distribution of heparan-sulfate proteoglycan in embryonic chicken retina and isolated inner limiting membrane. *Curr. Eye Res.* **13**:669–677.

Chaitin, M. H., and Brun-Zinkernagel, A. M., 1998, Immunolocalization of CD44 in the dystrophic rat retina. *Exp. Eye Res.* **67**:283–292.

Chaitin, M. H., Wortham, H. S., and Brunzinkernagel, A. M., 1994, Immunocytochemical localization of CD44 in the mouse retina. *Exp. Eye Res.* **58**:359–365.

Chakrabarti, S., Sima, A. A., Lee, J., Brachet, P., and Dicou, E., 1990, Nerve growth factor (NGF), proNGF and NGF-like immunoreactivity in BB rat retina. *Brain Res.* **523**:11–15.

Chan, C. C., Palestine, A. G., Nussenblatt, R. B., Roberge, F. G., and Benezra, N., 1985, Anti-retinal auto-antibodies in Vogt-Koyanagi-Harada Syndrome, Behcet's disease and sympathetic ophthalmia. *Ophthalmology* **92**:1025–1028.

Chan, C. C., Roberge, F. G., Ni, M., Ahang, W., and Nussenblatt, R. B., 1991, Injury of Müller cells increases the incidence of experimental autoimmmune uveoretinitis. *Clin. Imm. Immunopath.* **59**:201–207.

Chan-Ling, T., and Stone, J., 1991, Factors determining the migration of astrocytes into the developing retina: Migration does not depend on intact axons or patent vessels. *J. Comp. Neurol.* **303**:375–386.

Chao, T. I., Pannicke, T., Reichelt, W., and Reichenbach, A., 1993, Na^+ channels are expressed by mammalian retinal glial (Müller) cells. *NeuroReport* **4**:575–578.

Chao, T. I., Henke, A., Reichelt, W., Eberhardt, W., Reinhardt-Maelicke, S., and Reichenbach, A., 1994a, Three distinct types of voltage-dependent K^+ channels are expressed by Müller (glial) cells of the rabbit retina. *Pflügers Arch.* **426**:51–60.

Chao, T. I., Skachkov, S. N., Eberhardt, W., and Reichenbach, A., 1994b, Na^+ channels of Müller (glial) cells isolated from retinae of various mammalian species including man. *Glia* **10**:173–185.

Chao, T. I., Grosche, J., Friedrich, K. J., Biedermann, B., Francke, M., Pannicke, T., Reichelt, W., Wulst, M., Muhle, C., Pritz-Hohmeier, S., Kuhrt, H., Faude, F., Drommer, W., Kasper, M., Buse, E., and Reichenbach, A., 1997, Comparative studies on mammalian Müller (retinal glial) cells. *J. Neurocytol.* **26**:439–454.

Chen, H.-S. V., and Lipton, S. A., 1997, Mechanism of memantine block of NMDA-activated channels in rat retinal ganglion cells: Uncompetitive antagonism. *J. Physiol. (Lond.)* **499**:27–46.

Chen, H.-S. V., Pellegrini, J. W., Aggarwal, S. K., Lei, S. Z., Warach, S., Jensen, F. E., and Lipton, S. A., 1992, Open-channel block of N-methyl-D-aspartate (NMDA) responses by memen-tine: Therapeutic advantage against NMDA receptor-mediated neurotoxicity. *J. Neurosci.* **12**:4427–4436.

Chen, Q., Olney, J. W., Lukasiewicz, P. D., Almli, T., and Romano, C., 1998, Ca^{2+}-independent excitotoxic neurodegeneration in isolated retina, an intact neural net: A role for Cl^- and inhibitory transmitters. *Mol. Pharmacol.* **53**:564–572.

Chen, S. T., Garey, L. J., and Jen, L. S., 1994, Bcl_2 proto-oncogene protein immunoreactivity in normally developing and axotomized rat retinas. *Neurosci. Lett.* **172**:11–14.

Chesler, M., 1990, The regulation and modulation of pH in the nervous system. *Prog. Neurobiol.* **34**:401–427.

Chiu, S. Y., and Wilson, G. F., 1989, The role of potassium channels in Schwann cell proliferation in Wallerian degeneration of explant rabbit sciatic nerves. *J. Physiol. (Lond.)* **408**:199–222.

Choi, D. W., 1988, Calcium-mediated neurotoxicity: Relationship to specific channel types and role in ischemic damage. *Trends Neurosci.* **11**:465–469.

Choi, D. W., and Rothman, S. M., 1990, The role of glutamate neurotoxicity in hypoxic-ischemic neuronal death. *Annu. Rev. Neurosci.* **13**:171–182.

Clark, B., and Mobbs, P., 1992, Transmitter-operated channels in rabbit retinal astrocytes studied in situ by whole-cell patch clamping. *J. Neurosci.* **12**:664–673.

Clark, D. D., Nicklas, W. J., and Berl, S., 1970, Tricarboxylic acid-cycle metabolism in brain: Effect of fluoroacetate and fluorocitrate on the labeling of glutamate, aspartate, glu-tamine and γ-aminobutyrate. *Biochem. J.* **120**:345–351.

Clark, J. A., Deutch, A. Y., Gallipoli, P. Z., and Amara, S., 1992, Functional expression and CNS distribution of a β-alanine-sensitive neuronal GABA transporter. *Neuron* **9**:337–348.

Clarke, D. D., and Sokoloff, L., 1994, Circulation and energy metabolism of the brain, in: *Basic Neurochemistry: Molecular, Cellular, and Medical Aspects*, 5th edition (G. J. Siegel, B. W. Agranoff, R. W. Albers, and P. B. Molinoff, eds.). Raven Press, New York, pp. 645–680.

Clarke, K., and Geisert, Jr., E. E., 1998, The target of the antiproliferative antibody (TAPA) in the normal and injured rat retina. *Mol. Vision* **4**:3–13.

Cogan, D. G., and Kuwabara, T., 1959, Tetrazolium studies on the retina. II. Substrate depen-dent patterns. *J. Histochem. Cytochem.* **7**:334–341.

Cohen, A. I., 1965a, New details of the ultrastructure of the outer segments and ciliary connectives of the rods of human and Macaque retinas. *Anat. Rec.* **152**:63–79.

Cohen, A. I., 1965b, A possible cytological basis for the "R" membrane in the vertebrate eye. *Nature* **205**:1222–1223.

Cohen, A. I., 1967, An electron microscopic study of the modification by monosodium gluta-mate of the retinas of normal and "rodless" mice. *Am. J. Anat.* **120**:319–356.

Cohen, A. I., 1989, Interphotoreceptor contacts at the inner segment level in primate retinas. *Brain Res.* **490**:200–203.

Cohen-Armon, M., and Sokolovsky, M., 1993, Evidence for involvement of the voltage-dependent Na^+ channel gating in depolarization-induced activation of G-proteins. *J. Biol. Chem.* **268**:9824–9838.

Coles, J. A., 1996, Glial cells and the supply of substrates of energy metabolism to neurons, in: *Neuroglia* (H. Kettenman and B. R. Ransom, eds.) Oxford University Press, Oxford, pp. 793–804.

Condon, G. P., Browstein, S., Wang, N. S., Kearns, J. A., and Ewing, C. C., 1986, Congenital hereditary (juvenile X-linked) retinoschisis: Histopathologic and ultrastructural findings in three eyes. *Arch. Ophthalmol.* **104**:576–583.

Conner, J. D., Detwiler, P. B., and Sarthy, P. V., 1985, Ionic and electrophysiological properties of retinal Müller (glial) cells of the turtle. *J. Physiol.* **362**:79–92.

Connor, J. A., and Stevens, C. F., 1971, Voltage clamp studies of a transient outward membrane current in gastropod neural somata. *J. Physiol.* (Lond.) **213**:21–30.

Cooper, A. J. L., and Plum, F., 1987, Biochemistry and physiology of brain ammonia. *Physiol. Rev.* **67**:440–519.

Cornell-Bell, A. H., and Finkbeiner, S. M., 1991, Ca^{2+} waves in astrocytes. *Cell Calcium* **12**: 185–204.

Cornell-Bell, A. H., Finkbeiner, S. M., Cooper, M. S., and Smith, S. J., 1990, Glutamate induces calcium waves in cultured astrocytes: Long-range glial signaling. *Science* **247**:470–473.

Corvalan, V., Cole, R., de Vellis, J., and Hagiwara, S., 1990, Neuronal modulation of calcium channel activity in cultured rat astrocytes. *Proc. Natl. Acad. Sci. USA* **87**:4345–4348.

Craft, J. L., Sang, D. N., Dryja, T. P., Brockhurst, R. J., and Robinson, N. L., 1985, Glial cell component in retinoblastoma. *Exp. Eye Res.* **40**:647–659.

Cremer, H., Chazal, G., Goridis, C., and Represa, A., 1997, NCAM is essential for axonal growth and fasciculation in the hippocampus. *Mol. Cell. Neurosci.* **8**:323–335.

Crittenden, S. L., Pratt, R. S., Cook, J. H., Balsamo, J., and Lilien, J., 1987, Immunologically unique and common domains within a family of proteins related to the retina Ca^{2+}-dependent cell-adhesion molecule, NCALCAM. *Development* **101**:729–740.

Cubells, J. F., Walkley, S. U., and Makman, M. H., 1988, The effects of gabaculine on the distribution of GABA-like immunoreactivity in the rat retina. *Brain Res.* **458**:82–90.

Cudeiro, J., and Rivadulla, C., 1999, Sight and insight—on the physiological role of nitric oxide in the visual system. *Trends Neurosci.* **22**:109–116.

Cunha-Vaz, J. G., and Maurice, D. M., 1967, The active transport of fluorescein by the retinal vessels and the retina. *J. Physiol.* **191**:467–486.

Cunha-Vaz, J. G., Shakib, M., and Ashton, N., 1966, Studies on the permeability of the blood-retinal barrier. I. On the existence, development, and site of a blood-retinal barrier. *Br. J. Ophthalmol.* **50**:441–453.

Cuthbertson, R. A., Lang, R. A., and Coghlan, J. P., 1990, Macrophage products IL-1α, TNFα and bFGF may mediate multiple cytopathic effects in the developing eyes of GM-CSF transgenic mice. *Exp. Eye Res.* **51**:335–344.

Dabin, I., and Barnstable, C., 1995, Rat retinal Müller cells express Thy-1 following neuronal cell death. *Glia* **14**:23–32.

Dalessandri, L., Ranscht, B., Winterhalter, K. H., and Vaughan, L., 1995, Contactin/F11 and tenascin-C coexpression in the chick retina correlates with formation of the synaptic plexiform layers. *Curr. Eye Res.* **14**:911–926.

Dani, J. W., Chernjavsky, A., and Smith, S. J., 1992, Neuronal activity triggers calcium waves in hippocampal astrocyte networks. *Neuron* **8**:429–440.

Darius, S., Wolf, G., Huang, P. L., and Fishman, M. C., 1995, Localization of NADPH-diaphorase oxide synthase in the rat retina: An electron microscopic study. *Brain Res.* **690**:231–235.

Das, A., Pansky, B., Budd, G. C., and Kollarits, C. R., 1984, Immunocytochemistry of mouse and human retina with antisera to insulin and S-100 protein. *Curr. Eye Res.* **3**:1397–1403.

Das, A., Pansky, B., and Budd, G. C., 1987, Demonstration of insulin-specific mRNA in cultured rat retinal glial cells. *Invest. Ophthalmol. Vis. Sci.* **28**:1800–1810.

Das, S. R., Bharadwaj, N., Kjeldbye, H., and Gouras, P., 1992, Müller cells of chicken retina synthesize 11–cis-retinol. *Biochem. J.* **285**:907–913.

Davidson, M., Nasisse, M., and Krnegay, J., 1990, Intermediate filament complement of the normal and gliotic canine retina. *J. Comp. Path.* **103**:125–134.

Dawson, T. M., Bredt, D. S., Fotuhi, M., Hwang, P. M., and Snyder, S. H., 1991a, Nitric oxide synthase and neuronal NADPH diaphorase are identical in brain and peripheral tissues. *Proc. Natl. Acad. Sci. USA* **88**:7797–7801.

De Jong, P. T., Zrenner, E., van Meel, G. J., Keunen, J. E., and van Norren, D., 1991, Mizuo phenomenon in X-linked retinoschisis: Pathogenesis of the Mizuo phenomenon. *Arch. Ophthalmol.* **109**:1104–1108.

De Leeuw, A. M., Gaur, V. P., Saari, J. C., and Milam, A. H., 1990, Immunolocalization of cellular retinol-, retinaldehyde-, and retinoic acid-binding proteins in rat retina during pre-and post-natal development. *J. Neurocytol.* **19**:253–264.

DeFeudis, F. V., 1980, Binding studies with muscimol: Relation to γ-aminobutyrate receptors. *Neurosci.* **5**:675–688.

Del Priore, L. V., Lewis, A., Tan, S., Carley, W. W., and Webb, W. W., 1987, Fluorescence light microscopy of F-actin in retinal rods and glial cells. *Invest. Ophthalmol. Vis. Sci.* **28**:632–639.

Demyanenko, G. P., Tsai, A. Y., and Maness, P. F., 1999, Abnormalities in neuronal process extension, hippocampal development, and the ventricular system of L1 knockout mice. *J. Neurosci.* **19**:4907–4920.

Dermietzel, R., 1973, Visualization by freeze-fracturing of regular structures in glial cell membranes. *Naturwissenschaften* **60**:208–209.

Dermietzel, R., 1974, Junctions in the central nervous system of the cat. III. Gap junctions and membrane-associated orthogonal particle complexes (MOPC) in astrocytic membranes. *Cell Tissue Res.* **149**:121–135.

Dermietzel, R., Hertzberg, E. L., Kessler, J. A., and Spray, D. C., 1991, Gap junctions between cultured astrocytes: Immunocytochemical, molecular, and electrophysiological analysis. *J. Neurosci.* **11**:1421–1432.

Derouiche, A., and Rauen, T., 1995, Coincidence of L-glutamate/L-aspartate transporter (GLAST) and glutamine synthetase (GS) immunoreactions in retinal glia: Evidence for coupling of GLAST and GS in transmitter clearance. *J. Neurosci. Res.* **42**:131–143.

DeVries, S. H., and Schwartz, E. A., 1989, Modulation of an electrical synapse between solitary pairs of catfish horizontal cells by dopamine and second messengers. *J. Physiol.* **414**:351–375.

Di Scala-Guenod, D., Mouginot, D., and Strosser, M.-T., 1994, Increase of intracellular calcium induced by oxytocin in hypothalamic cultured astrocytes. *Glia* **11**:269–276.

Dick, E., and Miller, R. F., 1978, Light-evoked potassium activity in mudpuppy retina: Its relationship to the b-wave of the electroretinogram. *Brain Res.* **154**:388–394.

Dingledine, R., and McBain, C. J., 1993, Excitatory amino acid transmitters, in: *Basic Neurochemistry*, 5th edition (G. J. Siegel, B. W. Agranoff, R. W. Albers, and P. B. Molinoff, eds.), Raven Press, New York, pp. 367–387.

Distler, C., and Dreher, Z., 1996, Glial cells of the monkey retina. II. Müller cells. *Vis. Res.* **36**:2381–2394.

Dixon, R. G., and Eng, L. F., 1981, Glial fibrillary acidic protein in the retina of the albino rat: An immunoperoxidase study of paraffin-embedded tissue. *J. Comp. Neurol.* **195**:305–321.

Dorsky, R. I., Rapaport, D. H., and Harris, W. A., 1995, *Xotch* inhibits cell differentiation in the *Xenopus* retina. *Neuron* **14**:487–496.

Dowling, J. E., 1970, Organization of vertebrate retinas. *Invest. Ophthalmol. Vis. Sci.* **9**:655–680.

Dowling, J. E., 1987, *The Retina: An Approachable Part of the Brain*, Belknap Press, Cambridge, MA, pp. 282.

Dowling, J. E., and Ripps, H., 1972, Adaptation in skate photoreceptors. *J. Gen. Physiol.* **60**: 698–719.

Dowling, J. E., and Ripps, H., 1973, Effect of magnesium on horizontal cell activity in the skate retina. *Nature* **242**:101–103.

Dräger, U. C., 1983, Coexistence of neurofilaments and vimentin in a neurone of adult mouse retina. *Nature* **303**:169–172.

Dräger, U. C., and Edwards, D. L., 1983, Antibodies to intermediate filaments reveal abnormalities in retinas of mice with photoreceptor degeneration. *Invest. Ophthalmol. Vis. Sci.* **24**:S1115.

Draghun, A., Verdoorn, T. A., Ewert, M., Seeburg, P. H., and Sakmann, B., 1990, Functional and molecular distinction between recombinant rat $GABA_A$ receptor subtypes by Zn^{2+}. *Neuron* **5**:781–788.

Drazba, J., and Lemmon, V., 1991, The role of cell adhesion molecules in neurite outgrowth on Müller cells. *Dev. Biol.* **138**:82–93.

Dreher, Z., Wegner, M., and Stone, J., 1988, Müller cell endfeet at the inner surface of the retina: Light microscopy. *Visual Neurosci.* **1**:169–180.

Dreher, Z., Robinson, S. R., and Distler, C., 1992, Müller cells in vascular and avascular retinae: A survey of seven mammals. *J. Comp. Neurol.* **323**:59–80.

Drenckhahn, D., and Wagner, H.-J., 1985, Relation of retinomotor responses and contractile proteins in vertebrate retinas. *Eur. J. Cell Biol.* **37**:156–168.

Dreyer, E. B., Zurakowski, D., Schumer, R. A., Podos, S. M., and Lipton, S. A., 1996, Elevated glutamate levels in the vitreous body of humans and monkeys with glaucoma. *Arch. Ophthalmol.* **114**:299–305.

Dringen, R., Gebhardt, R., and Hemprecht, B., 1993a, Glycogen in astrocytes: Possible function as lactate supply for neighboring cells. *Brain Res.* **623**:208–214.

Dringen, R., Wiesinger, H., and Hamprecht, B., 1993b, Upake of L-lactate by cultured rat-brain neurons. *Neurosci. Lett.* **163**:5–7.

Duffy, M., Sun, Y., Wiggert, B., Duncan, T., Chader, G. J., and Ripps, H., 1993, Interphotoreceptor retinoid binding protein (IRBP) enhances rhodopsin regeneration in the experimentally detached retina. *Exp. Eye Res.* **57**:771–782.

Duke–Elder, S., ed., 1961, *System of Ophthalmology.* The Anatomy of the Visual System, Volume 2 (S. Duke–Elder and K. C. Wybar, eds.), C.V. Mosby, St. Louis, MO.

Durlu, Y. K., Ishiguro, S., Yoshida, A., Mito, T., Tsuchiya, M., and Tamai, M., 1990, Response of Müller cells following experimental lensectomy-vitreoctomy. *Graefés Arch. Clin. Exp. Ophthalmol.* **228**:44–48.

Eagle, Jr., R. C., 1984, Mechanisms of maculopathy. *Ophthalmology* **91**:613–625.

Eagle, Jr., R. C., 1999, *Eye Pathology: An Atlas and Basic Text,* W.B. Saunders Co., Philadelphia, PA, pp. 306.

Edelman, G. M., and Crossin, K. L., 1991, Cell adhesion molecules: Implications for a molecular histology. *Annu. Rev. Biochem.* **60**:155–190.

Edwards, R. B., Adler, A. J., Dev, S., and Claycomb, R. C., 1992, Synthesis of retinoic acid from retinol by cultured rabbit Müller cells. *Exp. Eye Res.* **54**:481–490.

Ehinger, B., 1977, Glial and neuronal uptake of GABA, glutamic acid, glutamine and glutathione in the rabbit retina. *Exp. Eye Res.* **25**:221–234.

Ehinger, B., Zucker, C. L., Bruun, A., and Adolph, A., 1994, In vivo staining of oligodendroglia in the rabbit retina. *Glia* **10**:40–48.

Eisenfeld, A. J., Bunt-Milam, A. H., and Sarthy, P. V., 1984, Müller cell expression of glial fibrillary acidic protein after genetic and experimental photoreceptor degeneration in the rat retina. *Invest. Ophthalmol. Vis. Sci.* **25**:1321–1328.

Eisenfeld, A. J., Bunt-Milam, A. H., and Saari, J. C., 1985, Localization of retinoid-binding proteins in developing rat retina. *Exp. Eye Res.* **41**:299–394.

Eisenfeld, A. J., Bunt-Milam, A. H., and Saari, J. C., 1987, Uveoretinitis in rabbits following immunization with interphotoreceptor retinoid-binding protein. *Exp. Eye Res.* **44**:425–438.

Ekstrom, P., Sanyal, S., Nafstrom, K., Chader, G. J., and van Veen, T., 1988, Accumulation of glial fibrillary acidic protein in Müller radial glia during retinal degeneration. *Invest. Ophthalmol. Vis. Sci.* **29**:1363–1371.

Eliasof, S., and Jahr, C. E., 1996, Retinal glial cell glutamate transporter is coupled to an anionic conductance. *Proc. Natl. Acad. Sci. USA* **93**:4153–4158.

Eliasof, S., and Werblin, F., 1993, Characterization of the glutamate transporter in retinal cones of the tiger salamander. *J. Neurosci.* **13**:402–411.

Eliasof, S., Arriza, J. L., Leighton, B. H., Kavanaugh, M. P., and Amara, S. G., 1998, Excitatory amino acid transporters of the salamander retina: Identification, localization, and function. *J. Neurosci.* **18**:698–712.

Elner, S. G., and Elner, V. M., 1997, The integrin superfamily and the eye. *Invest. Ophthalmol. Vis. Sci.* **37**:696–701.

Enkvist, M. O. K., Holopainen, I., and Åkerman, K. E. O., 1989, Glutamate receptor-linked changes in membrane potential and intracellular Ca^{2+} in primary rat astrocytes. *Glia* **2**: 397–402.

Erecinska, M., 1987, The neurotransmitter amino acid transport systems: A fresh outlook on an old problem. *Biochem. Pharmacol.* **36**:3547–3555.

Erickson, P. A., Fisher, S. K., Guerin, C. J., Anderson, D. H., and Kaska, D. D., 1987, Glial fibrillary acidic protein increases in Müller cells after retinal detachment. *Exp. Eye Res.* **44**:37–48.

Ezzeddine, D. Z., Yang, X., DeChiara, T., Yancopoulos, G., and Cepko, C. L., 1997, Postmitotic cells fated to become rod photoreceptors can be respecified by CNTF treatment of the retina. *Development* **124**:1055–1067.

Faber, D. S., 1969, Analysis of the slow trans-retinal potentials in response to light. Ph.D. dissertation, State University of New York, Buffalo, N.Y.

Fadool, J. M., and Linser, P. J., 1993, 5A11 antigen is a cell recognition molecule which is involved in neuronal-glial interactions in avian neural retina. *Dev. Dynamics* **196**:252–262.

Fain, G. L., Gold, G. H., and Dowling, J. E., 1976, Receptor coupling in the toad retina. *Cold Spr. Harbor Symp. Quant. Biol.* **40**:547–561.

Fairman, W. A., Vandenberg, R. J., Arriza, J. L., Kavanaugh, M. P., and Amara, S. G., 1995, An excitatory amino-acid transporter with properties of a ligand-gated chloride channel. *Nature* **375**:599–603.

Fakler, B., and Puppersberg, J. P., 1996, Functional and molecular diversity classifies the family of inward-rectifier K^+ channels. *Cell. Physiol. Biochem.* **6**:195–209.

Faude, F., Biedermann, B., Pannicke, T., Francke, M., Wiedemann, P., Reichenbach, A. and Reichelt, W., 1995, GABA and glutamate uptake currents in human Müller cells. *Invest. Ophthalmol. Vis. Sci.* (suppl.)**36**:926.

Fesenko, E. E., Kolesnikov, S. S., and Lyubarsky, A. L., 1985, Induction by cyclic GMP of cationic conductance in plasma membrane of retinal rod outer segment. *Nature* **313**:310–313.

Fine, B. S., 1961, Limiting membranes of the sensory retina and pigment epithelium: An electron microscopic study. *Arch. Ophthalmol.* **66**:847–860.

Fine, B. S., and Brucker, A. J., 1981, Macular edema and cystoid macular edema. *Am. J. Ophthalmol.* **92**:466–481.

Finkbeiner, S. M., 1993, Glial calcium. *Glia* **9**:83–104.

Fiordalisi, J. J., and Maness, P. F., 1999, *Rek*, an Axl/Tyro3–related receptor-class tyrosine kinase, is expressed in Müller glia and differentiating neural subpopulations in the avian retina. *Exp. Eye Res.* **68**:201–210.

Fisher, S. K., and Anderson, D. H., 1994, Cellular effects of detachment on the neural retina and the retinal pigment epithelium, in: *Retina*, Volume 3 (S. J. Ryan, ed.), C.V. Mosby, St. Louis, MO, pp. 2035–2061.

Fisher, S. K., Erickson, P. A., Lewis, G. P., and Anderson, D. H., 1991, Intraretinal proliferation induced by retinal detachment. *Invest. Ophthalmol. Vis. Sci.* **23**:1739–1748.

Fitzgerald, M. E., Vana, B. A., and Reiner, A., 1990, Evidence for retinal pathology following interruption of neural regulation of choroidal blood flow: Müller cells express GFAP following lesions of the nucleus of Edinger-Westphal in pigeons. *Curr. Eye Res.* **9**:583–598.

Fonnum, F., Johnsen, A., and Hassel, B., 1997, Use of fluorocitrate and fluoroacetate in the study of brain metabolism, Glia. 21:106–113.

Foos, R. Y., 1972, Vitreoretinal juncture: Topographical variations. *Invest. Ophthalmol.* **11**:801–808.

Foos, R. Y., 1974, Posterior peripheral retinal tears. *Ann. Ophthalmol.* **6**:679–687.

Friedenwald, J. S., and Chan, E., 1932, Pathogenesis of Retinitis Pigmentosa with a note on the phagocytic activity of Müller fibers. *Arch. Ophthalmol.* **8**:173–181.

Frishman, L. J., and Steinberg, R. H., 1989, Light-evoked increases in $[K^+]_o$ in proximal portion of the dark-adapted cat retina. *J. Neurophysiol.* **61**:1233–1243.

Fuchs, E., 1995, Keratins and the skin. *Annu. Rev. Cell Dev. Biol.* **11**:123–153.

Fuchs, E., and Cleveland, D. W., 1998, A structural scaffolding of intermediate filaments in health and disease. *Science* **279**:514–519.

Furakawa, T., and Hanawa, I., 1955, Effects of some common cations on electretinogram of toad. *Jpn. J. Physiol.* **5**:289–300.

Gainer, H., 1978, Intercellular transfer of proteins from glial cells to axons. *Trends Neurosci.* **1**:93–96.

Gallemore, R. P., Hughes, B. A., and Miller, S. S., 1997, Retinal pigment epithelial transport mechanisms and their contributions to the electroretinogram. *Prog. Ret. Eye Res.* **16**:509–566.

Gallo, V., and Russell, J. T., 1995, Excitatory amino acid receptors in glia: Different subtypes for distinct functions? *J. Neurosci. Res.* **42**:1–8.

Gally, J. A., Montague, P. R., Reeke, G. N., and Edelman, G. M., 1990, The NO hypothesis: Possible effects of short-lived, rapidly diffusible signal in the development and function of the nervous system. *Proc. Natl. Acad. Sci. USA* **87**:3547–3551.

Gao, J. M., and Hollyfield, J. G., 1992, Basic fibroblast growth factor (bFGF) immunolocalization in the rodent outer retina demonstrated with an anti-rodent bFGF antibody. *Brain Res.* **585**:355–360.

Gardner-Medwin, A. R., 1981, Possible roles of vertebrate neuroglia in potassium dynamics, spreading depression and migraine. *J. Exp. Biol.* **95**:11–127.

Gardner-Medwin, A. R., 1983a, A study of the mechanisms by which potassium moves through brain tissue in the rat. *J. Physiol. (Lond.)* **335**:353–374.

Gardner-Medwin, A. R., 1983b, Analysis of potassium dynamics in mammalian brain tissue. *J. Physiol. (Lond.)* **335**:393–426.

Garthwaite, J., 1991, Glutamate, nitric oxide and cell-cell signalling in the nervous system. *Trends Neurosci.* **14**:60–67.

Garthwaite, J., Charles, S. L., and Chess-Williams, R., 1988, Endothelium-derived relaxing factor release on activation of NMDA receptors suggests role as intercellular messenger in the brain. *Nature* **336**:385–388.

Gartner, S., and Henkind, P., 1982, Pathology of retinitis pigmentosa. *Ophthalmology* **89**:1425–1432.

Gass, J. D. M., 1999, Müller cell cone, an overlooked part of the anatomy of the fovea centralis. *Arch. Ophthalmol.* **117**:821–823.

Gass, J. D. M., and Norton, E. W. D., 1966, Cystoid macular edema and papilledema following cataract extraction: A fluorescein fundoscopic and angiographic study. *Arch. Ophthalmol.* **76**:82–86.

Gass, J. D. M., Anderson, D. R., and Davis, E. B., 1985, A clinical, fluorescein angiographic, and electron microscopic correlation of cystoid macular edema. *Am. J. Ophthalmol.* **100**:82–86.

Gaur, V. P., Eldred, W., and Sarthy, P. V., 1988, Distribution of Müller cells in the turtle retina: An immunocytochemical study. *J. Neurocytol.* **17**:683–692.

Gautron, S., Dos Santos, G., Pinto-Henrique, D., Koulakoff, A., Gros, F., and Berwald-Netter, Y., 1992, The glial voltage-gated sodium channel: Cell- and tissue-specific mRNA expression. *Proc. Natl. Acad. Sci. USA* **89**:7272–7276.

Geiger, B., and Ayola, O., 1992, Cadherins. *Annu. Rev. Cell Biol.* **8**:307–322.

Geiger, K., Howes, E., Gallina, M., Huang, X. J., Travis. G. H., and Sarvetnick, N., 1994, Transgenic mice expressing IFN-gamma in the retina develop inflammation of the eye and photoreceptor loss. *Invest. Ophthalmol. Vis. Sci.* **35**:2667–2681.

Germer, A., Biedermann, B., Wolburg, H., Schuck, J., Grosche, J., Kuhrt, H., Reichelt, W., Schousboe, A., Paasche, G., Mack, A. F., and Reichenbach, A., 1998a, Distribution of mitochondria within Müller cells. I. Correlation with retinal vascularization in different mammalian species. *J. Neurocytol.* **27**:329–345.

Germer, A., Schuck, J., Wolburg, H., Kuhrt, H., Mack, A. F., and Reichenbach, A., 1998b, Distribution of mitochondria within Müller cells. II. Post-natal development of the rabbit retinal periphery *in vivo* and *in vitro*: Dependence on oxygen supply. *J. Neurocytol.* **27**:347–359.

Gery, I., Mochizuki, M., and Nussenblatt, R. B., 1986, Retinal specific antigens and immunopathogenic processes they provoke. *Prog. Ret. Res.* **5**:75–109.

Giulian, D., Baker, T. J., Shih, L. N., and Lachman, L. B., 1986, Interleukin-1 of the central nervous system is produced by ameboid microglia. *J. Exp. Biol. Med.* **164**:594–604.

Glees, P., and Meller, K., 1968, Morphology of neuroglia, in: *The Structure and Function of Nervous Tissue*, Volume I (G. Bourne, ed.), Academic Press, New York, pp. 301–323.

Gold, G. H., and Dowling, J. E., 1979, Photoreceptor coupling in retina of toad, *Bufo marinus*. I. Anatomy. *J. Neurophysiol.* **42**:292–310.

Goldman, D. E., 1943, Potential, impedance and rectification in membranes. *J. Gen. Physiol.* **27**:37–60.

Goldman, S. S., 1988, Gluconeogenesis in the amphibian retina: Lactate is preferred to glutamate as the gluconeogenic precursor. *Biochem. J.* **254**:359–365.

Goldman, S. S., 1990, Evidence that the gluconeogenic pathway is combined to an enriched Müller cell fraction derived from the amphibian retina. *Exp. Eye Res.* **50**:213–218.

Goldstein, E. B., 1967, Early receptor potential of the isolated frog (*Rana pipiens*) retina. *Vis. Res.* **7**:837–845.

Goldstein, E. B., 1970, Cone pigment regeneration in the isolated frog retina. *Vis. Res.* **10**:1065–1068.

Goldstein, I. M., Ostwald, P., and Roth, S., 1996, Nitric oxide: A review of its role in retinal function and disease. *Vis. Res.* **36**:2979–2994.

Golgi, C., 1886, Sulla fine anatomia degli organi centrali del sisterna nervoso Milano: Hoepli.

Gomi, H., Yokoyama, T., Fujimoto, K., Ikeda, T., Katoh, A., Itoh, T., and Itohara, S., 1995, Mice devoid of the glial fibrillary acidic protein develop normally and are susceptible to scrapie prions. *Neuron* **14**:29–41.

Goodman, G., and Ripps, H., 1960, Electroretinography in the differential diagnosis of visual loss in children. *Arch. Ophthalmol.* **64**:221–235.

Gorovits, R., Ben Dror, I., Fox, L. E., Westphal, H. M., and Vardimon, L., 1994, Developmental changes in the expression and compartmentalization of the glucocorticoid receptor in embryonic retina. *Proc. Natl. Acad. Sci. USA* **91**:4786–4790.

Gorovits, R., Avidan, N., Avisar, N., Shaked, I., and Vardimon, L., 1997, Glutamine synthetase protects against neuronal degeneration in injured retinal tissue. *Proc. Natl. Acad. Sci. USA* **94**:7024–7029.

Gotow, T., and Hashimoto, P. H., 1988, Deep-etch structure of astrocytes at the superficial glia limitans with special emphasis on the internal and external organization of their plasma membranes. *J. Neurocytol.* **17**:399–413.

Gotow, T., and Hashimoto, P. H., 1989, Orthogonal arrays of particles in plasma membranes of Müller cells in the guinea pig retina. *Glia* **2**:273–285.

Gouras, P., 1958, Spreading depression of activity in amphibian retina. *Am. J. Physiol.* **195**:28–32.

Goureau, O., Lepoivre, M., and Courtois, Y., 1992, Lipopolysaccaride and cytokines induce a macrophage-type of nitric oxide synthase in bovine retinal pigmented epithelial cells. *Biochem. Biophys. Res. Comm.* **186**:854–859.

Goureau, O., Hicks, D., and Courtois, Y., 1994a, Human retinal pigmented epithelial cells produce nitric oxide in response to cytokines. *Biochem. Biophys. Res. Comm.* **198**:120–126.

Goureau, O., Hicks, D., Courtois, Y., and De Kozak, Y., 1994b, Induction and regulation of nitric oxide synthase in retinal Müller glial cells. *J. Neurochem.* **63**:310–317.

Goureau, O., Régnier-Ricard, F., and Courtois, Y., 1999, Requirement for nitric oxide in retinal neuronal cell death induced by activated Müller glial cells. *J. Neurochem.* **72**:2506–2515.

Graeber, M. B., Streit, W. J., and Kreutzberg, G. W., 1988, The microglial cytoskeleton: Vimentin is localized within activated cells in situ. *J. Neurocytol.* **17**:573–580.

Grafstein, B., 1956, Mechanism of spreading cortical potential. *J. Neurophysiol.* **19**:154–171.

Grant, G. B., and Dowling, J. E., 1995, A glutamate-activated chloride current in cone-driven ON bipolar cells of the white perch retina. *J. Neurosci.* **15**:3852–3862.

Graymore, C., 1970, Biochemistry of the retina, in: *Biochemistry of the Eye* (C. Graymore, ed.), Academic Press Inc., London, pp. 645–735.

Green, W. R., and Robertson, D. M., 1991. Pathological findings of photic retinopathy in the human eye. *Am. J. Ophthalmol.* **112**:520–527.

Griffith, O. W., and Meister, A., 1978, Differential inhibition of glutamine and gamma-glutamylcysteine synthetases by alpha-alkyl analogs of methionine sulfoximine that induce convulsions. *J. Biol. Chem.* **253**:2333–2338.

Grisar, T., Frere, J.-M., and Franck, G., 1979, Effects of K^+ ions on kinetic properties of the (Na^+, K^+)-ATPase (EC 3.6.1.3) of bulk isolated glial cells, perikarya and synaptosomes from rabbit brain cortex. *Brain Res.* **165**:87–103.

Grondona, J. M., Kastner, P., GansMüller, A., Decimo, D., Chambon, P., and Mark, M., 1996, Retinal dysplasia and degeneration in RARβ2/RARγ2 compound mutant mice. *Development* **122**:2173–2188.

Grosche, J., Hartig, W., and Reichenbach, A., 1995, Expression of glial fibrillary acidic protein (GFAP), glutamine synthetase (GS), and Bcl2 protooncogene protein by Müller (glial) cells in retinal light damage of rats. *Neurosci. Lett.* **185**:119–122.

Grossman, R., Fox, L. E., Gorovits, R., Ben Dror, I., Reisfeld, S., and Vardimon, L., 1994, Molecular basis for differential expression of glutamine synthetase in retina glia and neurons. *Mol. Brain Res.* **21**:312–320.

Grunwald G. B., 1996, Cadherin cell adhesion molecules in retinal development and pathology. *Prog. Ret. Eye Res.* **15**:363–392.

Guastella, J., Nelson, N., Nelson, H., Czyzyk, L., Keynan, S., Miedel, M. C., Davidson, N., Lester, H., and Kanner, B., 1990, Two pharmacologically distinct sodium- and chloride-coupled high affinity gamma-aminobutyric acid transporters are present in plasma membrane vesicles and reconstituted preparations from brain. *Science* **249**:1303–1306.

Guastella, J., Brecha, N., Weigmann, C., and Lester, H. A., 1992, Cloning expression and localization of a rat brain high affinity glycine transporter. *Proc. Natl. Acad. Sci. USA* **89**:7189–7193.

Guerin, C. J., Anderson, D. H., and Fisher, S. K., 1990a, Changes in intermediate filament immunolabeling occur in response to retinal detachment and reattachment in primates. *Invest. Ophthalmol. Vis. Sci.* **31**:1474–1482.

Guerin, C. J., Wolfshangen, R. W., Eifring, D. E., and Anderson, D. H., 1990b, Immunocytochemical identification of Müller glia as a component of human epiretinal membranes. *Invest. Ophthalmol. Vis. Sci.* **31**:1483–1491.

Gumbiner, B. M., 1996, Cell adhesion: The molecular basis of tissue architecture and morphogenesis. *Cell* **84**:345–357.

Gurevich, L., and Slaughter, M. M., 1993, Comparison of the waveforms of the ON bipolar neuron and the *b*-wave of the electroretinogram. *Vis. Res.* **33**:2431–2435.

Hageman, G. S., Kirhoff-Rempe, M. A., Lewis, G. P., Fisher, S. K., and Anderson, D. H., 1991, Sequestration of basic fibroblast growth factor in the primate retinal interphotoreceptor matrix. *Proc. Natl. Acad. Sci. USA* **88**:6706–6710.

Hagemann, G. S., and Johnson, L. V., 1984, Lectin-binding glycoproteins in the vertebrate vitreous body and inner limiting membrane: Tissue localization and biochemical characterization. *J. Cell Biol.* **99**:179a.

Hahn, J. S., Aizenman, E., and Lipton, S. A., 1988, Central mammalian neurons normally resistant to glutamate toxicity are made sensitive by elevated extracellular Ca^{2+}: Toxicity is blocked by the N-methyl-D-aspartate antagonist MK 801. *Proc. Natl. Acad. Sci. USA* **85**:6556–6560.

Halfter, W., and Schurer, B., 1994, A new heparan-sulfate proteoglycan in the extracellular matrix of the developing chick embryo. *Exp. Cell Res.* **214**:285–296.

Halfter, W., Diamantis, I., and Monard, D., 1988, Migratory behavior of cells on embryonic retina basal lamina. *Dev. Biol.* **130**:259–275.

Hall, H., and Schachner, M., 1998, Laminins and their ligands: Involvement of carbohydrates in formation of the extracellular matrix and in cell adhesion. *Trends Glycosci. Glyc.* **10**:361–382.

Hamberger, A. C., Chiang, G. H., Nylen, E. S., Scheff, S. W., and Cotman, C. W., 1979, Glutamate as a CNS transmitter. I. Evaluation of glucose and glutamine as precursors for the synthesis of preferentially released glutamate. *Brain Res.* **168**:513–530.

Hamill, O., Marty, A., Neher, E., Sakmann, B., and Sigworth, F. J., 1981, Improved patch-clamp techniques for high-resolution current recording from cells and cell-free membrane patches. *Pflügers Arch.* **391**:85–100.

Hamprecht, B., and Dringen, R., 1996, Energy metabolism, in: *Neuroglia* (H. Kettenman and B. R. Ransom, eds.), Oxford University Press, Oxford, pp. 473–487.

Hanani, M., Zamir, O., and Baluk, P., 1989, Glial cells in the guinea pig myenteric plexus are dye coupled. *Brain Res.* **497**:245–249.

Hanawa, I., Kuge, K., and Matsumura, K., 1968, Mechanism of the slow depressive potential production in the isolated frog retina. *Jpn. J. Physiol.* **18**:59–70.

Hanitzsch, R., Lichtenberger, T., and Mättig, W.-U., 1996, The influence of $MgCl_2$ and APB on the light-induced potassium changes and the ERG *b*-wave of the isolated superfused rat retina. *Vis. Res.* **36**:499–507.

Hansen, A. J., Lund-Andersen, H., and Crone, C., 1977, K^+-permeability of the blood-brain barrier, investigated by aid of a K^+-sensitive microelectrode. *Acta Physiol. Scand.* **101**:438–445.

Hansson, E., and Rönnbäck, L., 1991, Receptor regulation of the glutamate, GABA and taurine high-affinity uptake into astrocytes in primary culture. *Brain Res.* **548**:215–221.

Hansson, H. P. J., 1967, Histochemical demonstration of carbonic anhydrase activity. *Histochemie* **11**:112–128.

Harada, T., Harada, C., Watanabe, M., Inoue, Y., Sakagawa, T., Nakayama, N., Sasaki, S., Okuyama, S., Watase, K., Wada, K., and Tanaka, K., 1998, Functions of two glutamate transporters GLAST and GLT-1 in the retina. *Proc. Natl. Acad. Sci. USA* **95**:4663–4666.

Harris, G. S., and Yeung, J., 1976, Maculopathy of sex-linked juvenile retinoschisis. *Can. J. Ophthalmol.* **11**:1–10.

Harris, W. A., 1997, Cellular diversification in the vertebrate retina. *Curr. Opin. Genet. Cell Biol.* **7**:651– 658.

Härtig, W., Grosche, J., Distler, C., Grimm, D., El-Hifnawi, E., and Reichenbach, A., 1995, Alterations of Müller (glial) cells in dystrophic retinae of RCS rats. *J. Neurocytol.* **24**:507–517.

Hatten, M. E., 1990, Riding the glial monorail: A common mechanism for glial-guided neuronal migration in different regions of the developing brain. *Trends Neurosci.* **13**:179–184.

Hausman, R. E., Krishna Rao, A. S. M., Ren, Y., Sagar, G. D. V., & Shah, B. H., 1993, Retina cognin, cell signaling, and neuronal differentiation in the developing retina. *Dev. Dynamics* **196**:263–266.

He, S., Hofman, F., Lopez, P., Ryan, S. J., and Hinton, D. R., 1997, Thrombospondin-1 localization in human retina and choroidal neovascular membranes. *Invest. Ophthalmol. Vis. Sci.* **38**:S353.

Heegard, S., 1994, Structure of the human vitreo-retinal border region. *Ophthalmologica* **208**:82–91.

Heegard, S., 1997, Morphology of the vitreoretinal border region. *Acta Ophthalmol. Scand.* **75**: 7–31.

Heegard, S., Jensen, O. A., and Praue, J. U., 1986, Structure and composition of the inner limiting membranes of the retina. SEM on frozen resin-cracked and enzyme-digested retinas of *Macaca mulatta. Graefés Arch. Klin. Exp. Ophthalmol.* **224**:355–360.

Heinemann, U., and Lux, H. D., 1977, Ceiling of stimulus induced rises in extracellular potassium concentration in the cerebral cortex of cat. *Brain Res.* **120**:231–249.

Henn, F. A., and Hamberger, A., 1971, Glial cell function: Uptake of transmitter substances. *Proc. Natl. Acad. Sci. USA* **68**:2686–2690.

Henn, F. A., Goldstein, M. N., and Hamberger, A., 1974, Uptake of neurotransmitter candidate glutamate by glia. *Nature* **249**:663–664.

Henschel, D. S., and Miller, R. F., 1992, Catecholamine effects on dissociated tiger salamander Müller (glial) cells. *Brain Res.* **57**:208–214.

Hertz, L., Yu, A. C. H., Svenneby, G., Kvamme, E., Fosmark, H., and Schousboe, A., 1980, Absence of a preferential glutamine uptake into neurons—an indication of a net transfer of TCA constituents from nerve endings to astrocytes. *Neurosci. Lett.* **16**:103–109.

Hibbs, J. B., Taintor, R. R., Vavrin, Z., and Rachlin, E. M., 1988, Nitric oxide: A cytotoxic activated macrophage effector molecule. *Biochem. Biophys. Res. Comm.* **157**:87–94.

Hickey, W. F., and Kimura, H., 1988, Perivascular glial cells of the CNS are bone marrow-derived and present antigen in vivo. *Science* **239**:290–292.

Hicks, D., and Courtois, Y., 1990, The growth and behavior of rat retinal Müller cells in vitro 1. An improved method for isolation and culture. *Exp. Eye Res.* **51**:119–129.

Hille, B., 1992, *Ionic Channels of Excitable Membranes*, 2nd edition, Sinauer Associates, Sunderland, MA, p. 607.

Hinds, J. W., and Hinds, P. L., 1978, Early development of amacrine cells in the mouse retina: An electronmicroscopic, serial section analysis. *J. Comp. Neurol.* **179**:277–300.

Hiscott, P., Waller, H. A., Grierson, I., Butler, M. G., and Scott, D., 1992, Local production of fibronectin by ectopic human retinal cells. *Cell Tiss. Res.* **267**:185–192.

Hjelmeland, L. M., and Harvey, A. K., 1988, Gliosis of the mammalian retina: Migration and proliferation of retinal glia. *Prog. Ret. Res.* **7**:259–281.

Hock, P. A., and Marmor, M. F., 1983, Variability of the human c-wave. *Doc. Ophthalmol.* **37**:151–157.

Hodgkin, A. L., and Horowicz, P., 1959, The influence of potassium and chloride ions on the membrane potential of single muscle fibers. *J. Physiol. (Lond.)* **148**:127–160.

Hof, P. R., Pascale, E., and Magistretti, P. J., 1988, K^+ at concentrations reached in the extracellular space during neuronal activity promotes a Ca^{2+}-dependent glycogen hydrolysis in mouse cerebral cortex. *J. Neurosci.* **8**:1922–1928.

Hoff, M., and Gouras, P., 1973, Tolerance of mammalian retina to circulatory arrest. *Doc. Ophthalmol. Proc. Series* **2**:57–63.

Hofman, F. M., and Hinton, D. R., 1992, Tumor necrosis factor-alpha in the retina in acquired immune deficiency syndrome. *Invest. Ophthalmol. Vis. Sci.* **33**:1829–1835.

Hofmann, S., Friedlander, D. R., Choung, C.-M., Grument, M., and Edelman, G. M., 1986, Differential contributions of Ng-CAM and N-CAM to cell adhesion in differential neural regions. *J. Cell Biol.* **103**:145–158.

Hogan, M. J., and Feeney, L., 1963, The ultrastructure of the retinal blood vessels. 1. The large vessels. *J. Ultrastructure Res.* **9**:10–28.

Hogan, M. J., Alvarado, J. A., and Weddell, J. E., 1971, *Histology of the Human Eye*, W.B. Saunders, Philadelphia, p. 687

Holländer, H., Makarov, F., Dreher, Z., van Driel, D., Chan-Ling, T., and Stone, J., 1991, Structure of the macroglia of the retina: Sharing and division of labour between astrocytes and Müller cells. *J. Comp. Neurol.* **313**:587–603.

Holt, C. E., Bertsch, T. W., Ellis, H. M., and Harris, W. A., 1988, Cellular determination in the *Xenopus* retina is independent of lineage and birth date. *Neuron* **1**:15–26.

Honda, S., Yamamoto, M., and Saito, N., 1995, Immunocytochemical localization of three subtypes of GABA transporter in rat retina. *Mol. Brain Res.* **33**:319–325.

Hood, D. C., and Birch, D. G., 1990, A quantitative measure of the electrical activity of human rod photoreceptors using electroretinography. *Vis. Neurosci.* **5**:379–387.

Hope, B. T., Michael, G. J., Knigge, K. M., and Vincent, S. R., 1991, Neuronal NADPH diaphorase is a nitric oxide synthase. *Proc. Natl. Acad. Sci. USA* **88**:2811–2814.

Hoth, M., and Penner, R., 1992, Depletion of intracellular calcium stores activates a calcium current in mast cells. *Nature* **355**:353–356.

Howell, G. A., Welch, M. G., and Frederickson, C. J., 1984, Stimulation-induced uptake and release of zinc in hippocampal slices. *Nature* **308**:736–738.

Huang, P. L., Dawson, P. M., Bredt, D. S., Snyder, S. H., and Fishman, M. C., 1993, Targeted disruption of the neuronal nitric oxide synthase gene. *Cell* **75**:1273–1286.

Huck, S., Grass, F., and Hortnagl, H., 1984, The glutamate analog α-aminoadipic aid is taken up by astrocytes before exerting its gliotoxic effect in vitro. *J. Neurosci.* **10**:2650–2657.

Hughes, B. A., and Steinberg, R. H., 1990, Voltage-dependent currents in isolated cells of the frog retinal pigment epithelium. *J. Physiol. (Lond.)* **428**:273–297.

Hume, D. A., Perry, V., and Gordon, S., 1983, Immunohistochemical localization of a macrophage-specific antigen in developing mouse retina: Phagocytosis of dying neurons and differentiation of microglial cells to form a regular array in the plexiform layers. *J. Cell Biol.* **97**:253–257.

Hume, R. I., Dingledine, R., and Heinemann, S. F., 1991, Identification of a site in glutamate receptor subunits that controls calcium permeability. *Science* **253**:1028–1031.

Humphrey, M. F., Constable, I. J., Chu, Y., and Wiffen, S., 1993, A quantitative study of the lateral spread of Müller cell responses to retinal lesions in the rabbit. *J. Comp. Neurol.* **334**: 545–558.

Hunter, D. D., Murphy, M. D., Olsson, C. V., and Brunken, W. J., 1992, S-laminin expression in adult and developing retinae: A potential cue for photoreceptor morphogenesis. *Neuron* **8**:399–413.

Hutchinson, B. T., and Kuwabara, T., 1962, Phosphorylase and uridine diphosphoglucose glycogen synthetase in the retina. *Arch. Ophthalmol.* **68**:538–545.

Huxlin, K. R., 1995, NADPH-diaphorase expression in neurons and glia of the normal adult rat retina. *Brain Res.* **692**:195–206.

Huxlin, K. R., Sefton, A. J., and Furby, J. H., 1992, The origin and development of retinal astrocytes in the mouse. *J. Neurocytol.* **21**:530–544.

Huxlin, K. R., Dreher, Z., Schulz, M., and Dreher, B., 1995, Glial reactivity in the retina of adult rats. *Glia* **15**:105–118.

Hyatt, G. A., and Dowling, J. E., 1997, Retinoic acid: A key molecule for eye and photoreceptor development. *Invest. Ophthalmol. Vis. Sci.* **38**:1471–1475.

Hyatt, G. A., Schmitt, E. A., Fadool, J. M., and Dowling, J. E., 1996, Retinoic acid alters photoreceptor development in vivo. *Proc. Natl. Acad. Sci. USA* **93**:13298–13303.

Hyde, J. C., and Robinson, N., 1974, Localization of sites of GABA catabolism in the retina. *Nature* **248**:432–443.

Hydén, H., 1961, Satellite cells in the nervous system. *Sci. Am.* **205**:62–70.

Hydén, H., and Lange, P., 1960, Differences in the metabolism of oligodendroglia and nerve cells in the vestibular area, in: *Fourth International Neurochemical Symposium*, Pergamon Press, Oxford, pp. 190–199.

Hydén, H., and Lange, P. W., 1962, A kinetic study of the neuron-glia relationship. *J. Cell Biol.* **13**:233–237.

Hynes, R. O., and Lander, A. D., 1992, Contact and adhesive specificities in the associations, migrations, and targeting of cells and axons. *Cell* **68**:303–322.

Ikeda, T., and Puro, D. G., 1994, Nerve growth factor: A mitogenic signal for retinal Müller glial cells. *Brain Res.* **649**:260–264.

Ikeda, T., and Puro, D. G., 1995, Regulation of retinal glial cell proliferation by antiproliferative molecules. *Exp. Eye Res.* **60**:435–444.

Ikeda, T., Homma, Y., Nisida, K., Hirase, K., Sotozona, C., Kinoshira, S., and Puro, D. G., 1998, Expression of transforming growth factor-betas and their receptors by human retinal glial cells. *Curr. Eye Res.* **17**:546–550.

Inatani, M., Tanihara, H., Honjo, M., Hangai, M., Kresse, H., and Honda, Y., 1999, Expression of proteoglycan decorin in neural retina. *Invest. Ophthalmol. Vis. Sci.* **40**:1783–1791.

Ingraham, C. A., Cooke, M. P., Chuang, Y.-N., Perlmutter, R. M., and Maness, P. F., 1992, Cell type and developmental regulation of the *fyn* proto-oncogene in neural retina. *Oncogene* **7**:95–100.

Inuzuka, H., Miyatani, S., and Takeichi, M., 1991, R-cadherin—a novel Ca^{2+}-dependent cell cell-adhesion molecule expressed in the retina. *Neuron* **7**:69–79.

Ishida, A. T., and Fain, G. L., 1981, D-aspartate potentiates the effects of L-glutamate on horizontal cells in goldfish retina. *Proc. Natl. Acad. Sci. USA* **78**:5890–5894.

Ishigooka, H., Hirata, A., Kitaoka T., and Ueno, S., 1989, Cytochemical studies on pathological Müller cells after argon laser photocoagulation. *Invest. Ophthalmol. Vis. Sci.* **30**:509–520.

Ishii, M., Horio, Y., Tada, Y., Hibino, H., Inanobe, A., Ito, M., Yamada, M., Gotow, T., Uchiyama, Y., and Kurachi, Y., 1997, Expression and clustered distribution of an inwardly rectifying potassium channel, K_{AB}-2/Kir4.1, on mammalian retinal Müller cell membrane: Their regulation by insulin and laminin signals. *J. Neurosci.* **17**:7725–7735.

Ishimoto, S.-I., Zhang, J., Gullapalli, V. K., Pararajasegaram, G., and Rao, N. A., 1999, Antigen-presenting cells in experimental autoimmune uveoretinitis. *Exp. Eye Res.* **67**:539–548.

Izumi, Y., Benz, A. M., Zorumski, C. F., and Olney, J. W., 1994, Effects of lactate and pyruvate on glucose deprivation in rat hippocampal slices. *NeuroReport* **5**:617–620.

Izumi, Y., Benz, A. M., Kirby, C. O., Labruyere, J., Zorumski, C. F., Price, M. T., and Olney, J. W., 1995, An ex vivo rat retinal preparation for excitotoxicity studies. *J. Neurosci. Meth.* **60**:215–219.

Izumi, Y., Kirby, C. O., Benz, A. M., Olney, J. W., and Zorumski, C. F., 1999, Müller cell swelling, glutamate uptake, and excitotoxic neurodegeneration in the isolated rat retina. *Glia* **25**:379–389.

Jacobiec, F. A., 1982, *Ocular Anatomy, Embryology, and Teratology*, Harper and Row, Philadelphia.

Jakobiec, F. A., Brodie, S. E., Haik, B., and Iwamoto, T., 1983, Giant cell astrocytoma of the retina: A tumor of possible Müller cell origin. *Ophthalmology* **90**:1565–1576.

Jampol, L. M., 1994, Macular edema, in: *Retina*, Volume 2 (S. J. Ryan, ed.), C.V. Mosby, St. Louis, MO, pp. 999–1007.

Jan, L. Y., and Jan, Y. N., 1990, How might the diversity of potassium channels be generated? *Trends Neurosci.* **13**:415–418.

Jan, L. Y., and Jan, Y. N., 1994, Potassium channels and their evolving gates. *Nature* **371**:119–122.

Janzer, R. C., and Raff, M. C., 1987, Astrocytes induce blood-brain barrier properties in endothelial cells. *Nature* **325**:253–257.

Jasoni, C. L., and Reh, T. A., 1996, Temporal and spatial pattern of MASH-1 expression in the developing rat retina demonstrates progenitor cell heterogeneity. *J. Comp. Neurol.* **369**:319–327.

Jensen, A. M., and Chiu, S. Y., 1990, Fluorescence measurement of changes in intracellular calcium induced by excitatory amino acids in cultured cortical astrocytes. *J. Neurosci.* **10**:1165–1175.

Jensen, A. M., and Chiu, S. Y., 1991, Differential intracellular calcium responses to glutamate in type 1 and type 2 cultured brain astrocytes. *J. Neurosci.* **11**:1674–1684.

Jensen, A. M., and Wallace, V. A., 1997, Expression of Sonic hedgehog and its putative role as a precursor cell mitogen in the developing mouse retina. *Development* **124**:363–371.

Jeon, C.-J., Strettoi, E., and Masland, R. H., 1998, The major cell populations of the mouse retina. *J. Neurosci.* **18**:8936–8946.

Jerdan, J. A., Kao, L., and Glaser, B. M., 1986, The inner limiting membrane: A modified basement membrane? *Invest. Ophthalmol. Vis. Sci.* **27**:S230.

Johanson, N. F., 1977, Retinal glycogen content during ischemia. *Graefés Arch. Klin. Exp. Ophthalmol.* **203**:271–282.

Johns, P. R., 1982, Formation of photoreceptors in larval and adult goldfish. *J. Neurosci.* **2**:178–198.

Johnson, J., Chen, T. K., Rickman, D. W., Evans, C. and Brecha, N. C., 1996, Multiple gamma-aminobutyric acid plasma membrane transporters (GAT-1, GAT-2, GAT-3) in the rat retina. *J. Comp. Neurol.* **375**:212–224.

Johnson, W. B., Ruppe, M. D., Rockenstein, E. M., Price, J., Sarthy, V. P., Verderber, L. C., and Mucke, L., 1995, Indicator expression directed by regulatory sequences of the glial fibrillary acidic protein (GFAP) gene: In vivo comparison of distinct GFAP-lacZ transgenes. *Glia* **13**:174–184.

Johnston, G. A. R., Curtis, D. R., Groat, W. C., and Duggan, A. W., 1968, Central action of ibotenic acid and muscimol. *Biochem. Pharmacol.* **17**:2488–2489.

Jones, G. J., Crouch, R. K., Wiggert, B., Cornwall, M. C., and Chader, G. J., 1989, Retinoid requirements for recovery of sensitivity after visual-pigment bleaching in isolated photoreceptors. *Proc. Natl. Acad. Sci. USA* **86**:9606–9610.

Kahn, M. A., Ellison, J. A., Speight, J., and DeVellis, J., 1995, CNTF regulation of astrogliosis and the activation of microglia in the developing rat central nervous system. *Brain Res.* **685**:55–67.

Kaila, K., Panula, P., Karhunen, T., and Heinonen, E., 1991, Fall in intracellular pH mediated by $GABA_A$ receptors in cultured rat astrocytes. *Neurosci. Lett.* **126**:9–12.

Kalloniatis, M., Tomisich, G., and Marc, R. E., 1994, Neurochemical signatures revealed by glutamine labeling in the chicken retina. *Vis. Neurosci.* **11**:793–804.

Kanai, Y., and Hediger, M. A., 1992, Primary structure and functional characterization of a high affinity glutamate transporter. *Nature* **360**:467–471.

Kanner, B. I., 1983, Bioenergetics of neurotransmitter transport. *Biochim. Biophys. Acta* **726**:293–316.

Kanner, B. I., 1994, Sodium-coupled neurotransmitter transport: Structure, function and regulation. *J. Exp. Biol.* **196**:237–249.

Kanner, B. I., and Bendahan, A., 1982, Binding order of substrates to the sodium and potassium ion coupled L-glutamate transporter from rat brain. *Biochemistry* **21**:6327–6330.

Kanner, B. I., and Sharon, I., 1978, Active transport of L-glutamate by membrane vesicles isolated from rat brain. *Biochemistry* **17**:3949–3953.

Karschin, A., Wässle, H., and Schnitzer, J., 1986, Immunocytochemical studies on astroglia of the cat retina under normal and pathological conditions. *J. Comp. Neurol.* **249**:564–576.

Karwoski, C. J., and Proenza, L. M., 1977, Relationship between Müller cell responses, a local transretinal potential, and potassium flux. *J. Neurophysiol.* **40**:244–259.

Karwoski, C. J., and Proenza, L. M., 1978, Light-evoked changes in extracellular potassium concentration in mudpuppy retina. *Brain Res.* **142**:515–530.

Karwoski, C. J., Newman, E. A., Shimazaki, H., and Proenza, L. M., 1985, Light-evoked increases in extracellular K^+ in the plexiform layers of amphibian retinas. *J. Gen. Physiol.* **86**:189–213.

Karwoski, C. J., Lu, H.-K., and Newman, E. A., 1989, Spatial buffering of light-evoked potassium increases by retinal Müller (glial) cells. *Science* **244**:578–580.

Karwoski, C. J., Xu, X., and Yu, H. (1996). Current-source density analysis of the electroretinogram of the frog: Methodological issues and origin of components. *J. Opt. Soc. Am.* **13**:549–556.

Kato, S., Sugawara, K., Matsukawa, T., and Negishi, K., 1990, Gliotoxic effects of alpha-aminoadipic acid isomers on the carp retina: A long-term observation. *Neurosci.* **36**: 145–153.

Kato, S., Ishita, K., Sugawara, K., and Mawatari, K., 1993, Cystine/glutamate antiporter expression in retina Müller glial cells: Implications for DL-alpha-amino adipate toxicity. *Neurosci.* **57**:473–482.

Katz, B. J., Zhian, X., Zheng, J., and Oakley, B., II, 1992, PII component of the toad electroretinogram. *J. Neurophysiol.* **68**:333–341.

Kayyem, J. F., Roman, J. M., Vonboxberg, Y., Schwarz, U., and Dryer, W. J., 1992, Bravo/Nr-CAM is closely related to the cell adhesion molecules L1 and Ng-CAM and has a similar heterodimer structure. *J. Cell Biol.* **118**:1259–1270.

Keirstead, S. A., and Miller, R. F., 1997, Metabotropic glutamate receptor agonists evoke calcium waves in isolated Müller cells. *Glia* **21**:194–203.

Kelley, M. W., Turner, J. K., and Reh, T. A., 1994, Retinoic acid promotes differentiation of photoreceptors in vitro. *Development* **120**:2091–2102.

Kellner, U., Kraus, H., Heimann, H., Helbig, H., Bornfeld, N., and Foerster, M. H., 1998, Electrophysiological evaluation of visual loss in Müller cell sheen dystrophy. *Br. J. Ophthalmol.* **82**:650–654.

Kelly, M. M., Phanhthourath, C., Brees, D. K., McCabe, C. F., and Cole, G. J., 1995, Molecular characterization of EAP-300: A high molecular weight, embryonic polypeptide containing an aminoacid repeat comprised of multiple leucine-zipper motifs. *Dev. Brain Res.* **85**:31–47.

Kettenmann, H., and Ransom, B. R. (eds.), 1995, *Neuroglia,* Oxford University Press: Oxford.

Keynan, S., and Kanner, B. I., 1988, Gamma-aminobutyric acid transport in reconstituted preparations from rat brain: Coupled sodium and chloride fluxes. *Biochem.* **27**:12–17.

Kim, I.-B., Lee, E.-J., Kim, K.-Y., Ju, W.-K., Oh, S.-J., Joo, C.-K., and Chun, M.-H., 1999, Immunocytochemical localization of nitric oxide synthase in the mammalian retina. *Neurosci. Lett.* **267**:193–196.

Kim, M. K., Chan, C. C., Belfort, R., Farah, M., Burnier, M. P., Nussenblatt, R. B., Kuwabara, T., and Palestine, A. G., 1987, Histopathologic and immunopathologic features of subretinal fibrosis and uveitis syndrome. *Am. J. Ophthalmol.* **104**:15–23.

Kimelberg, H. K., Biddlecome, S., and Bourke, R. S., 1979, STS-inhibitable Cl^- transport and Na^+ dependent H^+ production in primary astroglial cultures. *Brain Res.* **173**:111–124.

Kimelberg, H. K., Jalonen, T., and Walz, W., 1993, Regulation of brain microenvironment: Transmitters and ions, in: *Astrocytes: Pharmacology and Function* (S. Murphy, ed.), Academic Press, San Diego, CA, pp. 193–228.

Kirsch, M., Fuhrmann, S., Wiese, A., and Hofman, H. D., 1996, CNTF exerts opposite effects on in vitro development of rat and chick photoreceptors. *NeuroReport* **7**:697–700.

Kish, P. E., Fischer-Bovenkerk, C., and Ueda, T., 1989, Active transport of γ-aminobutyric acid and glycine into synaptic vesicles. *Proc. Natl. Acad. Sci. USA* **86**:3877–3881.

Kline, R. P., Ripps, H., and Dowling, J. E., 1978, Generation of b-wave currents in the skate retina. *Proc. Natl. Acad. Sci. USA* **75**:5727–5731.

Kline, R. P., Ripps, H., and Dowling, J. E., 1985, Light-induced potassium fluxes in the skate retina. *Neuroscience.* **14**:225–235.

Kljavin, I. J., and Reh, T. A., 1991, Müller cells are a preferred substrate for in vitro neurite extension by rod photoreceptor cells. *J. Neurosci.* **11**:2985–2991.

Kljavin, I. J., Lagenaur, C., Bixby, J. L., and Reh, T. A., 1994, Cell-adhesion molecules regulating neurite growth from amacrine and rod photoreceptor cells. *J. Neurosci.* **14**:5035–5049.

Knapp, A. G., and Schiller, P. H., 1984, The contribution of on-bipolar cells to the electroretinogram of rabbits and monkeys. *Vis. Res.* **24**:1841–1846.

Kobayashi, S., Morgans, C. W., Casey, J. R., and Kopito, R. R., 1994, AE3 anion exchanger isoforms in the vertebrate retina: Developmental regulation and differential expression in neurons and glia. *J. Neurosci.* **14**:6266–6279.

Kofuji, P., Ceelen, P., Zahs, K. R., Surbeck, L. W., Lester, H., and Newman, E. H., 2000, Genetic inactivation of an inwardly rectifying potassium channel (K_{ir}4.1 subunit) in mice: Phenotypic impact in retina. *J. Neurosci.* **20**:5733–5740.

Koh, S. W., and Roberge, F. C., 1989, VIP modulation of cultured glial cells of the rat retina. *Curr. Eye Res.* **8**:1207–1210.

Kohner, E. M., 1994, Retinal ischemia, in: *Retina,* Volume 2 (S. J. Ryan, ed.), C.V. Mosby, St. Louis, MO, pp. 1009–1018.

Kohno, T., Sorgente, N., Patterson, R., and Ryan, S. J., 1983, Fibronectin and laminin distribution in bovine eye. *Jpn. J. Ophthalmol.* **27**:496–505.

Kohno, T., Sorgente, N., Ishibashi, T., Goodnight, R., and Ryan, S. J., 1987, Immunofluorescent studies of fibronectin and laminin in the human eye. *Invest. Ophthalmol. Vis. Sci.* **28**:506–514.

Kolliker, A., 1854, Mikroskopische Anatomie oder Gewebelehre des Menschen. 2, pt.2, 605–648.

Kondo, H., Takahashi, H., and Takahashi, Y., 1984, Immunohistochemical study of S-100 protein in the postnatal development of Müller cells and astrocytes in the rat retina. *Cell Tissue. Res.* **238**:503–508.

Kono, T., Kohno, T., and Inomata, H., 1995, Epiretinal membrane formation: Light and electron microsopic study in an experimental rabbit model. *Arch. Ophthalmol.* **113**:359–363.

Kopito, R. R., 1990, Molecular biology of the anion exchanger gene family. *Int. Rev. Cytol.* **123**:177–199.

Korhonen, E., and Korhonen, L. K., 1965, Histochemical demonstration of carbonic anhydrase activity in the eyes of rat and mouse. *Acta Ophthalmol.* **43**:475–581.

Korte, G. E., Hageman, G. S., Pratt, D. V., Glusman, S., Marko, M., and Ophir, A., 1992, Changes in Müller cell plasma membrane specializations during subretinal scar formation in the rabbit. *Exp. Eye Res.* **5**:155–162.

Koshibu, A., 1978, Ultrastructural studies on absorption of experimentally produced subretinal hemorrhage. *Acta Ophthalmol. Jpn.* **82**:471.

Kraig, R. P., and Cooper, A. J. L., 1987, Bicarbonate and ammonia changes in brain during spreading depression. *Can. J. Physiol. Pharmacol.* **65**:1099–1104.

Kriegler, S., and Chiu, S. Y., 1993, Calcium signaling of glial cells along mamalian axons. *J. Neurosci.* **13**:4229–4245.

Kroger, S., Hortoin, S. E., and Honig, L. S., 1995, The developing avian retina expresses agrin isoforms during synaptogenesis. *J. Neurobiol.* **29**:165–182.

Kuffler, S. W., 1967, The Ferrier Lecture. Neuroglial cells: Physiological properties and a potassium mediated effect of neuronal activity on the glial membrane potential. *Proc. Roy. Soc. B* **168**:1–21.

Kuffler, S. W., and Potter, D. D., 1964, Glia in the leech central nervous system. Physiological properties and neuron–glia relationship. *J. Neurophysiol.* **27**:290–320.

Kuffler, S. W., Nicholls, J. G., and Orkand, R. K., 1966, Physiological properties of glial cells in the central nervous system of amphibia. *J. Neurophysiol.* **29**:768–787.

Kugler, P., 1993, Enzymes involved in glutaminergic and GABAergic neurotransmission. *Int. Rev. Cytol.* **147**:285–336.

Kugler, P., and Drenckhahn, D., 1996, Astrocytes and Bergmann glia as an important site of nitric oxide synthase I. *Glia* **16**:165–173.

Kuhar, M. J., and Murrin, L. C., 1978, Sodium-dependent, high affinity choline uptake. *J. Neurochem.* **30**:15–21.

Kurenny, D. E., Moroz, L. L., Turner, R. W., Sharkey, K. A., and Barnes, S., 1994, Modulation of ion channels in rod photoreceptors by nitric oxide. *Neuron* **13**:315– 324.

Kurenny, D. E., Thurlow, G. A., Turner, R. W., Moroz, L. L., Sharkey, K. A., and Barnes, S. B., 1995, Nitric oxide synthase in tiger salamander retina. *J. Comp. Neurol.* **361**:525–536.

Kusaka, S., and Puro, D. G., 1997, Intracellular ATP activates inwardly rectifying K^+ channels in human and monkey retinal Müller (glial) cells. *J. Physiol. (Lond.)* **500**:593–604.

Kuwabara, T., 1965, Some aspects of retinal metabolism revealed by histochemistry. Müller cells in the pathological condition, in: *Biochemistry of the Retina* (C. N. Graymore, ed.), Academic Press, New York, pp. 93–97.

Kuwabara, T., and Cogan, D. G., 1959a, Tetrazolium studies in the retina: I. Introduction and technique. *J. Histochem. Cytochem.* **7**:329–333.

Kuwabara, T., and Cogan, D. G., 1959b, Tetrazolium studies in the retina: II. Substrate dependent patterns. *J. Histochem. Cytochem.* **7**:334–341.

Kuwabara, T., and Cogan, D. G., 1960, Tetrazolium studies in the retina: III. Activity of metabolic intermediates and miscellaneous substrates. *J. Histochem. Cytochem.* **8**:214–224.

Kuwabara, T., and Cogan, D. G., 1961, Retinal glycogen. *Arch. Ophthalmol.* **66**:680–688.

Kuwabara, T., and Weidman, T. A., 1974, Development of the prenatal rat retina. *Invest. Ophthalmol.* **13**:725–739.

Kuwabara, T., Cogan, D. G., Futterman, S., and Kinoshita, J. H., 1957, Dehydrogenases in the retina and Müller's fibers. *J. Histochem. Cytochem.* **7**:67–68.

Lam, D. M. K., 1975, Synaptic chemistry of identified cells in the vertebrate retina. *Cold Spring Harbor Symp. Quant. Biol.* **40**:571–579.

Lam, D. M. K., and Steinman, L., 1971, The uptake of γ-^{3}H aminobutyric acid in the goldfish retina. *Proc. Natl. Acad. Sci. USA* **68**:2777–2781.

Lamb, T. D., 1976, Spatial properties of horizontal cell responses in the turtle retina. *J. Physiol.* **263**:239–255.

Lamb, T. D., and Simon, E. J., 1976, The relation between intercellular coupling and electrical noise in turtle photoreceptors. *J. Physiol.* **263**:257–286.

Landis, D. M. D., 1994, The early reactions of non-neuronal cells to brain injury. *Annu. Rev. Neurosci.* **17**:133–151.

Landis, D. M. D., and Reese, T. S., 1974, Arrays of particles in freeze-fractured astrocytic membranes. *J. Cell Biol.* **60**:316–320.

Landis, D. M. D., and Reese, T. S., 1981a, Membrane structure in mammalian astrocytes: A review of freeze-fracture studies on adult, developing, reactive and cultured astrocytes. *J. Exp. Biol.* **95**:35–48.

Landis, D. M. D., and Reese, T. S., 1981b, Astrocyte membrane structure: Changes after circulatory arrest. *J. Cell Biol.* **88**:660–663.

Larrabee, M. G., 1983, Lactate uptake and release in the presence of glucose by sympathetic ganglia of chicken embryos and by neuronal and nonneuronal cultures prepared from these ganglia. *J. Neurochem.* **40**:1237–1250.

Larrabee, M. G., 1992, Extracellular intermediates of glucose metabolism: Fluxes of endogenous lactate and alanine through extracellular pools in embryonic sympathetic ganglia. *J. Neurochem.* **59**:1041–1052.

Lasek, R. J., Gainer, H., and Przybylski, R. J., 1974, Transfer of newly synthesized proteins from Schwann cells to the squid giant axon. *Proc. Natl. Acad. Sci. USA* **71**:1188–1192.

Laties, A. M., 1983, New light on the Müller cell. *Trans. Ophthalmol. Soc. U.K.* **103**:380–384.

Leão, A. A. P., 1944, Spreading depression of activity in the cerebral cortex. *J. Neurophysiol.* **7**:359–390.

Lee, J.-M., Zipfel, G. J., and Choi, D. W., 1999, The changing landscape of ischaemic brain injury mechanisms. *Nature* **399**:A7–A14.

Lee, S. H., Kim, W. T., Cornell-Bell, A. H., and Sontheimer, H., 1994, Astrocytes exhibit regional specificity in gap junction coupling. *Glia* **11**:315–325.

Lee, W.-H., Javedan, S., and Bundy, C. A., 1992, Coordinate expression of insulin-like growth factor system components by neuron and neuroglia during retinal and cerebellar development. *J. Neurosci.* **12**:4737–4744.

Lehre, K. P., Davanger, S., and Danbolt, N. C., 1997, Localization of the glutamate transporter protein GLAST in rat retina. *Brain Res.* **744**:129–137.

Lemmon, V., 1986, Localization of a filamin-like protein in glia of the chick central nervous system. *J. Neurosci.* **6**:43–51.

Lemmon, V., and Reiser, G., 1983, The developmental distribution of vimentin in the chick retina. *Dev. Brain Res.* **11**:191–197.

Lessell, S., and Kuwabara, T., 1964, Phosphatase histochemistry of the eye. *Arch. Ophthalmol.* **71**:851–860.

Letourneau, P. C., Condic, M. L., and Snow, D. M., 1994, Interactions of developing neurons with the extracellular matrix. *J. Neurosci.* **14**:915–928.

Levi, G., and Raiteri, M., 1993, Carrier-mediated release of neurotransmitters. *Trends Neurosci.* **16**:415–419.

Levine, E. M., Roelink, H., Turner, J., and Reh, T. A., 1997, Sonic hedgehog promotes rod photoreceptor differentiation in mammalian retinal cells *in vitro*. *J. Neurosci.* **17**:6277–6288.

Lewis, G. P., Kaska, D. D., Vaughan, D. K., and Fisher, S. K., 1988, An immunocytochemical study of cat retinal Müller cells in culture. *Exp. Eye Res.* **47**:855–868.

Lewis, G. P., Erickson, P. A., Guerin, C. J., Anderson, D. H., and Fisher, S. K., 1989, Changes in the expression of specific Müller cell-proteins during long-term retinal-detachment. *Exp. Eye Res.* **49**:93–111.

Lewis, G. P., Erickson, P. A., Guerin, C. J., Anderson, D. H., and Fisher, S. K., 1992, Basic fibroblast growth factor: A potential regulator of proliferation and intermediate filament expression in the retina. *J. Neurosci.* **12**:3968–3978.

Lewis, G. P., Guerin, C. J., Anderson, D. H., Matsumoto, B., and Fisher, S. K., 1994, Rapid changes in the expression of glial cell proteins caused by experimental retinal detachment. *Am. J. Ophthalmol.* **118**:368–376.

Li, F., Cao, W., Steinberg, R. H., and La Vail, M. M., 1999, Basic FGF-induced down-regulation of IGF-I mRNA in cultured rat Müller cells. *Exp. Eye Res.* **68**:19–27.

Li, L., Sheedlo, H. J., and Turner, J. E., 1993, Müller cell expression of glial fibrillary acidic protein (GFAP) in RPE-transplanted retinas of RCS dystrophic rats. *Curr. Eye Res.* **12**: 841–849.

Li, W. W. Y., Hui, B. S. W., Tsang, D. S. C., and Yew, D. T., 1987, Optic nerve transection and tectal removal affect phagocytic activity of the pigment epithelium in goldfish. *Acta Histochem.* **81**:149–153.

Li, Y.-C., Beard, D., Hayes, S., and Young, A. P., 1995, A transcriptional enhancer of the glutamine synthetase gene that is selective for retinal Müller glial cells. *J. Mol. Neurosci.* **6**:169–183.

Li, Z. Y., Kljavin, I. J., and Milam, A. H., 1995, Rod photoreceptor neurite sprouting in retinitis pigmentosa. *J. Neurosci.* **15**:5429–5438.

Libby, R. T., Xu, Y., Selfors, L. M., Brunken, W. J., and Hunter, D. D., 1997, Identification of the cellular source of laminin beta 2 in adult and developing vertebrate retinae. *J. Comp. Neurol.* **389**:655–667.

Liedtke, W., Edelmann, W., Bieri, P. L., Chiu, F.-C., Cowan, N. J., Kucherlapati, R., and Raine, C. S., 1997, GFAP is necessary for the integrity of CNS white matter architecture and long-term maintenance of myelination. *Neuron* **17**:607–615.

Liepe, B. A., Stone, C., Koistinaho, J., and Copenhagen, D. R., 1994, Nitric oxide synthase in Müller cells and neurons of salamander and fish retina. *J. Neurosci.* **14**:7641–7654.

Lillien, L., 1995, Changes in retinal cell fate induced by overexpression of EGF receptor. *Nature* **377**:158–162.

Linden, R., Cavalcante, L. A., and Barradas, P. C., 1986, Mononuclear phagocytes in the retina of developing rats. *Histochemistry* **85**:335–339.

Lindsey, R. M., 1986, Reactive gliosis, in: *Astrocytes*, Volume 3 (S. Fedoroff and A. Vernadakis, eds.), Academic Press, Orlando, FL, pp. 231–262.

Ling, E. A., 1981, The origin and nature of microglia. *Adv. Cell. Neurobiol.* **2**:33–82.

Ling, E. A., 1982, A light-microscopic demonstration of amoeboid microglia and microglial cells in the retina of rats of various ages. *Archivum Histologicum Japonicum* **45**:37–44.

Ling, T., and Stone, J., 1988, The development of astrocytes in the cat retina: Evidence of migration from the optic nerve. *Dev. Brain Res.* **44**:73–85.

Ling, T., Mitrofanis, J., and Stone, J., 1989, Origin of retinal astrocytes in the rat: Evidence for migration from the optic nerve. *J. Comp. Neurol.* **286**:345–352.

Linser, P., and Moscona, A. A., 1979, Induction of glutamine synthetase in embryonic neural retina: Localization in Müller fibers and dependence on cell interactions. *Proc. Natl. Acad. Sci. USA* **76**:6476–6480.

Linser, P., and Moscona, A. A., 1981a, Carbonic anhydrase C in the neural retina: Transition from generalized to a glia-specific cell localization during embryonic development. *Proc. Natl. Acad. Sci. USA* **78**:7190–7194.

Linser, P., and Moscona, A. A., 1981b, Developmental changes in the distribution of S-100 in avian neural retina. *Dev. Neurosci.* **4**:433–441.

Linser, P., and Moscona, A. A., 1984, Variable CA II compartmentalization in vertebrate retina. *Ann. NY Acad. Sci.* **429**:430–446.

Linser, P. J., 1987, Neuronal-glial interactions in retina development. *Am. Zool.* **27**:161–169.

Linser, P. J., 1988, Plasticity of retinal glioblast cells: Neuronal contact regulates phenotypic maturation of embryonic Müller cells, in: *Cell Interactions in Visual Development*, (S. R. Hilfer and J. B. Sheffield, eds.), Springer-Verlag, New York, pp. 29–46.

Linser, P. J., Sorrentino, M., and Moscona, A. A., 1984, Cellular compartmentalization of carbonic anhydrase and glutamine synthetase in developing and mature mouse neural retina. *Dev. Brain Res.* **13**:65–71.

Liu, J. J., Xue, Y. M., Agarwal, N., and Roque, R. S., 1999, Human Müller cells express VEGF183, a novel spliced variant of vascular endothelial growth factor. *Invest. Ophthalmol. Vis. Sci.* **40**:752–759.

Loeffler K. U., Li, Z. L., Fishman, G. A., and Tso, M. O., 1992, Dominantly inherited cystoid macular edema—a histopathologic study. *Ophthalmology* **99**: 1385–1392.

López, T., López-Colomé, A. M., and Ortega, A. (1994). AMPA/KA receptor expression in radial glia. *NeuroReport* *5*:504–506.

López-Colomé, A. M., and Romo-de-Vivar, M., 1991, Excitatory amino acid receptors in primary cultures of glial cells from the retina. *Glia* **4**:431–439.

Lowenstein, C. J., and Snyder, S. H., 1992, Nitric oxide, a novel biologic messenger. *Cell* **70**: 705–707.

Lowenstein, C. J., Glatt, C. S., Bredt, D. S., and Snyder, S. H., 1992, Cloned and expressed macrophage nitric oxide synthase contrasts with the brain enzyme. *Proc. Natl. Acad. Sci. USA* **89**:6711–6715.

Lucas, D. R., and Newhouse, J. P., 1957, The toxic effect of sodium L-glutamate on the inner layers of the retina. *Arch. Ophthalmol.* **58**:193–201.

Lund Karlsen, R., 1978, The toxic effect of sodium glutamate and DL-alpha-aminoadipic acid on rat retina: Changes in high affinity uptake of putative transmitters. *J. Neurochem.* **31**:1055–1061.

Lund Karlsen, R., Pedersen, O. O., Schousboe, A., and Langeland, A., 1982, Toxic effects of DL-alpha-amino adipic acid on Müller cells from rats in vivo and cultured cerebral astrocytes. *Exp. Eye Res.* **35**:305–311.

Luskin, M. B., Parnavelas, J. G., and Barfield, J. A., 1993, Neurons, astrocytes, and oligodendrocytes of the rat cerebral cortex originate from separate progenitor cells: An ultrastructural analysis of clonally related cells. *J. Neurosci.* **13**:1730–1750.

Luten-Drecoll, E., Lonnerholm, G., and Eichhorn, M., 1983, Carbonic anhydrase distribution in the human and monkey eye by light and electronmicroscopy. *Graefés Arch. Clin. Exp. Ophthalmol.* **220**:285–291.

MacVicar, B. A., 1984, Voltage-dependent calcium channels in glial cells. *Science* **226**:1345–1347.

MacVicar, B. A., Tse, F. W. Y., Crichton, S. A., and Kettenmann, H., 1989, GABA-activated Cl channels in astrocytes of hippocampal slices. *J. Neurosci.* **9**:3577–3583.

Madeperla, S. A., Geiger, G. L., Funata, M., de la Cruz, Z., and Green, W. R., 1994, Clinicopathological correlation of a macular hole treated by cortical vitreous peeling and gas tamponade. *Ophthalmology* **101**:682–686.

Madigan, M. C., Penfold, P. L., Provis, J. M., Balind, T. K., and Billson, F. A., 1994, Intermediate filament expression in human retinal macroglia. Histopathological changes associated with age-related macular degeneration. *Retina* **14**:65–74.

Magalhaes, M. M., and Coimbra, A., 1970, Electron microscope autoradiographic study of glycogen synthesis in rabbit retina. *J. Cell Biol.* **47**:263–275.

Mager, S., Kleinberger-Doron, N., Keshet, G. I., Davidson, N., Kanner, B. I., and Lester, H. A., 1996, Ion binding and permeation at the GABA transporter GAT1. *J. Neurosci.* **16**:5405–5414.

Magistretti, P. J., Hof, P. R., and Martin, J. L., 1986, Adenosine stimulates glycogenolysis in mouse cerebral cortex: A posible coupling mechanism between neuronal activity and energy metabolism. *J. Neurosci.* **6**:2558–2562.

Magistretti, P. J., Pellerin, L., Rothman, D. L., and Shulman, R. G., 1999, Energy on demand. *Science* **283**:496–497.

Malandro, M. S., and Kilberg, M. S., 1996, Molecular biology of mammalian amino acid transporters. *Annu. Rev. Biochem.* **65**:305–336.

Malchow, R. P., and Ripps, H., 1990, Effects of γ-aminobutyric acid on skate retinal horizontal cells: Evidence for an electrogenic uptake mechanism. *Proc. Natl. Acad. Sci. USA* **87**:8945–8949.

Malchow, R. P., Qian, H., and Ripps, H., 1989, γ-Aminobutyric acid (GABA)-induced currents of skate Müller (glial) cells are mediated by neuronal-like $GABA_a$ receptors. *Proc. Natl. Acad. Sci. USA* **86**:4326–4330.

Mano, T., and Puro, D. G., 1990, Phagocytosis by human retinal glial cells in culture. *Invest. Ophthalmol. Vis. Sci.* **31**:1047–1055.

Mantych, G. J., Hageman, G. S., and Devaskar, S. U., 1993, Characterization of glucose transporter isoforms in the adult and developing human eye. *Endocrinology* **133**:600–607.

Marc, R. E., Liu, W.-L. S., and Müller, J. F., 1988, Gap junctions in the inner plexiform layer of the goldfish retina. *Vis. Res.* **28**:9–24.

Maren, T. H., 1967, Carbonic anhydrase: Chemistry, physiology and inhibition. *Physiol. Rev.* **47**:597–781.

Marshall, J., and Voaden, M., 1974a, An autoradiographic study of the cells accumulating ^{3}H γ-aminobutyric acid in the isolated retinas of pigeons and chickens. *Invest. Ophthalmol.* **13**:606–607.

Marshall, J., and Voaden, M., 1974b, An investigation of the cells incorporating [^{3}H]GABA and [^{3}H]glycine in the isolated retina of the rat. *Exp. Eye Res.* **18**:367–370.

Marshall, J., and Voaden, M., 1975, Autoradiographic identification of the cells accumulating ^{3}H γ-aminobutyric acid in mammalian retinae: A species comparison. *Vis. Res.* **15**:459–461.

Martin, D. L., 1976, Carrier-mediated transport and removal of GABA from synaptic regions, in: *GABA in Nervous System Function* (E. Roberts, T. N. Chase, and D. B. Tower, eds.), Raven Press, New York, pp. 347–386.

Martin, D. L., 1992, Synthesis and release of neuroactive substances by glial cells. *Glia* **5**:81–94.

Martin, D. L., and Waniewski, R. A., 1996, Precursor synthesis and neurotransmitter uptake by astrocytes as targets of neurotoxicants, in: *The Role of Glia in Neurotoxicity* (M. Aschner and H. K. Kimelberg, eds.), CRC Press, Boca Raton, FL, pp. 335–357.

Martins-Ferreira, H., and De Oliveira Castro, G., 1966, Light-scattering changes accompanying spreading depression in isolated retina. *J. Neurophysiol.* **29**:715–726.

Mascarelli, F., Tassin, J., and Courtois, Y., 1991, Effects of FGF's on adult bovine Müller cells: Proliferation, binding and internalization. *Growth Factors* **4**:81–95.

Masland, R. H., and Ames, III, A., 1975, Dissociation of field potential from neuronal activity in the isolated retina: Failure of the b-wave with normal ganglion cell response. *J. Neurobiol.* **6**:305–312.

Massey, S. C., 1990. Cell types using glutamate as a neurotransmitter in the vertebrate retina. *Prog. Ret. Res.* **9**:399–425.

Matsuda, H., Saigusa, A., and Irisawa, H., 1987, Ohmic conductance through the inwardly rectifying K channel and blocking by internal Mg^{2+}. *Nature* **325**:156–159.

Matsunaga, M., Hotta, K., and Takeichi, M., 1988, Role of N-cadherin cell adhesion molecules in the histogenesis of neural retina. *Neuron* **1**:289–295.

Matsuura, T., Miller, W. H., and Tomita, T., 1978, Cone-specific c-wave in the turtle retina. *Vis. Res.* **18**:767–775.

McAdams B. D., and McLoon, S. C., 1995, Expression of chondroitin sulfate and keratan sulfate proteoglycans in the path of growing retinal axons in the developing chick. *J. Comp. Neurol.* **352**:594–606.

McCabe, C. F., and Cole, G. J., 1992, Expression of the barrier-associated proteins EAP-300 and clausterin in the developing central nervous system. *Dev. Brain Res.* **70**:9–24.

McCaffrey, P., Tempst, P., Lara, G., and Drager, U. C., 1991, Aldehyde dehydrogenase is a positional marker in the retina. *Development* **112**:693–702.

McCall, M. A., Gregg, R. G., Behringer, R. R., Brenner, M., Delaney, C. L., Galbreath, E. J., Zhang, C. L., Pearce, R. A., Chiu, S. Y., and Messing, A., 1996, Targeted deletion in astrocyte intermediate filament gene (*Gfap*) alters neuronal physiology. *Proc. Natl. Acad. Sci. USA* **93**:6361–6366.

McDonald, H. R., Schatz, H., and Johnson, R. N., 1994, Introduction to epiretinal membranes, in: *Retina*, Volume 2 (S. J. Ryan, ed.), C.V. Mosby, St. Louis, MO, pp. 1819–1825.

McFarlane, S., Zuber, M. E., and Holt, C. E., 1998, A role for the fibroblast growth factor receptor in cell fate decisions in the developing vertebrate retina. *Development* **125**:3967–3975.

McKenna M. C., Tildon, J. T., Stevenson, J. H., Boatright, R., and Huang, S., 1993, Regulation of energy-metabolism in synaptic terminals and cultured rat-brain astrocytes—differences revealed using aminooxyacetate. *Dev. Neurosci.* **15**:320–329.

McLoon, S. C., McLoon, L. K., Palm, S. L., and Furcht, L. T., 1988, Transient expression of laminin in the optic nerve of the developing rat. *J. Neurosci.* **8**:1981–1990.

Meister, A., 1974, Glutamine synthetase of mammals, in: *The Enzymes*, Volume 10 (P. D. Boyer, ed.), Academic Press, New York, pp. 699–754.

Meldrum, B., and Garthwaite, J., 1990, Excitatory amino acid neurotoxicity and neurodegenerative disease. *Trends Pharmacol. Sci.* **11**:379–387.

Meller, K., and Tetzlaff, W., 1976, Scanning electron microscopic studies in the development of the chick retina. *Cell Tiss. Res.* **170**:145–159.

Memo, M., Ricardi, F., Trabucchi, M., and Spano, P. F., 1981, Localization of GABA and dopamine receptor sites in retinal glial cells using D, L-alpha-aminoadipic acid. *Adv. Biomed. Pharmacol.* **26**:41–51.

Merrill, J. E., Murphy, S. P., Mitrovic, B., Mackenzie-Graham, A., Dopp, J. C., Ding, M., Griscavage, J., Ignarro, L. J., and Lowenstein, C. J., 1997, Inducible nitric oxide synthase and nitric oxide production by oligodendrocytes. *J. Neurosci. Res.* **48**:372–384.

Meyer-Franke, A., Kaplan, M. R., Pfrieger, F. W., and Barres, B. A., 1995, Characterization of the signaling interactions that promote the survival and growth of developing retinal ganglion cells in culture. *Neuron* **15**:805–819.

Mignery, G. A., Sudhof, T. C., Takei, K., and DeCamilli, P., 1989, Putative receptor for inositol 1,4,5–trisphosphate similar to ryanodine receptor. *Nature* **342**:192–195.

Miller, B., Miller, H., and Ryan, S. J., 1986, Experimental epiretinal proliferation induced by intravitreal red blood cells. *Am. J. Ophthalmol.* **102**:188–195.

Miller, R. F., 1973, Role of K^+ in generation of b-wave of electroretinogram. *J. Neurophysiol.* **36**:28–38.

Miller, R. F., and Dowling, J. E., 1970, Intracellular responses of the Müller (glial) cells of mudpuppy retina: Their relation to b-wave of the electroretinogram. *J. Neurophysiol.* **33**: 323–341.

Miller, S. S., Steinberg, R. H., and Oakley, III, B., 1978, The electrogenic sodium pump of the frog retinal pigment epithelium. *J. Membrane Biol.* **44**:259–279.

Milner, P. M., 1958, Note on a possible correspondence between the scotomas of migraine and spreading depression of Leão. *Electroenceph. Clin. Neurophysiol.* **10**:705.

Minelli, A., DeBiasi, S., Brecha, N. C., Zuccarello, L. B., and Conti, F., 1996, GAT-3, a high-affinity GABA plasma membrane transporter, is localized to astrocytic processes, and it is not confined to the vicinity of GABAergic synapses in the cerebral cortex. *J. Neurosci.* **16**: 6255–6264.

Miyachi, E.-I., and Nishikawa, C., 1994, Blocking effect of L-arginine on retinal gap junctions by activating guanylate cyclase via generation of nitric oxide. *Biogenic Amines* **10**:459–464.

Miyachi, E.-I., Miyakawa, A., and Murakami, M., 1991, Modulation of electrical coupling between horizontal cells by intracellular messengers. *Neurosci. Res. Suppl.* **15**:S41–S49.

Mizuo, G., 1913, A new discovery in dark adaptation in Oguchi's disease. *Acta Soc. Ophthalmol. Jpn.* **17**:1148–1150.

Mobbs, P., Brew, H., and Attwell, D., 1988, A quantitative analysis of glial cell coupling in the retina of the axolotl (*Ambystoma mexicanum*). *Brain Res.* **460**:235–245.

Molnar, M. L., Stefansson, K., Marton, L. S., Tripathi, R. C., and Molnar, G. K., 1984, Distribution of S-100 protein and glial fibrillary acidic protein in normal and gliotic human retina. *Exp. Eye Res.* **38**:27–34.

Moncada, S., Palmer, R. M. J., and Higgs, E. A., 1989, Biosynthesis of nitric oxide from L-arginine: A pathway for the regulation of cell function and communication. *Biochem. Pharmacol.* **38**:1709–1715.

Moncada, S., Palmer, R. M. J., and Higgs, E. A., 1991, Nitric oxide: Physiology, pathophysiology, and pharmacology. *Pharmacol. Rev.* **43**:109–142.

Mondon, H., Hamard, H., Girard, P., and Bregeat, P., 1970, Retinal gliosis associated with Coat's disease. *Bull. Soc. Ophthalmol. France* **70**:881–883.

Moore, C. L., and Gruberg, E. R., 1974, The distribution of succinic semialdehyde dehydrogenase in the brain and retina of tiger salamander. *Brain Res.* **67**:467–478.

Mori, S., Miller, W. H., and Tomita, T., 1976a, Microelectrode study of spreading depression (SD) in frog retina—Müller cell activity and [K^+] during SD. *Jpn. J. Physiol.* **26**:203–217.

Mori, S., Miller, W. H., and Tomita, T., 1976b, Müller cell function during spreading depression in frog retina. *Proc. Natl. Acad. Sci. USA* **73**:1351–1354.

Morissette, N., and Carbonetto, S., 1995, Laminin β2 chain (M Chain) is found within the pathway of avian and murine retinal projections. *J. Neurosci.* **15**:8067–8082.

Moromoto, A., Matsuda, S., Uryu, K., Fujita, H., Okumura, N., and Sakanaka, M., 1993, Light- and electronmicroscopic localization of basic fibroblast growth factor in adult rat retina. *Okajimas, Folia Anatomia Japonica* **70**:7–12.

Morris, J. E., Yanagishita, M., and Hascall, V. C., 1987, Proteoglycans synthesized by embryonic chick retina in culture: Composition and compartmentalization. *Arch. Biochem. Biophys.* **258**:206–218.

Moscona, A. A., 1987, Role of cell contact in phenotype stability and modification—studies on retina cells. *Am. Zool.* **27**:137–144.

Moscona, A. A., and Linser, P. J., 1983, Developmental and experimental changes in retina glia cells: Cell interactions and control of phenotype expression and stability. *Curr. Top. Dev. Biol.* **18**:155–188.

Moscona, A. A., and Vardimon, L., 1988, Cell contacts and gene regulation in embryonic retina, in: *Molecular Biology of the Eye: Genes, Vision, and Ocular Disease* (J. Piatigorsky, T. Shinohara, and P. S. Zelenka, eds.), Alan R. Liss, New York, pp. 259–268.

Moscona, A. A., Meyerson, P., Linser, P., and Moscona, M., 1980, Regulatory aspects of the induction of glutamine synthetase in embryonic neural retina, in: *Glutamine: Metabolism, Enzymology and Regulation* (J. Mora and R. Palacios, eds.), Academic Press, New York, pp. 299–313.

Moscona, M., and Moscona, A. A., 1979, The development of inducibility for glutamine synthetase in embryonic neural retina: Inhibition of BrdU. *Differentiation* **13**:165–172.

Mudhar, H. S., Pollock, R. A., Wang, C., Stiles, C., and Richardson, W. D., 1993, PDGF and its receptors in the developing rodent retina and optic nerve. *Development* **118**:539–552.

Müller, A., Villain, M., and Bonne, C., 1997, The release of amino acids from ischemic retina. *Exp. Eye Res.* **64**:291–293.

Müller, A., Maurin, L., and Bonne, C., 1998, Free radicals and glutamate uptake in the retina. *Gen. Pharmacol.* **30**:315–318.

Müller, H., 1851, Zur Histologie der Netzhaut. *Zeitschrift für Wissenschaft und Zoologie* **3**:234–237.

Müller, T., Möller, T., Berger, T., Schnitzer, J., and Kettenmann, H., 1992, Calcium entry through kainate receptors and resulting potassium-channel blockade in Bergmann glial cells. *Science* **256**:1563–1566.

Müller, T., Grosche, J., Ohlemeyer, C., and Kettenmann, H., 1993, NMDA-activated currents in Bergmann glial cells. *NeuroReport* **4**:671–674.

Müller, T., Möller, T., Neuhaus, J., and Kettenmann, H., 1996, Electrical coupling among Bergmann glial cells and its modulation by glutamate receptor activation. *Glia* **17**:274–284.

Murabe, Y., Ibata, Y., and Sano, Y., 1982, Morphological studies on neuroglia. IV. Proliferative response of nonneuronal elments in the hippocampus of the rat to kainic acid-induced lesions. *Cell Tissue Res.* **222**:223–226.

Murakami, M., Fukuyama, K., Hubbard, S., Matsuzawa, K., Dirks, P. B., and Rutka, J. T., 1996, Inducible expression of glial fibrillary acidic protein in HT-1080 human fibrosarcoma cells. *Cell Growth Diff.* **7**:1697–1703.

Murphy, S., and Pierce, B., 1987, Functional receptors for neurotransmitters on astroglial cells. *Neuroscience* **22**:381–394.

Murphy, T. H., Blatter, L. A., Wier, W. G., and Baraban, J. M., 1993, Rapid communication between neurons and astrocytes in primary cortical cultures. *J. Neurosci.* **13**:2672–2679.

Musser, G. L., and Rosen, S., 1973, Localization of carbonic anhydrase activity in the vertebrate retina. *Exp. Eye Res.* **15**:105–119.

Mutsuga, N., Schuette, W. H., and Lewis, D. V., 1976, The contribution of local blood flow to the rapid clearance of potassium from the cortical extracellular space. *Brain Res.* **116**:431–436.

Naka, K.-I., and Rushton, W. A. H., 1967, The generation and spread of S-potentials in fish (cyprinidae). *J. Physiol.* **192**:437–461.

Napoli, J. L., Posch, K. P., Fiorella, P. D., and Boerman, M. H. E. M., 1991, Physiological occurrence, biosynthesis and metabolism of retinoic acid: Evidence for roles of cellular retinol-binding protein (CRBP) and cellular retinoic acid-binding protein (CRABP) in the pathway of retinoic acid homeosynthesis. *Biomed. Pharmacother.* **45**:131–143.

Nathan, C. F., 1992, Nitric oxide as a secretory product of mammalian cells. *FASEB J.* **6**:3051–3064.

Nawy, S., and Jahr, C. E., 1990, Suppression by glutamate of cGMP-activated conductance in retinal bipolar cells. *Nature* **346**:269–271.

Nawy, S., and Jahr, C. E., 1991, cGMP-gated conductance in retinal bipolar cells is suppressed by the photoreceptor transmitter. *Neuron* **7**:677–683.

Neal, M. J., and Iverson, L. L., 1972, Autoradiographic localization of ^{3}H-GABA in rat retina. *Nature New Biol.* **235**:217–218.

Neal, M. J., and White, R. D., 1971, Uptake of ^{14}C-glutamate in rat retina. *Br. J. Pharmacol.* **43**:442–443.

Neal, M. J., and Starr, M. S., 1973, Effects of inhibitors of γ-aminobutyrate aminotransferase on the accumulation of ^{3}H-γ-aminobutyric acid by the retina. *Br. J. Pharmacol.* **47**:543–555.

Neal, M. J., Cunningham, J. R., Shah, M. A., and Yazulla, S., 1989, Immunocytochemical evidence that vigabatrin in rats causes GABA accumulation in glial cells of the retina. *Neurosci. Lett.* **98**:29–32.

Nedergaard, M., 1994, Direct signaling from astrocytes to neurons in cultures of mammalian brain cells. *Science* **263**:1768–1771.

Nedergaard, M., Cooper, A. J. L., and Goldman, S. A., 1995, Gap junctions are required for the propagation of spreading depression. *J. Neurobiol.* **28**:433–444.

Neophytou, C., Vernallis, A. B., Smith, A., and Raff, M. C., 1997, Müller cell-derived leukemia inhibitory factor arrests rod photoreceptor differentiation at a postmitotic pre-rod stage of development. *Development* **124**:2345–2354.

Neugebauer, K. M., Tomaselli, K. J., Lillien, J., and Reichardt, L. F., 1988, N-cadherin, NCAM, and integrins promote retinal neurite outgrowth on astrocytes *in vitro*. *J. Cell Biol.* **107**:1177–1187.

Neugebauer, K. M., Emmett, C. J., Venstrom, K. A., and Reichardt, L. F., 1991, Vitronectin and thrombospondin promote retinal neurite outgrowth: Developmental regulation and role of integrins. *Neuron* **6**:345–358.

Newman, E., and Reichenbach, A., 1996, The Müller cell: A functional element of the retina. *Trends Neurosci.* **19**:307–312.

Newman, E. A., 1980, Current source-density analysis of the b-wave of frog retina. *J. Neurophysiol.* **43**:1355–1366.

Newman, E. A., 1984, Regional specialization of retinal glial cell membrane. *Nature* **309**: 155–157.

Newman, E. A., 1985a, Membrane physiology of retinal glial (Müller) cells. *J. Neurosci.* **5**:2225–2239.

Newman, E. A., 1985b, Voltage-dependent calcium and potassium channels in retinal glial cells. *Nature* **317**:809–811.

Newman, E. A., 1986, The Müller cell, in: *Astrocytes*, Volume 1 (S. Federoff and A. Vernadakis, eds.), Academic Press, Orlando, FL, pp. 149–171.

Newman, E. A., 1987, Distribution of potassium conductance in mammalian Müller (glial) cells: A comparative study. *J. Neurosci.* **7**:2423–2432.

Newman, E. A., 1991, Sodium-bicarbonate co-transport in retinal Müller (glial) cells of the salamander. *J. Neurosci.* **11**:3972–3983.

Newman, E. A., 1993, Inward-rectifying potassium channels in retinal glial (Müller) cells. *J. Neurosci.* **13**:3333–3345.

Newman, E. A., 1996, Acid flux from retinal glial cells generated by sodium bicarbonate cotransport. *J. Neurosci.* **16**:159–168.

Newman, E. A., and Odette, L. L., 1984, Model of electroretinogram b-wave generation: A test of the K^+ hypothesis. *J. Neurophysiol.* **51**:164–182.

Newman, E. A., and Astion, M. L., 1991, Localization and stoichiometry of electrogenic bicarbonate cotransport in retinal glial cells. *Glia* **4**:424–428.

Newman, E. A., and Zahs, K. R., 1997, Calcium waves in retinal glial cells. *Science* **275**:844–847.

Newman, E. A., Frambach, D. A., and Odette, L. L., 1984, Control of extracellular potassium levels by retinal glial cell K^+ siphoning. *Science* **225**:1174–1175.

Newsome, D. A., and Hewitt, A. T., 1985, Laminin detection in normal and retinitis pigmentosa human retina. *Exp. Eye Res.* **4**:925–932.

Nicholls, D., and Attwell, D., 1990, The release and uptake of excitatory amino acids. *Trends Pharmacol. Sci.* **11**:462–468.

Nicholson, C., 1980, Dynamics of the brain cell microenvironment. *Neurosci. Res. Program Bull.* **18**:177–322.

Nicholson, C., and Freeman, J. A., 1975, Theory of current source-density analysis and determination of conductivity tensor for anuran cerebellum. *J. Neurophysiol.* **38**:356–368.

Nicholson, C., and Kraig, R. P., 1981, The behavior of extracellular ions during spreading depression, in: *The Application of Ion-Selective Electrodes* (T. Zeuthen, ed.), Elsevier/North Holland, Amsterdam, pp. 217–238.

Niederkorn, J. Y., 1990, Immune privilege and immune regulation in the eye. *Adv. Immunol.* **48**:191–226.

Niemeyer, G., 1997, Glucose concentration and retinal function. *Clin. Neurosci.* **4**:327–335.

Niemeyer, G., Rungger-Brandle, E., and Lansel, N., 1997, Glycogen content related to function in the cat retina *in vitro*. *Invest. Ophthalmol. Vis. Sci.* **38**:S888.

Nilius, B., and Reichenbach, A., 1988, Efficient K^+ buffering by mammalian retinal glial cells is due to cooperation of specialized ion channels. *Pflügers Arch.* **411**:654–660.

Nilsson, M., Hannson, E., and Rönnbäck, L., 1992, Agonist evoked Ca^{2+} calcium transients in primary astroglial cultures—modulatory effects of valproic acid. *Glia* **5**:201–209.

Nilsson, M., Eriksson, P. S., Rönnbäck, L., and Hansson, E., 1993, GABA induces Ca^{2+} transients in astrocytes. *Neurosci.* **54**:605–614.

Nishizono, H., Murata, Y., Tanaka, M., Soji, T., and Herbert, D. C., 1993, Evidence that Müller cells can phagocytize egg-lecithin-coated silicone particles. *Tissue Cell. Res.* **25**: 305–310.

Noell, W. K., 1953, Experimentally induced toxic effects on structure and function of visual cells and pigment epithelium. *Am. J. Ophthalmol.* **36**:103–116.

Nomura, Y., and Kitamura, Y., 1993, Inducible nitric oxide synthase in glial cells. *Neurosci. Res.* **18**:103–107.

Norenberg, M. D., and Martinez-Hernandez. A., 1979, Fine structural localization of glutamine synthetase in astrocytes of rat brain. *Brain Res.* **161**:303–310.

Norfstrom, K., and Nilsson, S. E., 1986, Progressive retinal atrophy in the Abyssinian cat: Electron microscopy. *Invest. Ophthalmol. Vis. Sci.* **27**:1569–1576.

Nork, T. M., Ghobrial, M. W., Peyman, G. A., and Tso, M. O. M., 1986, Massive retinal gliosis: A reactive proliferation of Müller cells. *Arch. Ophthalmol.* **104**:1383–1389.

Nork, T. M., Wallow, I. H., Sramek, S. J., and Anderson, G., 1987, Müller's cell involvement in proliferative diabetic retinopathy. *Arch. Ophthalmol.* **105**:1424–1429.

Nork, T. M., Wallow, I. H., Sramek, S. H., Stevens, T. S., and De Venecia, G., 1990, Immunocytochemical study of an eye with proliferative vitreoretinopathy and retinal tacks. *Retina* **10**:78–85.

Nowak, L., Ascher, P., and Berwald-Netter, Y., 1987, Ionic channels in mouse astrocytes in culture. *J. Neurosci.* **7**:101–109.

Oakley, B. H., and Wen, R., 1989, Extracellular pH in the isolated retina of the toad in darkness and during illumination. *J. Physiol.* **419**:353–378.

Oakley, II, B., 1977, Potassium and the photoreceptor-dependent pigment epithelial hyperpolarization. *J. Gen. Physiol.* **70**:405–425.

Oakley, II, B., and Green, D. G., 1976, Correlation of light-induced changes in retinal extracellular potassium concentration with c-wave of the electroretinogram. *J. Neurophysiol.* **39**:1117–1133.

Oakley, II, B., Flaming, D. G., and Brown, K. T., 1979, Effects of the rod receptor potential upon retinal extracellular potassium concentration. *J. Gen. Physiol.* **74**:713–737.

Oakley, II, B., Katz, B. J., Xu, Z., and Zheng, J., 1992, Spatial buffering of extracellular potassium by Müller (glial) cells in the toad retina. *Exp. Eye Res.* **55**:539–550.

Ogden T. E., 1978, Nerve fiber layer astrocytes of the primate retina: Morphology, distribution, and density. *Invest. Ophthalmol. Vis. Sci.* **17**:499–510.

Ohira, A., and de Juan, Jr., E., 1990, Characterization of glia involvement in proliferative diabetic retinopathy. *Ophthalmolgica* **201**:187–195.

Ohira, A., Oshima, K., and Kikichi, M., 1984, Localization of glial fibrillary acidic (GFA) protein in the human Müller cell. *Nippon Ganka Zasshi.* **88**:1068–1074.

Ohki, K., Yoshida, K., Yamakawa, A., Harada, T., Mitsuda, H., and Imaki, J., 1995, Jun-B gene expression in rat retinal cells following focal retinal injury. *Curr. Eye Res.* **14**:1021–1024.

Okada, M., Matsumura, M., Yamakawa, R., Shirakawa, H., Yoshimura, N., and Ogino, N., 1987, Immunoelectron microscopic study on glial fibrillary acidic protein (GFAP)-containing cells in preretinal proliferative tissue. *Nippon Ganka Zasshi-Acta Societatis Opthalmologicaae Japonicae* **91**:657–664.

Okada, M., Matsumura, M., Ogina, N., and Honda, Y., 1990, Müller cells in detached human retina express glial fibrillary acidic protein and vimentin. *Graefés Arch. Klin. Exp. Ophthalmol.* **228**:467–474.

Olney, J. W., 1978, Neurotoxicity of excitatory amino acids, in: *Kainic Acid as a Tool in Neurobiology* (E. G. McGeer, P. L. McGeer, and J. W. Olney, eds.), Raven Press, New York, pp. 95–121.

Olney, J. W., Ho, O. L., and Rhee, V., 1971, Cytotoxic effects of acidic and sulphur containing amino acids on the infant mouse central nervous system. *Exp. Brain Res.* **14**:61–76.

Ong, D. E., 1994, Cellular-transport and metabolism of vitamin-A—Roles of the cellular retinoid-binding proteins. *Nutritional Rev.* **52**:24–31.

Opremcak, E. M., 1994, Cellular cooperation, in: *Principles and Practice of Ophthalmology* (D. M. Albert and F. A. Jakobiec, eds.), W.B. Saunders, Philadelphia, PA, pp. 771–776.

Orkand, R. K., 1980, Extracellular potassium accumulation in the nervous system. *Fed. Proc.* **39**:1515–1518.

Orkand, R. K., Nicholls, J. G., and Kuffler, S. W., 1966, The effect of nerve impulses on the membrane potential of glial cells of the central nervous system of amphibia. *J. Neurophysiol.* **29**:788–806.

Osborne, N. N., 1989, [H^3]-Glycogen hydrolysis elicitated by adenosine in rabbit retina-involvement of A2–receptors. *Neurochem. Int.* **14**:419–422.

Osborne, N. N., Block, F., and Sontag, K. H., 1991, Reduction of ocular blood flow results in glial fibrillary acidic protein (GFAP) expression in rat retinal Müller cells. *Vis. Neurosci.* **7**:637–639.

Osborne, N. N., Barnett, N. L., and Herrera, A. J., 1993, NADPH diaphorase localization and nitric oxide synthase activity in the retina and anterior uvea of the rabbit eye. *Brain Res.* **610**:194–198.

Otori, Y., Shimada, S., Tanaka, K., Ishimoto, I., Tano, Y. T., and Tohyama, M., 1994, Marked increase in glutamate-aspartate transporter (GLAST/GluT-1) mRNA following transient retinal ischemia. *Mol. Brain Res.* **27**:310–314.

Ozaki, S., Johnson, L. V., Mullins, R. F., Hageman, G. S., and Anderson, D. H., 1999, The human retina and retinal pigment epithelium are abundant sources of vitronectin mRNA. *Biochem. Biophys. Res. Comm.* **258**:524–529.

Pahuja, S. L., and Reid, T. W., 1985, Bovine retinal glutamine synthetase. 2. Regulation and properties on the basis of glutamine synthetase and glutamyl transferase reactions. *Exp. Eye Res.* **40**:75–83.

Pahuja, S. L., Mullins, B. T., and Reid, T. W., 1985, Bovine retinal glutamine synthetase. 1. Purification, characterization and immunologiical properties. *Exp. Eye Res.* **40**:61–74.

Pandey, S., Blanks, J. C., Spee, C., Jiang, M. S., and Fong, H. K. W., 1994, Cytoplasmic retinal localization of an evolutionary homolog of the visual pigments. *Exp. Eye Res.* **58**:605–613.

Parpura, V., Basarsky, T. A., Liu, F., Jeftinija, K., Jeftinija, S., and Haydon, P. G., 1994, Glutamate-mediated astrocyte-neuron signaling. *Nature* **369**:744–747.

Parpura, V., Liu, F., Jeftinija, K. V., Haydon, P. G., Jeftinija, S. D., 1995, Neuroligand-evoked calcium-dependent release of excitatory amino acids from Schwann cells. *J. Neurosci.* **15**:5831–5839.

Pasti, L., Volterra, A., Pozzan, T., and Carmignoto, G., 1997, Intracellular calcium oscillations in astrocytes: A highly plastic, bidirectional form of communication between neurons and astrocytes *in situ*. *J. Neurosci.* **17**:7817–7830.

Paulsen, R. E., and Fonnum, F., 1990, Using microdialysis to evaluate the role of astrocytes for transmitter amino acid inactivation and precursor production in the rat brain, in: *Differentiation and Functions of Glial Cells* (G. Levi, ed.), Wiley-Liss, New York, pp. 301–302.

Peachey, N. S., Green, D. J., and Ripps, H., 1993, Ocular ischemia and the effects of allopurinol on functional recovery in the retina of the arterially perfused cat eye. *Invest. Ophthalmol. Vis. Sci.* **34**:58–65.

Pearce, B., Albrecht, J., Morrow, C., and Murphy, S., 1986, Astrocyte glutamate receptor activation promotes inositol phospholipid turnover and calcium flux. *Neurosci. Lett.* **72**: 335–340.

Pedersen, O. Ø., and Lund Karlsen, R., 1979, Destruction of Müller cells in the adult rat by intravitreal injection of D,L-α-aminoadipic acid: An electron microscopic study. *Exp. Eye Res.* **28**:569–575.

Pedler, C., 1960, The inner limiting membrane of the retina. *Br. J. Ophthal.* **45**:423–438.

Pedler, C., 1962, The radial fibres of the retina. *Doc. Ophthal.* **16**:208–220.

Pedler, C., 1963, The fine structure of the radial fibers in the reptile retina. *Exp. Eye Res.* **2**:296–303.

Pekny, M., Leveen, P., Pekna, M., Eliasson, C., Berthold, C.-H., Westermark, B., and Betsholtz, C., 1995, Mice lacking glial fibrillary acidic protein display astrocytes devoid of intermediate filaments but develop and reproduce normally. *EMBO J.* **14**:1590–1598.

Pellerin, L., and Magistretti, P. J., 1994, Glutamate uptake into astrocytes stimulates aerobic glycolysis: A mechanism coupling neuronal activity to glucose utilization. *Proc. Natl. Acad. Sci. USA* **91**:10625–10629.

Penn, J. S., Thum, L. A., Rhem, M. N., and Dell, S. J., 1988, Effects of oxygen rearing on the electroretinogram and GFA-protein in the rat. *Invest. Ophthalmol. Vis. Sci.* **29**:1623–1630.

Penn, R. D., and Hagins, W. A., 1969, Signal transmission along retinal rods and the origin of the electroretinographic a-wave. *Nature* **223**:201–204.

Penner, R., Fasolato, C., and Hoth, M., 1993, Calcium influx and its control by calcium release. *Curr. Opin. Neurobiol.* **3**:368–374.

Pentreath, V. W., and Kai-Kai, M. A., 1982, Significance of the potassium signal from neurones to glial cells. *Nature* **295**:59–61.

Perez, M.-T. R., and Caminos, E., 1995 Expression of brain-derived neurotrophic factor and of its functional receptor in neonatal and adult rat retina. *Neurosci. Lett.* **183**:96–99.

Perez, R. G., and Halfter, W., 1993, Tenascin in the developing chick visual system—distribution and potential role as a modulator of retinal axon growth. *Dev. Biol.* **156**: 278–292.

Pérez-Clausell, J., and Danscher, G., 1985, Intravesicular localization of zinc in rat telencephalic boutons: A histochemical study. *Brain Res.* **337**:91–98.

Pfeiffer, B., Grosche, J., Reichenbach, A., and Hemprecht, B., 1994, Immunocytochemical demonstration of glycogen-phosphorylase in Müller (glial) cells of the mammalian retina. *Glia* **12**:62–67.

Pfrieger, F. W., and Barres, B. A., 1996, New views on synapse-glia interactions. *Curr. Opin. Neurobiol.* **6**:615–621.

Pfrieger, F. W., and Barres, B. A., 1997, Synaptic efficacy enhanced by glial cells *in vitro. Science* **277**:1684–1687.

Piccolino, M., Neyton, J., and Gerschenfeld, H. M., 1984, Decrease of gap junction permeability induced by dopamine and cyclic adenosine 3′:5′-monophosphate in horizontal cells of turtle retina. *J. Neurosci.* **4**:2477–2488.

Pines, G., Danbolt, N. C., Bjoras, M., Zhang, Y., Bendahan, A., Eide, L., Koepsell, H., Storm-Mathisen, J., Seeberg, E., and Kanner, B., 1992, Cloning and expression of a rat brain L-glutamate transporter. *Nature* **360**:464–467.

Poitry-Yamate, C., Poitry, S., and Tsacopoulos, M., 1995, Lactate released by Müller glial cells is metabolized by photoreceptors from mammalian retina. *J. Neurosci.* **15**:5179–5191.

Poitry-Yamate, C. L., and Tsacopoulos, M., 1991, Glial (Müller) cells take up and phosphorylate [^{3}H]-deoxy-D-glucose in mammalian retina. *Neurosci. Lett.* **122**:241–244.

Poitry-Yamate, C. L., and Tsacopoulos, M., 1992, Glucose metabolism in freshly isolated Müller glial cells from a mammalian retina. *J. Comp. Neurol.* **320**:257–266.

Polk, T. D., Gass, J. D. M., Green, W. R., Novak, M. A., and Johnson, M. W., 1997, Familial internal membrane dystrophy: A new sheen retinal dystrophy. *Arch. Ophthalmol.* **115**: 878–885.

Porter, J. T., and McCarthy, K. D., 1996, Hippocampal astrocytes in situ respond to glutamate released from synaptic terminals. *J. Neurosci.* **16**:5073–5081.

Pottek, M., Schultz, K., and Weiler, R., 1997, Effects of nitric oxide on the horizontal cell network and dopamine release in the carp retina. *Vis. Res.* **37**:1091–1102.

Potts, R. A., Dreher, B., and Bennett, M.R., 1982, The loss of ganglion cells in the developing retina of the rat. *Dev. Brain Res.* **3**:481–486.

Pourcho, R. G., Goebel, D. J., and McReynolds, J. S., 1984, Autoradiographic studies of [^{3}H]-glycine, [^{3}H]-GABA, and [^{3}H]-muscimol uptake in the mudpuppy retina. *Exp. Eye Res.* **39**:69–81.

Pow, D. V., and Barnett, N. L., 1999, Changing patterns of spatial buffering of glutamate in developing rat retinae are mediated by the Müller cell glutamate transporter GLAST. *Cell Tissue Res.* **297**:57–66.

Pow, D. V., and Robinson, S. R., 1994, Glutamate in some retinal neurons is derived solely from glia. *Neurosci.* **60**:355–366.

Powell, S. K., and Kleinman, H. K., 1997, Neuronal laminins and their cellular receptors. *Int. J. Biochem. Cell B* **29**:401–414.

Prada, F. A., Espinar, A., Chmielewski, C. E., Dorado, M. E., and Genis-Galvez, J. M., 1989a, Regional adaptation of Müller cells in the chick retina: A Golgi and electron microscopic study. *Histol. Histopathol.* **4**:309–315.

Prada, F. A., Mahalhaes, M. M., Coimbra, A., and Genis-Galvez, J. M., 1989b, Morphological differentiation of the Müller cell: Golgi and electron microscopic study in the chick retina. *J. Morphol.* **201**:11–22.

Provis, J. M., Penfold, P. L., Edwards, A. J., and van Driel, D., 1995, Human retinal microglia: Expression of immune markers and relationship to the *glia limitans*. *Glia* **14**:243–256.

Pull, I., and McIlwain, H., 1972, Adenine derivatives as neurohumoral agents in the brain. The quantities liberated on excitation of superfused cerebral tissues. *Biochem. J.* **130**:975–981.

Puro, D. G., 1991a, A calcium-activated, calcium-permeable ion channel in human retinal glial cells: Modulation by basic fibroblast growth factor. *Brain Res.* **548**:329–333.

Puro, D. G., 1991b, Stretch-activated channels in human retinal Müller cells. *Glia* **4**:456–460.

Puro, D. G., 1994a, Calcium channels of human retinal glial cells, in: *Methods in Neurosciences*, Volume 19 (T. Narahashi, ed.), Academic Press, Inc., San Diego, pp. 68–81.

Puro, D. G., 1994b, Growth factors and Müller cells. *Prog. Ret. Res.* **15**:89–101.

Puro, D. G., and Mano, T., 1991a, Modulation of calcium channels in human retinal glial cells by basic fibroblast growth factor: A possible role in retinal pathobiology. *J. Neurosci.* **11**:1873–1880.

Puro, D. G., and Mano, T., 1991b, Modulation of calcium channels in human retinal glial cells: Modulation by basic fibroblast growth factor. *Brain Res.* **548**:329–333.

Puro, D. G., Roberge, F., and Chan, C. C., 1989, Retinal glial cell proliferation and ion channels: A possible link. *Invest. Ophthalmol. Vis. Sci.* **30**:521–529.

Puro, D. G., Mano, T., Chan, C. C., Fukuda, M., and Shimada, H., 1990, Thrombin stimulates the proliferation of human retinal glial cells. *Graefés Arch. Klin. Exp. Ophthalmol.* **282**:169–173.

Puro, D. G., Yuan, J. P., and Sucher, N. J., 1996, Activation of NMDA receptor-channels in human retinal Müller glial cells inhibits inward-rectifying potassium currents. *Vis. Neurosci.* **13**:319–326.

Purves, D., and Lichtman, J. W., 1985, *Principles of Neural Development.* Sinauer Associates, Sunderland, MA.

Putney, Jr., J. W., 1993, Excitement about calcium signaling in inexcitable cells. *Science* **262**:676–678.

Qian, H., and Dowling, J. E., 1993, Novel GABA responses from rod-driven retinal horizontal cells. *Nature* **361**:162–164.

Qian, H., and Dowling, J. E., 1995, $GABA_A$ and $GABA_C$ receptors on hybrid bass retinal bipolar cells. *J. Neurophysiol.* **74**:1920–1928.

Qian, H., and Ripps, H., 1992, Receptive-field properties of rod-driven horizontal cells in the skate retina. *J. Gen. Physiol.* **100**:457–478.

Qian, H., and Ripps, H., 1999, GABA- and glutamate-induced responses of Müller cells from the baboon retina. *Invest. Ophthalmol. Vis. Sci.* **40**(suppl.):S235.

Qian, H., Malchow, R. P., and Ripps, H., 1993, The effects of lowered extracellular sodium on γ-aminobutyric acid (GABA)-induced currents of Müller (glial) cells of the skate retina. *Cell. Mol. Neurobiol.* **13**:147–158.

Qian, H., Malchow, R. P., Chappell, R. L., and Ripps, H., 1996, Zinc enhances ionic currents induced in skate Müller (glial) cells by the inhibitory neurotransmitter GABA. *Proc. Roy. Soc. Lond. B* **263**:791–796.

Qian, H., Li, L., Chappell, R. L., and Ripps, H., 1997, GABA receptors of bipolar cells from the skate retina: Actions of zinc on GABA-mediated membrane currents. *J. Neurophysiol.* **78**:2402–2412.

Quandt, F. N., and MacVicar, B. A., 1986, Calcium activated potassium channels in cultured astrocytes. *Neuroscience* **19**:29–41.

Quigley, H. A., 1977, Gap junctions between optic nerve head astrocytes. *Invest. Ophthalmol. Vis. Sci.* **16**:582–585.

Raff, M. C., 1989, Glial cell diversification in the rat optic nerve. *Science* **243**:1450–1455.

Rager, G., 1979, The cellular origin of the b-wave in the electroretinogram—a developmental approach. *J. Comp. Neurol.* **188**:225–244.

Rakic, P., 1978, Neuronal migration and contact guidance in primate telencephelon. *Postgrad. Med. J.* **54**:25–40.

Ramirez, J. M., Trivino, A., Ramirez, A. I., Salazar, J. J., and Garcia-Sanchez, J., 1996, Structural specializations of human retinal glial cells. *Vis. Res.* **36**:2029–2036.

Rao, N. A., Tso, M. O. M., and Rosenthal, A. R., 1976, Chalcosis in the human eye. *Arch. Ophthalmol.* **94**:1379–1384.

Rao, N. A., Brown, C. J., and Marak, G. E., 1986, Ultrastructural analysis of experimental allergic uveitis in rabbit. *Ophthalmic Res.* **18**:15–20.

Rasmussen, K.-E., 1973, A morphometric study of the Müller cells, their nuclei and mitochondria, in the rat retina. *J. Ultrastruct. Res.* **44**:96–112.

Rasmussen, K.-E., 1974, The Müller cell: A comparative study of rod and cone retinas with and without retinal vessels. *Exp. Eye Res.* **19**:243–257.

Rasmussen, K.-E., 1975, A morphometric study of the Müller cell in rod and cone retinas with and without retinal vessels. *Exp. Eye Res.* **20**:151–166.

Rauen, T., and Kanner, B. I., 1994, Localization of the glutamate transporter GLT-1 in rat and macaque monkey retinae. *Neurosci. Lett.* **169**:137–140.

Rauen, T., Rothstein, J. D., and Wassle, H., 1996, Differential expression of three glutamate transporter subtypes in the rat retina. *Cell Tissue Res.* **286**:325–336.

Raviola, E., and Gilula, N. B., 1973, Gap junctions between photoreceptor cells in the vertebrate retina. *Proc. Natl. Acad. Sci. USA* **70**:1677–1681.

Raymond, P. A., and Rivland, P. K., 1987, Germinal cells in the goldfish retina produce rod photoreceptors. *Dev. Biol.* **122**:120–138.

Reese, B. E., Harvey, A. R., and Tan, S.-S., 1995, Radial and tangential dispersion patterns in the mouse retina are cell-class specific. *Proc. Natl. Acad. Sci. USA* **92**:2494–2498.

Reh, T., 1987, Cell-specific regulation of neuronal production in the larval frog retina. *J. Neurosci.* **7**:3317–3324.

Reh, T. A., and Tully, T., 1986, Regulation of tyrosine hydroxylase-containing amacrine cell number in larval frog retina. *Dev. Biol.* **114**:463–469.

Reichardt, L. F., and Tomaselli, K. J., 1991, Extracellular molecules and their receptors: Functions in neural development. *Annu. Rev. Neurosci.* **14**:531–570.

Reichelt, W., Dettmer, D., Bruckner, G., Brust, P., and Eberhardt, W., 1989, Potassium as a signal for both proliferation and differentiation of rabbit retinal (Müller) glia growing in cell culture. *Cellular Signaling* **1**:187–194.

Reichelt, W., Müller, T., Pastor, A., Pannicke, T., Orkand, P. M., Kettenmann, H., and Schnitzer, J., 1993, Patch-clamp recordings from Müller (glial) cell endfeet in the intact isolated retina and acutely isolated Müller cells of mouse and guinea pig. *Neuroscience* **57**: 599–613.

Reichelt, W., Hernandez, M., Damian, R. T., Kisaalita, W. S., and Jordan, B. L., 1996, $GABA_A$ receptor currents recorded from Müller glial cells of the baboon (*Papio cynocephalus*) retina. *Neurosci. Lett.* **203**:159–162.

Reichelt, W., Hernandez, M., Damian, R. T., Kisaalita, W. S., and Jordan, B. L., 1997a, Voltage- and GABA-evoked currents from Müller cells of the baboon retina. *NeuroReport* **8**:541–544.

Reichelt, W., Stabel-Burow, J., Pannicke, T., Weichert, H., and Heinemann, U. 1997b, The glutathione level of retinal Müller glial cells is dependent on the high-affinity sodium-dependent uptake of glutamate. *Neuroscience* **77**:1213–1224.

Reichenbach, A., and Eberhardt, W., 1986, Intracellular recordings from isolated rabbit retinal Müller (glial) cells. *Pflügers Arch.* **407**:348–353.

Reichenbach, A., and Eberhardt, W., 1988, Cytotopographical specialization of enzymatically isolated rabbit retinal Müller (glial) cells: K^+ conductivity of the cell membrane. *Glia* **1**: 191–197.

Reichenbach, A., and Robinson, S. R., 1995, Phylogenetic constraints on retinal organization and development, in: *Progress in Retinal Eye Research*, Volume 15 (N. N. Osborne and G. J. Chader, eds.), Pergamon Press, Oxford, pp. 139–171.

Reichenbach, A., and Wohlrab, F., 1986, Morphometric parameters of Müller (glial) cells dependent on their topographic localization in the nonmyelinated part of the rabbit retina: A consideration of functional aspects of radial glia. *J. Neurocytol.* **15**:451–459.

Reichenbach, A., Dettmer, D., Reichelt, W., and Eberhardt, W., 1985, Na^+, K^+-activated adenosine triphosphatase of isolated Müller cells from the rabbit retina shows a potassium dependence similar to that of brain astrocytes. *Neurosci. Lett.* **59**:281–284.

Reichenbach, A., Nilius, B., and Eberhardt, W., 1986, Potassium accumulation by the glial membrane pump as revealed by membrane potential recordings from isolated rabbit retinal Müller cells. *Neurosci. Lett.* **63**:280–284.

Reichenbach, A., Schneider, H., Leibnitz, L., Reichelt, W., Schaaf, P., and Schumann, R., 1989, The structure of rabbit retinal Müller (glial) cells is adapted to the surrounding retinal layers. *Anat. Embryol.* **180**:71–79.

Reichenbach, A., Henke, A., Eberhardt, W., Reichelt, W., and Dettmer, D., 1992, K^+ ion regulation in retina. *Can. J. Physiol.* **70**:S239–S247.

Reichenbach, A., Stolzenburg, J.-U., Eberhardt, W., Chao, T. I., Dettmer, D., and Hertz, L., 1993, What do retinal Müller (glial) cells do for their neuronal "small siblings?" *J. Chem. Neuroanat.* **6**:201–213.

Reichenbach, A., Frömter, C., Engelmann, R., Wolburg, H., Kasper, M., and Schnitzer, J., 1995, Müller glial cells of the tree shrew retina. *J. Comp. Neurol.* **360**:257–270.

Reid, S. N. M., Akhmedov, N. B., Piriev, N. I., Kozak, C. A., Danciger, M., and Farber, D. B., 1999, The mouse X-linked juvenile retinoschisis cDNA: Expression in photoreceptors. *Gene* **227**:257–266.

Reier, P. J., 1996, Gliosis following CNS injury: The anatomy of astrocytic scars and their influences on axonal elongation, in: *Astrocytes,* Volume 3 (S. Fedoroff and A. Vernadakis, eds.), Academic Press, Orlando, FL, pp. 263–324.

Reiser, G., Donié, F., and Binmöller, F.-J., 1989, Serotonin regulates cytosolic calcium activity and membrane potential in a neuronal and in a glial cell line via $5HT_3$ and $5HT_2$ receptors by different mechanisms. *J. Cell Sci.* **93**:545–555.

Reisfeld, S., and Vardimon, L., 1994, Cell-to-cell contacts control the transcription activity of the glucocorticoid receptor. *Mol. Endocrinol.* **8**:1224–1233.

Rentsch, F. J., 1977, The ultrastructure of preretinal macular fibrosis. *Graefés Arch. Klin. Exp. Ophthalmol.* **203**:213–337.

Rhodes, R. H., 1984, Ultrastructure of Müller cells in the developing human retina. *Graefés Arch. Klin. Exp. Ophthalmol.* **221**:171–178.

Rich, K. A., Figueroa, S. L., Zhan, Y., and Blanks, J. C., 1995, Effects of Müller cell disruption on mouse photoreceptor cell development. *Exp. Eye Res.* **61**: 235–248.

Richter, W., Reichenbach, A., and Reichelt, W., 1990, Orthogonal arrays of intramembranous particles in the Müller cell and astrocyte endfoot membrane of rabbit retina. Postnatal development and adulthood. *J. Neurocytol.* **19**:127–139.

Rickman, D. W., and Brecha, N. C., 1995, Expression of the proto-oncogene, trk, receptors in the developing rat retina. *Vis. Neurosci.* **12**:215–222.

Ridderstrale, Y., Wistrand, P. J., and Brechue, W. F., 1994, Membrane-associated CA activity in the eye of the CAII-deficient mice. *Invest. Ophthalmol. Vis. Sci.* **35**:2577–2584.

Ridet J. L., Malhotra S. K., Privat, A., and Gage, F. H., 1997, Reactive astrocytes: Cellular and molecular cues to biological function. *Trends Neurosci.* **20**:570–577.

Riehl, R., Johnson, K., Bradley, R., Grunwald, G. B., Cornel, E., Lilienbaum, A., and Holt, C. E., 1996, Cadherin function is required for axon outgrowth in retinal ganglion cells *in vivo. Neuron* **17**:837–848.

Riepe, R. E., and Norenberg, M. D., 1977, Müller cell localization of glutamine synthetase in rat retina. *Nature* **268**: 654–655.

Ripps, H., 1982, Night blindness revisited: From man to molecules. The Proctor Lecture. *Invest. Ophthalmol. Vis. Sci.* **23**:588–609.

Ripps, H., and Witkovsky, P., 1985, Neuron-glia interaction in the brain and retina, in: *Progress in Retinal Research,* Volume 4 (N. N. Osborne and G. J. Chader, eds.), Pergamon Press, Elmsford, NY, pp. 181–219.

Ripps, H., Shakib, M., and MacDonald, E. D., 1974, Turnover of synaptic vesicles in photoreceptor terminals of the skate. *Biol. Bull.* **147**:495.

Ripps, H., Shakib, M., and MacDonald, E. D., 1976, Peroxidase uptake by photoreceptor terminals of the skate retina. *J. Cell Biol.* **70**:86–96.

Ripps, H., Mehaffey, III, L., and Siegel, I. M., 1981, "Rapid regeneration" in the cat retina. A case for spreading depression. *J. Gen. Physiol.* **77**:335–346.

Ripps, H., Mehaffey, III, L., Siegel, I. M., and Niemeyer, G., 1989, Vincristine-induced changes in the retina of the isolated arterially-perfused cat eye. *Exp. Eye Res.* **48**:771–790.

Ripps, H., Peachey, N. S., Xu, X., Nozell, S. E., Smith, S. B., and Liou, G. I., 2000, The rhodopsin cycle is preserved in IRBP "knockout" mice despite abnormalities in retinal structure and function. *Vis. Neurosci.* **17**:97–105.

Ritchie, J. M., 1987, Voltage-gated cation and anion channels in mammalian Schwann cells and astrocytes. *J. Physiol. (Paris)* **82**:248–257.

Ritchie, J. M., 1992, Voltage-gated ion channels in Schwann cells and glia. *Trends Neurosci.* **15**:345–350.

Roberge, F. C., Caspi, R. R., Chan, C. C., Kuwabara, T., and Nussenblatt, R. B., 1985, Longterm-culture of Müller cells from adult rats in the presence of activated lymphocytes/monocyte products. *Curr. Eye Res.* **4**:975–982.

Roberge, F. C., Caspi, R. R., and Nussenblatt, R. B., 1988, Glial retinal Müller cells produce IL-1 activity and have a dual effect on autoimmune T-helper lymphocytes. Antigen presentation manifested after removal of suppressive activity. *J. Immunology* **140**: 2193–2196.

Roberge, F. C., Caspi, R. R., Chan, C. C., and Nussenblatt, R. B., 1991, Inhibition of T-lymphocyte proliferation by retinal glial Müller cells: Reversal of inhibition by glucocorticoids. *J. Autoimmunity* **4**:307–314.

Robinson, S., 1992, Neuroglia: Brains as well as brawn? *Today's Life Science* **4**:18–23.

Robinson, S. R., 1991. Development of the mammalian retina, in: *Neuroanatomy of the Visual Pathways and Their Development* (B. Dreher, B. and S. R. Robinson, eds.), *Vision and Visual Dysfunction,* Volume 3 (J. R. Conly-Dillon, series ed.), Macmillan Press, London, UK, pp. 69–128.

Robinson, S. R., and Dreher, Z., 1989, Evidence for three morphological classes of astrocyte in the adult rabbit retina: Functional and developmental implications. *Neurosci. Lett.* **106**:261–268.

Robinson, S. R., and Dreher, Z., 1990, Müller cells in adult rabbit retinae: Morphology, distribution and implications for function and development. *J. Comp. Neurol.* **292**:178–192.

Robinson, S. R., Hampson, E. C. G. M., Munro, M. N., and Vaney, D. I., 1993, Unidirectional coupling of gap junctions between neuroglia. *Science* **262**:1072–1074.

Robson, J. G., and Frishman, L. J., 1995, Response linearity and kinetics of the cat retina: The bipolar cell component of the dark-adapted electroretinogram. *Vis. Neurosci.* **12**:837–850.

Rodrigues, M. M., Bardenstein, D., Wiggert, B., Lee, L., Fletcher, R. T., and Chader, G., 1987, Retinitis pigmentosa with sequential massive retinal gliosis. An immunohistochemical, biochemical and ultrastructural study. *Ophthalmology* **92**:180–186.

Rodriguez-Peralta, L., 1962, Experiments on the site of the blood-ocular barrier. *Anat. Rec.* **142**:273.

Romano, C., Chen, Q., and Olney, J. W., 1998, The intact isolated (ex vivo) retina as a model system for the study of excitotoxicity. *Prog. Ret. Eye Res.* **17**:465–483.

Roque, R. S., and Caldwell, R. B., 1990, Müller cell changes precede vascularization of the pigment epithelium in the dystrophic rat retina. *Glia* **3**:464–475.

Roque, R. S., Caldwell, R. B., and Behzadian, M. A., 1992, Cultured Müller cells have high levels of epidermal growth factor receptors. *Invest. Ophthalmol. Vis. Sci.* **33**:2587–2595.

Roque, R. S., Agarwal, N., Wordinger, R. J., Brun, A. M., Xue, Y. M., Huang, L. C., Nguyen, L. P., and Shay, J. W., 1997, Human papillomavirus-16 E6/E7 transfected retinal cell line expresses the Müller cell phenotype. *Exp. Eye Res.* **64**:519–527.

Rosenberg, P. A., and Aizenman, E., 1989, Hundred-fold increase in neuronal vulnerability to glutamate-toxicity in astrocyte-poor cultures of rat cerebral cortex. *Neurosci. Lett.* **103**:162–168.

Rosenberg, P. A., Amin, S., and Leitner, M, 1992, Glutamate uptake disguises neurotoxic potency of glutamate agonists in cerebral cortex in dissociated cell culture. *J. Neurochem.* **12**:56–61.

Rosenthal, A. R., and Appleton, B., 1975, Histochemical localization of intraocular copper foreign bodies. *Am. J. Ophthalmol.* **79**:613–625.

Ross, C. D., Bowers, M., and Godfrey, D. A., 1987, Distribution of glutaminase activity in retinal layers of rat and pig. *Brain Res.* **401**:168–172.

Rothman, S. M. and Olney, J. W., 1986, Glutamate and the pathophysiology of hypoxic-ischemic brain damage. *Ann. Neurol.* **19**:105–111.

Rothman, S. M., and Olney, J. W., 1987, Excitotoxicity and the NMDA receptor. *Trends Neurosci.* **10**:299–302.

Rothstein, J. D., Martin, L., Levey, A. I., Dykes-Hoberg, M., Jin, L. Wu, D., Nash, N., and Kuncl, R. W., 1994, Localization and neuronal and glial glutamate transporters. *Neuron* **13**: 713–725.

Rothstein, J. D., Dykes-Hoberg, M., Pardo, C. A., Bristol, L. A., Jin, L., Kuncl, R. W., Kanai, Y., Hediger, M. A., Wang, Y., Schielke, J. P., and Welty, D. F., 1996, Knockout of glutamate transporters reveals a major role for astroglial transport in excitotoxicity and clearance of glutamate. *Neuron* **16**:675–686.

Rubinson, K., 1990, The developing visual system and metamorphosis in the lamprey. *J. Neurobiol.* **21**:1123–1135.

Rudy, B., 1988, Diversity and ubiquity of K channels. *Neuroscience* **25**:729–749.

Ruiz, M., Egal, H., Sarthy, V., Qian, X., and Sarkar, H. K., 1994, Cloning, expression, and localization of a mouse γ-aminobutyric acid transporter. *Invest. Ophthalmol. Vis. Sci.* **35**: 4039–4048.

Rungger-Brändle, E., Messerli, J. M., Niemeyer, G., and Eppenberger, H. M., 1993, Confocal microscopy and computer-assisted image reconstruction of astrocytes in the mammalian retina. *Eur. J. Neurosci.* **5**:1093–1106.

Rungger-Brandle, E., Kolb, H., and Niemeyer, G., 1996, Histochemical demonstration of glycogen in neurons of the cat retina. *Invest. Ophthalmol. Vis. Sci.* **37**:702–715.

Russell, S. R., Shepherd, J. D., and Hageman, G. S., 1991, Distribution of glycoconjugates in the human retinal internal limiting membrane. *Invest. Ophthalmol. Vis. Sci.* **32**:1986–1995.

Rutishauser, U., Thierry, J.-P., Brackenbury, R., and Edelman, G. M., 1978, Adhesion among neural cells of the chick embryo. III. Relationship of the surface molecule CAM to cell adhesion and the development of histotypic patterns. *J. Cell Biol.* **79**: 371–381.

Saari, J. C., 1994, *Retinoids in Photosensitive Systems* (M. B. Sporn, A. B. Roberts, and D. S. Goodman, eds.), Raven Press, New York, pp. 351–385.

Saari, J. C., and Bredberg, D. L., 1982, Enzymatic reduction of 11–cis-retinal bound to cellular retinaldehyde binding protein. *Biochim. Biophys. Acta* **716**:266–272.

Saari, J. C., Futterman, S., and Bredberg, L., 1978, Cellular retinal- and retinoic acid-binding proteins of bovine retina. *J. Biol. Chem.* **253**: 6432–6436.

Saari, J. C., Bredberg, D. L., and Garwin, G. G., 1982, Identification of the endogenous retinoids associated with three cellular retinoid-binding proteins from bovine retina and retinal pigment epithelium. *J. Biol. Chem.* **257**:13329–13333.

Sagar, S. M., Edwards, R. H., and Sharp, F. R., 1991, Epidermal growth factor and transforming growth factor alpha induce c-fos gene expression in retinal Müller cells *in vivo*. *J. Neurosci. Res.* **29**: 549–559.

Sahel, J. A., Albert, D. M., and Lessell, S., 1990, Proliferation of retinal glia and excitatory amino acids. *Ophthalmologie* **4**:1–16.

Sahel, J. A., Albert, D. M., Lessell, S., Adler, H., and McGee, T. L., 1991, Mitogenic effects of excitatory amino acids in the adult rat retina. *Exp. Eye Res.* **53**:657–664.

Sandell, J. H., 1985, NADPH diaphorase cells in the the mammalian inner retina. *J. Comp. Neurol.* **238**:466–472.

Sandell, J. H., 1986, NADPH diaphorase histochemistry in the macaque striate cortex. *J. Comp. Neurol.* **251**:388–397.

Sarantis, M., and Attwell, D., 1990, Glutamate uptake in mammalian retinal glia is voltage- and potassium-dependent. *Brain Res.* **516**:322–325.

Sarthy, P. V., 1983, Release of [^{3}H]γ-aminobutyric acid from glial (Müller) cells of the rat retina: Effects of K^+, veratridine, and ethylenediamine. *J. Neurosci.* **3**:2494–2503.

Sarthy, P. V., 1990, Reactive gliosis in retinal degenerations, in: *Retinal Degenerations* (R. E. Anderson, J. G. Hollyfield, and M. M. LaVail, eds.), CRC Press, Boca Raton, FL, pp. 109–115.

Sarthy, P. V., and Fu, M., 1989, Transcriptional activation of an intermediate filament protein gene in mice with retinal dystrophy. *DNA J. Mol Cell. Biol.* **8**:437–446.

Sarthy, P. V., and Fu, M., 1990, Localization of laminin B1 mRNA in retinal ganglion cells by *in situ* hybridization. *J. Cell Biol.* **110**: 2099–2108.

Sarthy, P. V., and Lam, D. M. K., 1979, Biochemical studies of isolated glial (Müller) cells from the turtle retina. *J. Cell Biol.* **78**:675–684.

Sarthy, P. V., Curtis, B. M., and Catterall, W. A., 1983, Retrograde labeling, enrichment and characterization of retinal ganglion cells from the neonatal rat. *J. Neurosci.* **3**:2532–2544.

Sarthy, P. V., Hendrickson, A. E., and Wu, J.-Y., 1986, L-glutamate: A neurotransmitter candidate for cone photoreceptors in the monkey retina. *J. Neurosci.* **6**:637–643.

Sarthy, P. V., Fu, M., and Huang, J., 1991, Developmental expression of the glial fibrillary acidic protein (GFAP) gene in the mouse retina. *Cell. Mol. Neurobiol.* **11**:623–637.

Sarthy, V., 1993, Collagen IV mRNA expression during development of the mouse retina: An *in situ* hybridization study. *Invest. Ophthalmol. Vis. Sci.* **34**:145–152.

Sarthy, V., and Egal, H., 1995, Transient induction of the glial intermediate filament protein gene in Müller cells in the mouse retina. *DNA Cell Biol.* **14**:313–320.

Sarthy, V., Brodjian, S. J., and Smith, S., 1997, Ciliary neurotrophic factor (CNTF) induces reactive gliosis and GFAP gene expression in Müller cells. *Invest. Ophthalmol. Vis. Sci.* **38**: S590.

Sarthy, V., Brodjian, S., Dutt, K., Kennedy, B., French, R., and Crabb, J., 1998, Establishment and characterization of a retinal Müller cell line. *Invest. Ophthalmol. Vis. Sci.* **39**:212–216.

Sauer C. G., Gehrig, A., Warneke-Wittstock, R., Marquardt, A., Ewing, C. C., Gibson, A., Lorenz, B., Jurklies, B., and Weber, B. H. F., 1997, Positional cloning of the gene associated with X-linked juvenile retinoschisis. *Nat. Genet.* **17**:164–170.

Schabadasch, A. L., and Schabadasch, S. A., 1972, Localization and dynamic changes of glycogen in frog retina adapted to darkness or light. *Vis. Res.* **12**:1595–1604.

Schacher, S. M., Holtzman, E., and Hood, D. C., 1974, Uptake of horseradish peroxidase by frog photoreceptor synapses in the dark and the light. *Nature* **249**:261–263.

Shen, J., Petersen, K. F., Behar, K. L., Brown, P., Nixon, T. W., Mason, G. F., Petroff, O. A. C., Shulman, G. I., Shulman, R. G., and Rothman, D. I., 1999, Determination of the rate of the glutamine/glutamate cycle in the human brain by in vivo ^{13}CNMR. *Proc. Natl. Acad. Sci. USA* **96**:8235–8240.

Scherer, J., and Schnitzer, J., 1994, Growth factor effects on the proliferation of different retinal glial cells *in vitro*. *Dev. Brain Res.* **80**:209–221.

Schlosshauer, B., 1991, Neurothelin-Molecular characteristics and developmental regulation in the chick CNS. *Development* **113**:129–140.

Schlosshauser, B., and Herzog, K.-H., 1990, Neurothelin: An inducible cell surface glycoprotein of blood-brain barrier-specific endothelial cells and distinct neurons. *J. Cell Biol.* **110**: 1261–1274.

Schlosshauer, B., Grauer, D., Dutting, D., and Vanselow, J., 1991, Expression of a novel Müller glia specific antigen during development and after optic nerve lesion. *Development* **111**:789–799.

Schlosshauer, B., Steir, H., and Egert, U., 1993, Graded distribution of the neural 2A10 antigen in the developing chicken retina. *Dev. Brain Res.* **76**:13–22.

Schmidt, H. H. H. W., Lohmann, S. M., and Walter, U., 1993, The nitric oxide and cGMP signal transduction system: Regulation and mechanism of action. *Biochim. Biophys. Acta* **1178**:153–175.

Schmidt, K.-F., Nöll, G. N., and Yamamoto, Y., 1992, Sodium nitroprusside alters dark voltage and light responses in isolated retinal rods during whole-cell recording. *Vis. Neurosci.* **9**: 205–209.

Schmitz, Y., and Witkovsky, P., 1996, Glutamate release by the intact light-responsive photoreceptor layer of the *Xenopus* retina. *J. Neurosci. Meth.* **68**:55–60.

Schnitzer, J., 1987a, Immunocytochemical localization of S-100 protein in astrocytes and Müller cells in the rabbit retina. *Cell Tissue Res.* **248**:55–61.

Schnitzer, J., 1987b, Retinal astrocytes: Their restriction to vascularized parts of the mammalian retina. *Neurosci. Lett.* **78**:29–34.

Schnitzer, J., 1988a, Astrocytes in the guinea pig, horse, and monkey retina: Their occurrence coincides with the presence of blood vessels. *Glia* **1**:74–89.

Schnitzer, J., 1988b, Astrocytes in mammalian retina. *Prog. Retinal Res.* **7**:209–231.

Schoenherr C. J., Paquette, A., and Anderson, D. J., 1996, Identification of potential target genes for the neuron-restrictive silencer factor. *Proc. Natl. Acad. Sci. USA* **93**:9881–9886.

Schousboe, A., Krogsgaard-Larsen, P., Svenneby, G., and Hertz, L., 1978, Inhibition of the high-affinity, net uptake of GABA into cultured astrocytes by β-proline, nipecotic acid and other compounds. *Brain Res.* **153**:623–626.

Schousboe, A., Westergaard, N., Waagepetersen, H. S., Larsson, O. M., Bakken, I. J., and Sonnewald, U., 1997, Trafficking between glia and neurons of TCA cycle intermediates and related metabolites. *Glia* **21**:99–105.

Schultz, K., and Stell, W. K., 1996, Immunocytochemical localization of the high affinity glutamate transporter, EAACl, in the retina of representative vertebrate species. *Neurosci. Lett.* **211**:191–194.

Schuman, E. M., and Madison, D. V., 1991, A requirement for the intercellular messenger nitric oxide in long-term potentiation. *Science* **254**:1503–1506.

Schuman, E. M., and Madison, D. V., 1994, Nitric oxide and synaptic function. *Annu. Rev. Neurosci.* **17**:153–183.

Schurr, A., Miller, J. J., Payne, R. S., and Rigor, B. M., 1999, An increase in lactate output by brain tissue serves to meet the energy needs of glutamate-activated neurons. *J. Neurosci.* **19**:34–39.

Schwartz, E. A., 1987, Depolarization without calcium can release γ-aminobutyric acid from a retinal neuron. *Science* **238**:350–355.

Schwartz, E. A., 1993, L-glutamate conditionally modulates the K^+ current of Müller glial cells. *Neuron* **10**:1141–1149.

Schwartz, E. A., and Tachibana, M., 1990, Electrophysiology of glutamate and sodium co-transport in a glial cell of the salamander retina. *J. Physiol. (Lond.)* **426**:43–80.

Seal, R. P., and Amara, S. G., 1999, Excitatory amino acid transporters: A family in flux. *Annu. Rev. Pharmacol. Toxicol.* **39**:431–456.

Segal, R. A., and Greenberg, M. E., 1996, Intracellular signaling pathways activated by neurotrophic factors. *Annu. Rev. Neurosci.* **19**:463–489.

Semple-Rowland, S. L., 1991, Expression of glial fibrillary acidic protein by Müller cells in *rd* chick retina. *J. Comp. Neurol.* **305**:582–590.

Shank, R. P., and Aprison, M. H., 1981, Present status and significance of the glutamine cycle in neural tissues. *Life Sci.* **28**:837–842.

Shank, R. P., Bennett, G. S., Freytag, S. O., and Campbell, G. L., 1985, Pyruvate carboxylase: An astrocyte-specific enzyme implicated in the replenishment of amino acid neurotransmitter pools. *Brain Res.* **329**:364–367.

Shaw, G., and Weber, K., 1983, The structure and development of the rat retina: An immunofluorescence microscopical study using antibodies specific for intermediate filament proteins. *Eur. J. Cell Biol.* **30**:219–232.

Shay, J., and Ames, A., 1976, Retina subjected to components of ischemia *in vitro.* Selective vulnerability and minimum lethal exposure of neurons and glia to oxygen and/or glucose deprivation and to loss of exchange with incubating medium. *Arch. Neurol.* **33**:715–721.

Sheardown, M. J., 1993, The triggering of spreading depression in the chicken retina: A pharmacological study. *Brain Res.* **607**:189–194.

Sheardown, M. J., 1997, The triggering of spreading depression in the chick retina by nicotinic receptor agonists. *Eur. J. Pharmacol.* **337**:209–212.

Sheffield, J. B., and Li, H.-P., 1984, Interactions among cells of the developing neural retina in vitro. *Am. Zool.* **27**:145–159.

Shen, J., Petersen, K. F., Behar, K. L., Brown, P., Nixon, T. W., Mason, G. F., Petroff, O. A. C., Shulman, G. I., Shulman, R. G., and Rothman, D. L., 1999, Determination of the rate of the glutamine/glutamate cycle in the human brain by *in vivo* ^{13}CNMR. *Proc. Natl. Acad. Sci. USA* **96**:8235–8240.

Shiells, R. A., and Falk, G., 1990, Glutamate receptors of rod bipolar cells are linked to a cyclic GMP cascade via a G-protein. *Proc. R. Soc. Lond. B* **242**:91–94.

Shiells, R. A., and Falk, G., 1992, Retinal on-bipolar cells contain a nitric oxide-sensitive guanylate cyclase. *NeuroReport* **3**:845–848.

Shrager, P., Chiu, S. Y., and Ritchie, J. M., 1985, Voltage dependent sodium and potassium channels in mammalian cultured Schwann cells. *Proc. Natl. Acad. Sci. USA* **82**:948–952.

Sibson, N. R., Dhankar, A., Mason, G. F., Behar, K. L., Rothman, D. L., and Shulman, R. G., 1998a, Stoichiometric coupling of brain glucose metabolism and glutamatergic neuronal activity. *Proc. Natl. Acad. Sci. USA* **95**:316–321.

Sibson, N. R., Shen, J., Mason, G. F., Rothman, D. L., Behar, K. L., and Shulman, R. G., 1998b, Functional energy metabolism: In vivo C-13–NMR spectroscopy evidence for coupling of cerebral glucose consumption and glutamatergic neuronal activity. *Dev. Neurosci.* **20**: 321–330.

Sickel, W., 1972, Retinal metabolism in dark and light, in: *Handbook of Sensory Physiology, Physiology of Photoreceptor Organs*, Vol. VII/2 (M. G. F. Fuortes, ed.), Springer-Verlag, Berlin, pp. 667–727.

Sidman, R. L., 1961, Histogenesis of the mouse retina studied with thymidine-H^3, in: *The Structure of the Eye* (G. K. Smelser, ed.), Academic Press, NY, pp. 487–506.

Sieghart, W., 1992, $GABA_A$ receptors: Ligand-gated Cl^- ion channels modulated by multiple drug-binding sites. *Trends Pharmacol. Sci.* **13**:446–450.

Silver, J., and Robb, R. M., 1979, Studies on the development of the eyecup and optic nerve in normal and in mutants with congenital optic nerve aplasia. *Dev. Biol.* **68**:175–190.

Simmons, M. L., and Murphy, S., 1992, Induction of nitric oxide synthase in glial cells. *J. Neurochem.* **59**:897–905.

Sisk, D. R., and Kuwabara, T., 1985, Histologic changes in the inner retina of albino rats following intravitreal injection of monosodium L-glutamate. *Graefés Arch. Klin. Exp. Ophthalmol.* **223**:250–258.

Slaughter, M. M., and Bai, S.-H., 1989, Differential effects of baclofen on sustained and transient cells in the mudpuppy retina. *J. Neurophysiol.* **61**:374–381.

Small, R. K., Patel, P., and Watkins, B. A., 1991, Response of Müller cells to growth factors alters with time in culture. *Glia* **4**:469–483.

Small, R. K., Watkins, B. A., Munro, P. M., and Liu, D., 1993, Functional properties of retinal Müller cells following transplantation to the anterior eye chamber. *Glia* **7**:158–169.

Smith, Jr., G. N., Linsenmayer, T. F., and Newsome, D. A., 1976, Synthesis of type II collagen *in vitro* by embryonic chick neural retina tissue. *Proc. Natl. Acad. Sci. USA* **73**:4420–4423.

Smith, S. B., Brodjian, S., Patel, S., and Sarthy, V., 1997, Glial fibrillary acidic protein in the early stages of the photoreceptor cell degeneration of the mi^{vit}/mi^{vit} (vitiligo) mice. *Exp. Eye Res.* **64**:645–650.

Snow, D. M., Watanable, M., Letourneau, P. C., and Silver, J., 1991, A chondroitin sulfate proteoglycan may influence the direction of retinal ganglion cell outgrowth. *Development* **113**:1473–1485.

Snyder, S. H., and Bredt, D. S., 1992, Biological roles of nitric oxide. *Sci. Am.* **266**:68–77.

Somjen, G. G., 1979, Extracellular potassium in the mammalian central nervous system. *Annu. Rev. Physiol.* **41**:159–177.

Somjen, G. G., 1988, Nervenkett: Notes on the history of the concept of neuroglia. *Glia* **1**:2–9.

Somjen, G. G., Aitken, P. G., Czeh, G. L., Kerreras, O., Jing, J., and Young, J. N., 1992, Mechanisms of spreading depression: A review of recent findings and a hypothesis. *Can. J. Physiol. Pharmacol.* **70**:S248–S254.

Sontheimer, H., 1994, Voltage-dependent ion channels in glial cells. *Glia* **11**:156–172.

Sontheimer, H., and Waxman, S. G., 1992, Ion channels in spinal cord astrocytes *in vitro*. II.

Biophysical and pharmacological analysis of two Na^+ current types. *J. Neurophysiol.* **68**: 1001–1011.

Sosula, L., Beaumont, P., Hollows, F. C., Jonson, K. M., and Regtop, H. L., 1974, Glycogen accumulation in retinal neurons and glial cells of streptozotocin-diabetic rats. Quantitative electron microscopy. *Diabetes* **23**:221–231.

Sperling, H. G., Harcomb, E. S., and Johnson, C., 1982, Stimulus-controlled labeling of cones in the macaque monkey with ^{3}H-2–D-deoxyglucose, in: *The Structure of the Eye* (J. G. Hollyfield and E. A. Vidrio, eds.), Elsevier North Holland, Inc., New York, pp. 55–60.

Spitznas, M., 1970, Zur Feinstruktur der sog. Membrana limitans externa der menschlichen Retina. *Graefes Arch. Klin. Exp. Ophthalmol.* **80**:44–56.

Sramek, S. J., Wallow, I. H. L., Bindley, C., and Sterken, G., 1987, Fibronectin distribution in the rat eye: An immunocytochemical study. *Invest. Ophthalmol. Vis. Sci.* **28**:500–505.

Staehelin, L. A., and Hull, B. E., 1978, Junctions between living cells. *Sci. Am.* **288**:140–152.

Steinberg, R. H., 1969, Comparison of the intraretinal *b*-wave and d.c. component in the area centralis of cat retina. *Vis. Res.* **9**:317–331.

Steinberg, R. H., and Miller, S. S., 1979, Transport and membrane properties of the retinal pigment epithelium, in: *The Retinal Pigment Epithelium* (K. M. Zinn and M. F. Marmor, eds.), Harvard Univ. Press, Cambridge, MA, pp. 205–225.

Steinberg, R. H., Schmidt, R., and Brown, K. T., 1970, Intracellular responses to light from cat pigment epithelium: Origin of the electroretinogram c-wave. *Nature* **227**:728–730.

Steinberg, R. H., Oakley, II, B., and Niemeyer, G., 1980, Light-evoked changes in $[K^+]_o$ in retina of intact cat eye. *J. Neurophysiol.* **44**:897–921.

Steinhauser, C., 1993, Electrophysiologic characteristics of glial cells. *Hippocampus* **3**:113–123.

Steinhäuser, C., and Gallo, V., 1996, News on glutamate receptors in glial cells. *Trends Neurosci.* **19**:339–345.

Steinhauser, C., Kressen, K., Kuprijanova, E., Weber, M., and Seifert, G., 1994, Properties of voltage-activated Na^+ and K^+ currents in mouse hippocampal glial cells *in situ* and after acute isolation from tissue slices. *Pflügers Arch.* **428**:610–620.

Stenkamp, D. L., Gregory, J. K., and Adler, R., 1993, Retinoid effects in purified cultures of chick embryo retina neurons and photoreceptors. *Invest. Ophthalmol. Vis. Sci.* **34**:2424–2436.

Stirling, C. E., and Sarthy, P. V., 1985, Localization of the Na,K pump in turtle retina. *J. Neurocytol.* **14**:33–47.

Stockton, R. A., and Slaughter, M. M., 1989, B-wave of the electroretinogram. A reflection of ON bipolar cell activity. *J. Gen. Physiol.* **93**:101–122.

Stolzenburg, J. U., Haas, J., Hartig, W., Paulke, B. R., Wolburg, H., Reichelt, W., Chao, T. I., Wolff, J. R., and Reichenbach, A., 1992, Phagocytosis of latex beads by rabbit retinal Müller (glial) cells *in vitro, Journal fur Hirnforschung* **33**:557–564.

Stone, J., and Dreher, Z., 1987, Relationship between astrocytes, ganglion cells and vasculature of the retina. *J. Comp. Neurol.* **255**:35–49.

Stone, J., Itin, A., Alon, T., Peer, J., Gnessin, H., Chan-Ling, T., and Keshet, E., 1995a, Development of retinal vasculature is mediated by hypoxia-induced vascular endothelial growth-factor (VEGF) expression by neuroglia. *J. Neurosci.* **15**:4738–4747.

Stone, J., Makarov, F., and Hollander, H., 1995b, The glial ensheathment of the soma and axon hillock of retinal ganglion cells. *Vis. Neurosci.* **12**:273–279.

Storck, T., Schulte, S., Hofmann, K., and Stoffel, W., 1992, Structure, expression and functional analysis of a Na^+-dependent glutamate/aspartate transporter from rat brain. *Proc. Natl. Acad. Sci. USA* **89**:10955–10959.

Straznicky, C., and Gábriel, R., 1991, NADPH-diaphorase positive neurons in the retina of *Bufo marinus*: Selective staining of bipolar and amacrine cells. *Arch. Histol. Cytol.* **54**:213–220.

Streit, W. J., Graeber, M. B., and Kreutzberg, G. W., 1988, Functional plasticity of microglia: A review. *Glia* **1**:301–307.

Struyk, A. F., Canoll, P. D., Wolfgang, M. J., Rosen, C. L., D'Eustachio, P., and Salzer, J. L., 1995, Cloning of neurotrimin defines a new subfamily of differentially expressed neural cell adhesion molecules. *J. Neurosci.* **15**:2141–2156.

Sucher, N. J., Lei, S. Z., and Lipton, S. A., 1991, Calcium channel antagonists attenuate NMDA receptor-mediated neurotoxicity of retinal ganglion cells in culture. *Brain Res.* **297**:297–302.

Sugawara, K., Torigoe, K., Okoyama, S., Negishi, K., and Kato, S., 1990, Neurotoxic effects of L-alpha-aminoadipic acid. *Neuroscience* **36**:155–163.

Sugaya, E., Takato, M., and Noda, Y., 1975, Neuronal and glial activity during spreading depression in cerebral cortex of cat. *J. Neurophysiol.* **38**:822–841.

Sweadner, K. J., 1979, Two molecular forms of ($Na^+ + K^+$)-stimulated ATPase in brain. Separation and difference in affinity for strophanthidin. *J. Biol. Chem.* **254**:6060–6067.

Sykova, E., 1983, Extracellular K^+ accumulation in the central nervous system. *Prog. Biophys. Mol. Biol.* **42**:135–189.

Sypert, G. W., and Ward, Jr., A. A., 1974, Changes in extracellular potassium activity during neocortical propagated seizures. *Exp. Neurol.* **45**:19–41.

Szabo, C., 1996, Physiological and pathological roles of nitric oxide in the central nervous system. *Brain Res. Bull.* **41**:131–141.

Szamier, R. B., Ripps, H., and Chappell, R. L., 1981, Changes in ERG b-wave and Müller cell structure induced by alpha-aminoadipic acid. *Neurosci. Lett.* **21**:307–312.

Szatkowski, M., and Attwell, D., 1994, Triggering and execution of neuronal death in brain ischemia—2 phases of glutamate release by different mechanisms. *Trends Neurosci.* **17**:359–365.

Szatkowski, M., Barbour, B., and Attwell, D., 1990, Non-vesicular release of glutamate from glial cells by reversed electrogenic glutamate uptake. *Nature* **348**:443–446.

Szatkowski, M., Barbour, B., and Attwell, D., 1991, The potassium-dependence of excitatory amino acid transport: Resolution of a paradox. *Brain Res.* **555**:343–345.

Takahashi, J. B., Hoshimaru, M., Kikuchi, H., and Hatanaka, M., 1993, Developmental expression of trkB and low-affinity NGF receptor in the rat retina. *Neurosci. Lett.* **151**:174–177.

Takahashi, S., Driscoll, B. F., Law, M. J., and Sokoloff, L., 1995, Role of sodium and potassium in regulation of glucose metabolism in cultured astroglia. *Proc. Natl. Acad. Sci. USA* **92**:4616–4620.

Takeichi, M., 1991, Cadherin cell adhesion receptors as morphogenetic regulator. *Science* **251**: 1451–1455.

Takumi, T., Ishii, T., Horio, Y., Morishige, K.-I., Takahashi, N., Yamada, M., Yamashita, T., Kiyama, H., Sohmiya, K., Nakanishi, S., and Kurachi, Y., 1995, A novel ATP-dependent inward rectifier potassium channel expressed predominantly in glial cells. *J. Biol. Chem.* **270**:16339–16346.

Tanaka, K., Watase, K., Manabe, T., Yamada, K., Watanabe, M., Takahashi, K., Iwama, H., Nishikawa, T., Ichihara, N., Kikuchi, T., Okuyama, S., Kawashima, N., Hori, S., Takimoto, M., and Wada, K., 1997, Epilepsy and exacerbation of brain injury in mice lacking the glutamate transporter GLT-1. *Science* **276**:1699–1702.

Tanihara, H., Hangai, M., Sawaguchi, S., Abe, H., Kageyama, M., Nakazawa, F., Shirawa, E., and Honda, Y., 1997a, Up-regulation of glial fibrillary acidic protein in the retina of primate eyes with experimental glaucoma. *Arch. Ophthalmol.* **115**:752–756.

Tanihara, H., Inatani, M., and Honda, Y., 1997b, Growth factors and their receptors in the retina and pigment epithelium. *Prog. Ret. Eye Res.* **16**:271–301.

Tao-Cheng, J.-H., Nagy, Z., and Brightman, M. W., 1987, Tight junctions of brain endothelium *in vitro* are enhanced by astroglia. *J. Neurosci.* **7**:3293–3299.

Tashian, R. E., Hewett-Emmett, D., and Venta, J. P., 1991, Diversity and evolution on the carbonic anhydrase gene family, in: *Carbonic Anhydrase: From Biochemistry and Genetics to Physiology and Clinical Medicine* (F. Bortre, G. Gros, and B. Storey, eds.), Weinheim VCH, Weinheim, Germany, pp. 151–161.

Tassignon, M. J., Stempels, N., Nguyen-Legros, J., De Wilde, F., and Brihaye, M. 1991, The effect of wavelength on glial fibrillary acidic protein immunoreactivity in laser-induced lesions in rabbit retina. *Graefés Arch. Klin. Exp. Ophthalmol.* **229**:380–388.

Täumer, R., Wichmann, W., Rohde, N., and Röver, J., 1976, ERG of human without c-wave. *Graefés Arch. Klin. Exp. Ophthalmol.* **198**:275–289.

Tessier-Lavigne, M., and Goodman, C. S., 1996, The molecular biology of axon guidance. *Science* **274**:1123–1138.

Tholey, G., Roth-Schecheter, B. F., and Mandel, P., 1981, Activity and isoenzyme pattern of lactate dehydrogenase in neurons and astroblasts cultured from brains of chick embryos. *J. Neurochem.* **36**:77–81.

Tian, N., and Slaughter, M. M., 1995, Correlation of dynamic responses in the ON bipolar neuron and the b-wave of the electroretinogram. *Vis. Res.* **35**:1359–1364.

Tomita, T., and Yanagida, T., 1981, Origins of the ERG waves. *Vis. Res.* **21**:1703–1707.

Tonus, J. G., and Dickson, D. H., 1979, Neuroglial relationships at the external limiting membrane of the newt retina. *Exp. Eye Res.* **28**:93–110.

Tout, S., Ashwell, K., and Stone, J., 1988, The development of astrocytes in the albino rabbit retina and their relationship to retinal vasculature. *Neurosci. Lett.* **90**:191–210.

Tout, S., Chan-Ling, T., Holländer, H., and Stone, J., 1993, The role of Müller cells in the formation of the blood-retinal barrier. *Neuroscience* **55**:291–301.

Trifonov, Y. A., 1968, Study of synaptic transmission between the photoreceptor and the horizontal cell using electrical stimulation of the retina. *Biofizika* **13**:809–817.

Tsacopoulos, M., and Magistretti, P. J., 1996, Metabolic coupling between glia and neurons. *J. Neurosci.* **16**:877–885.

Tse, F. W., Fraser, D. D., Duffy, S., and MacVicar, B. A., 1992, Voltage-activated K^+ currents in acutely isolated hippocampal astrocytes. *J. Neurosci.* **12**:1781–1788.

Tso, M. O. M., 1982, Pathology of cystoid macular edema. *Ophthalmology* **89**:902–911.

Tsukamoto, Y., Taira, E., Yamate, J., Nakane, Y., Kajimura, K., Tsudzuki, M., Kiso, Y., Kotani, T., Miki, N., and Sakuma, S., 1997, Gicerin, a cell adhesion molecule, participates in the histogenesis of retina. *J. Neurobiol.* **33**:769–780.

Tsuyama, Y., Nöll, G. N., and Schmidt, K.-F., 1993, L-arginine and nicotinamide adenine dinucleotide phosphate alter dark voltage and accelerate light response recovery in isolated retinal rods of the frog (*Rana temporaria*). *Neurosci. Lett.* **149**:95–98.

Tucker, R. P., 1991, The distribution of tenascin and its transcript in the developing avian central nervous system. *J. Exp. Zool.* **259**:78–91.

Tuckwell, H. C., and Hermansen, C. L., 1981, Ion and transmitter movements during spreading cortical depression. *Int. J. Neuroscience* **12**:109–135.

Turner, D. L., and Cepko, C. L., 1987, A common progenitor for neurons and glia persists in rat retina late in development. *Nature* **328**:131–136.

Turner, D. L., Snyder, E. Y., and Cepko, C. L., 1990, Lineage-independent determination of cell type in the embryonic mouse retina. *Neuron* **4**:833–845.

Tyler, N. K., and Burns, M. S., 1991, Alterations in glial cell morphology and glial fibrillary acidic protein expression in urethane-induced retinopathy. *Invest. Ophthalmol. Vis. Sci.* **32**:2689–2695.

Tytell, M., Greenberg, S. G., and Lasek, R. J., 1986, Heat shock-like protein is transferred from glia to axon. *Brain Res.* **363**:161–164.

Uchihori, Y., and Puro, D. G., 1993a, Mitogenic and chemotactic effects of platelet-derived growth factor on human retinal glial cells. *Invest. Ophthalmol. Vis. Sci.* **32**:2689–2695.

Uchihori, Y., and Puro, D. G., 1993b, Glutamate as a neuron-glial signal for mitogenesis: Role of glial N-methyl-D-aspartate receptors. *Brain Res.* **613**:212–220.

Uga, S., 1974, Some structural features of the retinal Müllerian cells in the juxta-optic nerve region. *Exp. Eye Res.* **19**:105–115.

Uga, S., and Smelser, G. K., 1973a, Electron microscopic study of the development of retinal Müllerian cells. *Invest. Ophthalmol.* **12**:295–307.

Uga, S., and Smelser, G. K., 1973b, Comparative study of the fine structure of retinal Müller cells in various vertebrates. *Invest. Ophthalmol.* **12**:434–448.

Uhl, G. R., 1992, Neurotransmitter transporters (plus): A promising new gene family. *Trends Neurosci.* **15**:265–268.

Unanue, E. R., and Allen, P. M., 1987, The basis for the immunoregulatory role of macrophages and other accessory cells. *Science* **236**:551–557.

Usowicz, M. M., Gallo, V., and Cull-Candy, S. G., 1989, Multiple conductance channels in type-2 cerebellar astrocytes activated by excitatory amino acids. *Nature* **339**:380–383.

van den Berg, C. J., Krzalic, L. J., Mela, P., and Waelsch, H., 1969, Compartmentalization of glutamate metabolism in brain. Evidence for the existence of two different tricarboxylic acid cycles in brain. *Biochem. J.* **113**:281–290.

van Harreveld, A., 1978, Two mechanisms of spreading depression in the chicken retina. *J. Neurobiol.* **9**:419–431.

van Harreveld, A., and Fifkova, E., 1970, Glutamate release from the retina during spreading depression. *J. Neurobiol.* **2**:13–29.

Vaney, D. I., 1994, Patterns of neuronal coupling in the retina. *Prog. Retinal Eye Res.* **13**:301–355.

Vaney, D. I., and Young, H. M., 1988, GABA-like immunoreactivity in NADPH-diaphorase amacrine cells of the rabbit retina. *Brain Res.* **474**:380–385.

Vardimon, L., Fox, L. L., Degenstein, L., and Moscona, A. A., 1988. Cell contacts are required for induction by cortisol of glutamine synthetase gene-transcription in the retina. *Proc. Natl. Acad. Sci. USA* **85**:5981–5985.

Vardimon, L., Ben-Dror, I., Havazelet, N., and Fox, L. E., 1993, Molecular control of glutamine synthetase expression in the developing retina tissue. *Dev. Dynamics* **196**:276–282.

Vardimon, L., Ben-Dror, I., Avisar, N., Oren, A., and Shiftan, L., 1999, Glucocorticoid control of glial gene expression. *J. Neurobiol.* **40**:513–527.

Varki, A., 1994, Selectin ligands. *Proc. Natl. Acad. Sci. USA* **91**:7390–7397.

Vaughan, D. K., and Lasater, E. M., 1990, Glial and neuronal markers in bass retinal horizontal and Müller cells. *Brain Res.* **537**:131–140.

Velasco, I., Tapia, R., and Massieu, L., 1996, Inhibition of glutamate uptake induces progressive accumulation of extracellular glutamate and neuronal damage in rat cortical cultures. *J. Neurosci. Res.* **44**:551–561.

Verbavatz, J.-M., Ma, T., Gobin, R., and Verkman, A. S., 1997, Absence of orthogonal arrays in kidney, brain and muscle from transgenic knockout mice lacking water channel aquaporin-4. *J. Cell Sci.* **110**:2855–2860.

Verderber, L. C., Johnson, W., Mucke, L., and Sarthy, V., 1995, Differential regulation of a GFAP-lacZ transgene in retinal astrocytes and Müller cells. *Invest. Ophthalmol. Vis. Sci.* **36**:1137–1143.

Vincent, S. R., and Hope, B. T., 1992, Neurons that say NO. *Trends Neurosci.* **15**:108–113.

Virchow, R., 1846, Ueber das granulierte Ansehen der Wandungen der Gehirnventrikel. *Allg. Zeitsschr. Psychiatr.* **3**:242–250.

Virchow, R., 1856, *Gesammelte Abhandlungen zur Wissenschaftlichen Medicin*, Hamm, Frankfurt.

Voaden, M. J., 1978, The localization and metabolism of neuroactive amino acids in the retina, in: *Amino Acids as Chemical Transmitters* (F. Fonnum, ed.), Plenum Press, New York, pp. 257–274.

Voaden, M. J., and Starr, M. S., 1972, The efflux of radioactive GABA from rat retina *in vitro. Vis. Res.* **12**:559–566.

Voaden, M. J., Marshall, J., and Murani, N., 1974, The uptake of [^{3}H]γ-aminobutyric acid and [^{3}H]glycine by the isolated retina of the frog. *Brain Res.* **67**:115–132.

Vogel, D. A., and Green, D. G., 198, Potassium release and b-wave generation: A test of the Müller cell hypothesis. *Invest. Ophthalmol. Vis. Sci.* **19**:S39.

von der Mark, K., von der Mark, H., Timpl, R., and Trelstad, R. L., 1977, Immunofluorescent localization of collagen types I, II and III in the chick eye. *Dev. Biol.* **59**:75–85.

Vorwerk, C. K., Lipton, S. A., Zurakowski, D., Hyman, B. T., Sabel, B. A., and Dreyer, E. B., 1996, Chronic low-dose glutamate is toxic to retinal ganglion cells. Toxicity blocked by memantine. *Invest. Ophthalmol. Vis. Sci.* **37**:1618–1624.

Vrabec, F., 1970, Microglia in the monkey and rabbit retina. *J. Neuropath. Exp. Neurol.* **29**:217–224.

Vrabec, F., 1983, Early differentiation of the human retina. *Graefés Arch. Clin. Exp. Ophthalmol.* **220**:47–52.

Vyskocil, F., Kriz, N., and Bures, J., 1972, Potassium-selective microelectrodes used for measuring the extracellular brain potassium during spreading depression and anoxic depolarization in rats. *Brain Res.* **39**:255–259.

Wagner, E. C., and Raymond, P. A., 1991, Müller glial cells of the goldfish retina are phagocytic *in vitro* but not *in vivo*. *Exp. Eye Res.* **53**:583–589.

Wakakura, M., and Yamamoto, N., 1994, Cytosolic calcium transient increase through the AMPA/kainate receptor in cultured Müller cells. *Vis. Res.* **34**:1105–1109.

Wald, G., 1968, The molecular basis of visual excitation. *Nature* **219**:800–807.

Walz, W., and Hertz, L., 1982, Ouabain-sensitive and oubain-resistant net uptake of potassium into astrocytes and neurons in primary cultures. *J. Neurochem.* **39**:70–77.

Walz, W., and Hertz, L., 1983, Functional interactions between neurons and astrocytes. II. Potassium homeostasis at the cellular level. *Prog. Neurobiol.* **20**:133–183.

Wang, Y., Dudley, V. J., Sarkar, H. and Sarthy, V., 1999, Cloning and developmental expression of the glutamate transporter, GLAST, in the mouse retina. *Invest. Ophthalmol. Vis. Sci.* **40**: S236.

Wassilewa, P., Hockwin, O., and Korte, G., 1976, Glycogen concentration changes in retina, vitreous body and other eye tissues caused by disturbances of blood circulation. *Graefés Arch. Klin. Exp. Ophthalmol.* **199**:115–120.

Watanabe, T., and Raff, M. C., 1988, Retinal astrocytes are immigrants from the optic nerve. *Nature* **332**:834–837.

Watanabe, T., Mio, Y., Hoshino, F. B., Naamastsu, S., Hirosawa, K., and Nakahara, K., 1994, GLUT2 expression in the rat retina: Localization at the apical ends of Müller cells. *Brain Res.* **655**:128–134.

Weidman, T. A., 1975, Fine structure of the developing retina. *Int. Ophthalmol. Clin.* **15**:65–84.

Weikert, S., Fryer, D., Weih, M., Isaev, N., Busch, C., Schultze, J., Megow, D., and Dirnagl, U., 1997, Rapid Ca^{2+}-dependent NO-production from central nervous system cells in culture measured by NO-nitrite/ozone chemoluminescence. *Brain Res.* **748**:1–11.

Weiler, R., and Kewitz, B., 1993, The marker for nitric oxide synthase, NADPH-diaphorase, co-localizes with GABA in horizontal cells and cells of the inner retina in the carp retina. *Neurosci. Lett.* **158**:151–154.

Welinder, E., Textorius, O., and Nilsson, S. E. G., 1982, Effects of intravitreally injected DL-α-aminoadipic acid on the c-wave of the D.C.-recorded electroretinogram in albino rabbits. *Invest. Ophthalmol. Vis. Sci.* **23**:240–245.

Wen, R., and Oakley, III, B., 1990, K^+-evoked Müller cell depolarization generates b-wave of electroretinogram in toad retina. *Proc. Natl. Acad. Sci. USA* **87**:2117–2121.

Wen, R., Song, Y., Cheng, T., Matthes, M. T., Yasumura, D., LaVail, M. M., and Steinberg, R. H., 1995, Injury-induced upregulation of bFGF and CNTF messenger-RNAs in the rat retina. *J. Neurosci.* **15**:7377–7385.

Werblin, F. S., and Dowling, J. E., 1969, Organization of the retina of the mudpuppy, *Necturus maculosus*. II. Intracellular recording. *J. Neurophysiol.* **32**:339–355.

Westgaard, N., Sonnewald, U., and Schousboe, A., 1995, Metabolic trafficking between neurons and astrocytes: The glutamate/glutamine cycle revisited. *Dev. Neurosci.* **17**:203–211.

Wetts, R., and Fraser, S. E., 1988, Multipotent precursors can give rise to all major cell types of the frog retina. *Science* **239**:1142–1145.

Wetts, R., Serbedzija, G. N., and Fraser, S. E., 1989. Cell lineage analysis reveals multipotent precursors in the ciliary margin of the frog retina. *Dev. Biol.* **136**:254–263.

White, R. D., and Neal, M. J., 1976, The uptake of L-glutamate by the retina. *Brain Res.* **111**:79–93.

Wiesinger, H., Hamprecht, B., and Dringen, R., 1997, Metabolic pathways for glucose in astrocytes. *Glia* **21**:22–34.

Willbold, E., and Layer, P. G., 1992a, Formation of neuroblastic layers in chicken retinospheroids: The fiber layer of Chievitz secludes AchE-positive cells from mitotic cells. *Cell Tissue Res.* **268**:401–408.

Willbold, E., and Layer, P. G., 1992b, A hidden retinal regenerative capacity from the chick ciliary margin is reactivated *in vitro*, that is accompanied by down-regulation of butyrylcholinesterase. *Eur. J. Neurosci.* **4**:210–220.

Willbold, E., and Layer, P. G., 1998, Müller glia cells and their possible roles during retina differentiation *in vivo* and *in vitro*. *Histol. Histopathol.* **13**:531–552.

Willbold, E., Brummendorf, T., Rathjen, F. G., Schwarz, H., and Weiss, B., 1997, The neural cell recognition molecule F11 is expressed on Müller cells and Schwann cells *in vitro*. *J. Brain Res.* **38**:71–80.

Williams, D. S., Arikawa, K. and Paallysaho, T., 1990, Cytoskeletal components of the adherens junctions between the photoreceptors and the supportive Müller cells. *J. Comp. Neurol.* **295**:155–164.

Winkler, B., 1981, Glycolytic and oxidative metabolism in relation to retinal function. *J. Gen. Physiol.* **77**:667–692.

Winter, C. G., Saotome, Y., Levinson, S. W., and Hirsh, D., 1995, A role for CNTF as an inducer of reactive gliosis, the glial response to central nervous system injury. *Proc. Natl. Acad. Sci. USA* **92**:5865–5869.

Wistrand, P. J., Schenholm, M., and Lonnerholm, G., 1986, Carbonic anhydrase isozymes CAI and CAII in the human eye. *Invest. Ophthalmol. Vis. Sci.* **27**:419–428.

Witkovsky, P., and Yang, C. Y., 1982, Uptake and localization of ^{3}H-2–Deoxy-D-glucose by retinal photoreceptors. *J. Comp. Neurol.* **204**:105–116.

Witkovsky, P., Shakib, M., and Ripps, H., 1974, Interreceptoral junctions in the teleost retina. *Invest. Ophthalmol.* **13**:996–1009.

Witkovsky, P., Dudek, F. E., and Ripps, H., 1975, Slow PIII of the carp electroretinogram. *J. Gen Physiol.* **65**:119–134.

Wolburg, H., and Berg, K., 1987, Heterogeneity of Müller cell endfeet in the rabbit retina as revealed by freeze-fracturing. *Neurosci. Lett.* **82**:273–277.

Wolburg, H., and Berg, K., 1988, Distribution of orthogonal arrays of particles in the Müller cell membrane of the mouse retina. *Glia* **1**:246–252.

Wolburg, H., Reichelt, W., Stoltzenburg, J.-U., Richter, W., and Reichenbach, A., 1990, Rabbit retinal Müller cells in cell culture show gap and tight junctions which they do not express *in situ*. *Neurosci. Lett.* **111**:58–63.

Wolburg, H., Berg-von der Emde, K., and Naujoks-Manteuffel, C., 1992, Müller (glial) cells in the retina of urodeles and anurans reveal different morphology by means of freeze-fracturing. *Neurosci. Lett.* **138**:89–92.

Wolter, J. R., 1959. Glia of the human retina. *Am. J. Ophthalmol.* **48**:370–393.

Wolter, J. R., 1981, The histopathology of cystoid macular edema. *Graefés Arch. Klin. Exp. Ophthalmol.* **216**:85–101.

Wood, J., and Garthwaite, J., 1994, Models of the diffusional spread of nitric oxide: Implications for neural nitric oxide signalling and its pharmacological properties. *Neuropharmacol.* **33**:1235–1244.

Worrall, D. M., and Williams, D. C., 1994, Sodium ion-dependent transporters for neurotransmitters: A review of recent developments. *Biochem. J.* **297**:425–436.

Wu, S. M., Qiao, X., Noebels, J. L., and Yang, X. L., 1993, Localization and modulatory actions of zinc in vertebrate retina. *Vis. Res.* **33**:2611–2616.

Wyllie, D. J. A., Mathie, A., Symonds, C. J., and Cull-Candy, S. G., 1991, Activation of glutamate receptors and glutamate uptake in identified macroglial cells in rat cerebellar cultures. *J. Physiol.* **432**:235–258.

Xiao, M., Sastry, S. M., Li, Z.-Y., Possin, D. E., Chang, J. H., Klock, I. B., and Milam, A. H., 1998, Effects of retinal photocoagulation on photoreceptor basic fibroblast growth factor and survival. *Invest. Ophthalmol. Vis. Sci.* **39**:618–630.

Xie, Q.-W., Cho, H. J., Calaycay, J., Mumford R. A., Swiderek, K. M., Lee, T. D., Ding, A., Troso, T., and Nathan, C., 1992, Cloning and characterization of inducible nitric oxide synthase from mouse macrophages. *Science* **256**:225–228.

Xu, X., and Karwoski, C. J., 1994a, Current source density (CSD) analysis of retinal field potentials. I. Methodological considerations and depth profiles. *J. Neurophysiol.* **72**:84–95.

Xu, X., and Karwoski, C. J., 1994b, Current source density (CSD) analysis of retinal field potentials. II. Pharmacological analysis of the b-wave and M-wave. *J. Neurophysiol.* **72**:96–105.

Xu, X., and Karwoski, C. J., 1995, Current source density analysis of the electroretinographic d-wave of frog retina. *J. Neurophysiol.* **73**:2459–2469.

Xu, X., and Karwoski, C. J., 1997, The origin of slow PIII in frog retina: Current source density analysis in the eyecup and isolated retina. *Vis. Neurosci.* **14**:827–833.

Yamada, E., 1969, Some structural features of the fovea centralis in the human retina. *Arch. Ophthalmol.* **82**:151–159.

Yang, C. Y., Brecha, N. C., and Tsao, E., 1997, Immunocytochemical localization of gamma-aminobutyric acid plasma membrane transporters in the tiger salamander retina. *J. Comp. Neurol.* **389**:117–126.

Yang, Q. Z., and Hatton, G. I., 1999, Nitric oxide via cGMP-dependent mechanisms increases dye coupling and excitability of rat supraoptic nucleus neurons. *J. Neurosci.* **19**:4270–4279.

Yanoff, M., Rahn, E. K., and Zimmerman, L. E., 1968, Histopathology of juvenile retinoschisis. *Arch. Ophthalmol.* **79**:49–53.

Yanoff, M., Fine, B. S., Brucker, A. J., and Eagle, R. C., 1984, Pathology of human cystoid macular edema, *Sur. Ophthalmol.* **28**:505–511.

Yarowsky, P. J., Boyne, A. F., Wiewille, R., and Brookes, N., 1986, Effect of monensin on deoxyglucose uptake in cultured astrocytes: Energy metabolism is coupled to sodium entry. *J. Neurosci.* **6**:859–866.

Yazulla, S., 1985, Evoked efflux of [^{3}H]GABA from goldfish retina in the dark. *Brain Res.* **325**: 171–180.

Yazulla, S., 1986, GABAergic mechanisms in the retina. *Prog. Ret. Res.* **5**:1–51.

Yoshida, A., Ishiguro, S., and Tamai, M., 1993, Expression of glial fibrillary acidic protein in rabbit Müller cells after lensectomy-vitreoctomy. *Invest. Ophthalmol. Vis. Sci.* **34**:3154–3160.

Yoshida, K., Kuraki, Y., Ohki, K., Harada, T., Ohashi, T., Matsuda, H., and Imaki, J., 1995, C-fos gene expression in rat retinal cells after focal retinal injury. *Invest. Ophthalmol. Vis. Sci.* **36**:251–254.

Young, R. W., 1985, Cell differentiation in the retina of the mouse. *Anat. Rec.* **212**:199–205.

Zahs, K. R., and Newman, E. A., 1997, Asymmetric gap junctional coupling between glial cells in rat retina. *Glia* **20**:10–22.

Zeevalk, G. D., and Nicklas, W. J., 1994, Nitric oxide in retina: Relation to excitatory amino acids and excitotoxicity. *Exp. Eye Res.* **58**:343–350.

Zhang, H., Li, Y. C., and Young, A. P., 1993, Protein kinase A activation of glucocorticoid signaling in embryonic retina. *Proc. Natl. Acad. Sci. USA* **90**:3880–3884.

Index